Guide to Solutions for Inor

GUIDE TO SOLUTIONS FOR
INORGANIC CHEMISTRY

Third Edition

to accompany Inorganic Chemistry
by Shriver and Atkins

S. H. Strauss

Colorado State University

W. H. Freeman and Company

ISBN 0–7167–3438–9

Printed in the U.K.

First printing, 1999

TABLE OF CONTENTS

PREFACE

This book covers all of the Self-tests and Exercises in the textbook *Inorganic Chemistry*, 3rd edition, by Duward F. Shriver and Peter Atkins. It is written to help you study a fascinating subject. As you read each chapter in *Inorganic Chemistry* and work on a set of exercises, you should consult the short answers listed in the back of the textbook. That will allow you to check quickly the accuracy of your answers. Then, *Guide to Solutions* will help you advance the learning process several steps further, by presenting complete solutions that permit a detailed understanding of the questions and the short answers and, more importantly, by explicitly tying the exercises and solutions to principles developed in the text. The solutions include nearly all of the figures and drawings asked for in the exercises. They also include many other figures that will help you visualize new concepts. A special feature of *Guide to Solutions* is a Quiz at the end of each chapter consisting of ten questions.

Learning about inorganic chemistry will require more than a quick reading of the text and a glance at the short answers. There is no substitute for reading and re-reading the relevant material. One way to study is to outline each chapter and to copy your lecture notes within a few days of each lecture. That will alert you to concepts you may not yet understood fully. When studying for an exam or quiz, you may want to work through the exercises for a second time. If your solution is not as complete as the one in this book, you should review relevant sections of the textbook. Keep in mind that a good way to be sure that you understand something is to be able to explain it to someone else.

Steven H. Strauss
Fort Collins, CO
October 1998

ACKNOWLEDGMENT

I would like to thank Duward F. Shriver and Peter Atkins for insightful comments and valuable assistance during the preparation of the third edition of *Guide to Solutions*. I would also like to thank the many faculty members who used the first and second editions and wrote to me with their suggestions and comments and the many students in my classes during the past few years who brought to my attention many ambiguities, as well as mistakes, in the second edition. Last, but certainly not least, I would like to thank Mr. Brady Clapsaddle for his excellent proofreading of the entire manuscript. I probably failed to correct some of the mistakes Brady brought to my attention, and, for this reason, any remaining errors are mine alone.

1 Atomic structure

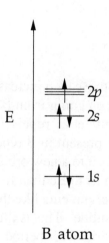

B atom

Chemists frequently use energy level diagrams like this one. You will find similar ones throughout the text (Figures 1.7 and 1.19 are examples). The vertical axis is an energy axis, with increasing energy from bottom to top. The short horizontal lines represent the energies of orbitals. Frequently, small arrows are used to represent electrons in those orbitals. In this diagram, the electron configuration of a ground-state boron atom is shown.

S1.1 **Neutron capture by $^{80}_{35}$Br?** You should use the same reasoning as in the example. Neutron capture by an atom involves an increase in mass number, in this case from 80 to 81. The atomic number remains the same, however; the atom in question is a bromine atom before *and* after the neutron capture. Any excess energy will appear as a photon in the γ-ray region of the electromagnetic spectrum. The balanced nuclear reaction is:

$$^{80}_{35}\text{Br} + ^{1}_{0}\text{n} \rightarrow ^{81}_{35}\text{Br} + \gamma$$

S1.2   **Probability of being close to the nucleus?**   There is no figure showing the radial distribution functions for $3p$ and $3d$ orbitals, so you must reason by analogy. In the Example, you saw that an electron in a p orbital has a smaller probability than in an s orbital of close approach to the nucleus because an electron in a p orbital has a greater angular momentum than in an s orbital. Study Section 1.5, *Atomic orbitals*, and you will see that, similarly, an electron in a d orbital has a greater angular momentum than in a p orbital. In other words, $l(d) > l(p) > l(s)$. Therefore, an electron in a p orbital has a greater probability than in a d orbital of close approach to the nucleus.

S1.3   **Relative changes in Z_{eff}?**   The configuration of the valence electrons, called the valence configuration, is as follows for the four atoms in question:

Li: $2s^1$ B: $2s^22p^1$

Be: $2s^2$ C: $2s^22p^2$

When an electron is added to the $2s$ orbital on going from Li to Be, Z_{eff} increases by 0.63, but when an electron is added to an empty p orbital on going from B to C, Z_{eff} increases by 0.72. The s electron already present in Li repels the incoming electron more strongly than the p electron already present in B repels the incoming p electron, because the incoming p electron goes into a new orbital. Therefore, Z_{eff} increases by a smaller amount on going from Li to Be than from B to C. However, extreme caution must be exercised with arguments like this, because the effects of electron–electron repulsions are very subtle. This is illustrated in period 3, where the effect is opposite to that just described for period 2.

S1.4   **Electron configuration of Ni and Ni^{2+}?**   Following the example, for an atom of Ni with $Z = 28$ the electron configuration is:

Ni: $1s^22s^22p^63s^23p^63d^84s^2$ or $[Ar]3d^84s^2$

Once again, the $4s$ electrons are listed last since the energy of the $4s$ orbital is higher than the energy of the $3d$ orbitals. Despite this ordering of the individual $3d$ and $4s$ energy levels for elements past Ca (see Figure 1.20), interelectronic repulsions prevent the configuration of a Ni atom from being $[Ar]3d^{10}$. For a Ni^{2+} ion, with two fewer electrons than a Ni atom but with the same Z as a Ni atom, interelectronic repulsions are less important: the higher energy $4s$ electrons are removed from Ni to form Ni^{2+}, and the electron configuration of the ion is:

Ni^{2+}: $1s^22s^22p^63s^23p^63d^8$ or $[Ar]3d^8$

S1.5 $I_1(Cl) < I_1(F)$? When considering questions like these, it is always best to begin by writing down the electron configurations of the atoms or ions in question. If you do this routinely, a confusing comparison may become more understandable. In this case the relevant configurations are:

$$F \quad 1s^22s^22p^5 \text{ or } [He]2s^22p^5$$

$$Cl \quad 1s^22s^22p^63s^23p^5 \text{ or } [Ne]3s^23p^5$$

The electron removed during the ionization process is a $2p$ electron for F and a $3p$ electron for Cl. The principal quantum number, n, is lower for the electron removed from F ($n = 2$ for a $2p$ electron), so this electron is bound more strongly by the F nucleus than a $3p$ electron in Cl is bound by its nucleus.

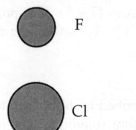

A general trend: within a group, *larger* atoms have *lower* ionization energies. There are only a few exceptions to this trend, and they are found in Groups 13 (IIIA) and 14 (IVA).

S1.6 $A_e(C) > A_e(N)$? The electron configurations of these two atoms are:

$$C: \quad [He]2s^22p^2 \qquad\qquad N: \quad [He]2s^22p^3$$

An additional electron can be added to the empty $2p$ orbital of C, and this is a favorable process ($A_e = 1.263$ eV). However, all of the $2p$ orbitals of N are already half occupied, so an additional electron added to N would experience sufficiently strong repulsions that the electron-gain process for N is unfavorable ($A_e = -0.07$ eV). This is despite the fact that the $2p$ Z_{eff} for N is larger than $2p$ Z_{eff} for C (see Table 1.3). This tells you that attraction to the nucleus is not the only force that determines electron affinities (or, for that matter, ionization energies). Interelectronic repulsions are also important.

1.1 **Nuclear reactions?** (a) $^{14}_{7}N + {}^{4}_{2}He$? You can tackle these questions easily if you do an accounting of protons and neutrons right away. The nuclear reactants, ^{14}N and ^{4}He, together contain 9 protons (7 + 2) and 9 neutrons. You are told that one of the products is ^{17}O, which contains 8 protons and 9 neutrons. (Remember, you can tell how many protons an atom contains by consulting a periodic table and noting its atomic number). Therefore, since one proton is not yet accounted for on the right hand side of the equation, the balanced equation is $^{14}_{7}N + ^{4}_{2}He \rightarrow ^{17}_{8}O + p + \gamma$.

(b) $^{12}_{6}C + p$? The atomic number of carbon is 6. If you add one proton to a carbon nucleus, the product must be the nucleus of the element with an atomic number of $6 + 1 = 7$, which of course is a nitrogen nucleus. Remember, the mass number will also increase whenever a proton, or a neutron, is added to a nucleus and no subsequent fission occurs. Therefore, the product is a ^{13}N nucleus and the balanced equation is $^{12}_{6}C + p \rightarrow ^{13}_{7}N + \gamma$.

(c) $^{14}_{7}N + ^{1}_{0}n$? Here you are told that the products are $^{3}_{1}H$ and $^{12}_{6}C$. You should confirm that the equation $^{14}_{7}N + ^{1}_{0}n \rightarrow ^{3}_{1}H + ^{12}_{6}C$ is indeed balanced. The reactants together contain 7 protons and 8 neutrons $(7 + 1)$. The products also contain 7 protons $(1 + 6)$ and 8 neutrons $(2 + 6)$. Therefore, the equation is balanced.

1.2 $^{22}_{10}Ne + \alpha$? Use the accounting procedure described in the answer to Exercise 1.1, above. Neon has atomic number 10 (therefore 10 protons) and magnesium has atomic number 12 (12 protons). In Section 1.1, *Nucleosynthesis of light elements*, you learned that an α particle is a helium-4 nucleus, ^{4}He. Therefore, the nuclear reactants together contain 12 protons $(10 + 2)$ and 14 neutrons $(12 + 2)$. One of the products is a $^{25}_{12}Mg$ nucleus, which contains 12 protons and 13 neutrons. Since one neutron is not yet accounted for, it must appear as a product, and the balanced nuclear equation is $^{22}_{10}Ne + \alpha \rightarrow ^{25}_{12}Mg + n$.

1.3 **Draw the periodic table?** See Figure 1.4 and the inside front cover of this book. You should start learning the names and positions of elements that you do not know. Start with the alkali metals and the alkaline earths. Then learn the elements in the *p* block. A blank periodic table can be found on the inside back cover of this book. You should make several photocopies of it and should test yourself from time to time, especially after studying each chapter.

1.4 $E(He^+)/E(Be^{3+})$? The ground-state energy of a hydrogenic ion, like He^+ or Be^{3+}, is defined as the orbital energy of its single electron, which is given by Equations 1.3 and 1.4:

$$E = -Z^2 m_e e^4 / 32\pi^2 (\varepsilon_0)^2 (h/2\pi)^2 n^2 \qquad (n = 1 \text{ for the ground state})$$

For the ratio $E(\text{He}^+)/E(\text{Be}^{3+})$, the constants can be ignored, and:

$$E(\text{He}^+)/E(\text{Be}^{3+}) = Z(\text{He}^+)^2/Z(\text{Be}^{3+})^2 = 2^2/4^2 = 0.25$$

1.5 $E(\textbf{H}, \textbf{\textit{n}} = \textbf{1}) - E(\textbf{H}, \textbf{\textit{n}} = \textbf{6})$? The expression for E given in Equations 1.3 and 1.4 (see above) can be used for a hydrogen atom as well as for hydrogenic ions. The ratio $E(\text{H}, n = 1)/E(\text{H}, n = 6)$ can be determined as follows:

$$E(\text{H}, n = 1)/E(\text{H}, n = 6) = (1/1^2)/(1/6^2) = 36$$

Therefore, $E(\text{H}, n = 6) = (E(\text{H}, n = 1))/36 = -0.378$ eV, and the difference:

$$E(\text{H}, n = 1) - E(\text{H}, n = 6) = -13.2 \text{ eV}$$

1.6   **Principal quantum number and its relation to l?** The principal quantum number, n, labels one of the shells of an atom. For a hydrogen atom or a hydrogenic ion, n alone determines the energy of all of the orbitals contained in a given shell (since there are n^2 orbitals in a shell, these would be n^2-fold degenerate). For a given value of n, the angular momentum quantum number, l, can assume all integer values from 0 to $n-1$.

1.7   **How many orbitals for a given value of n?** For the first shell ($n = 1$), there is only one orbital, the $1s$ orbital. For the second shell ($n = 2$), there are four orbitals, the $2s$ orbital and the three $2p$ orbitals. For $n = 3$, there are 9 orbitals, the $3s$ orbital, three $3p$ orbitals, and five $3d$ orbitals. The progression of the number of orbitals so far is 1, 4, 9, which is the same as n^2 (for example, $n^2 = 1$ for $n = 1$, $n^2 = 4$ for $n = 2$, etc.). As a further verification, consider the fourth shell ($n = 4$), which, according to the analysis so far, should contain $4^2 = 16$ orbitals. Does it? Yes; the fourth shell contains the $4s$ orbital, three $4p$ orbitals, five $4d$ orbitals, and seven $4f$ orbitals, and $1 + 3 + 5 + 7 = 16$.

1.8   **Radial wavefunctions vs. radial distribution functions vs. angular wavefunctions?** The plots of ψ vs. r shown in Figures 1.9 and 1.10 are plots of the radial parts of the total wavefunctions for the indicated orbitals. Notice that the plot of $\psi(2s)$ vs. r takes on both positive (small r) and negative (large r) values, requiring that for some value of r the wavefunction $\psi(2s) = 0$ (i.e. the wavefunction has a node at this value of r; for a hydrogen atom or a hydrogenic ion, $\psi(2s) = 0$ when $r = 2a_0/Z$). Notice also that the plot of $\psi(2p)$ vs. r is positive for all values of r. Although a $2p$ orbital does have a node, it is not due to the radial wavefunction (the radial part of the total wavefunction). The plot of $4\pi r^2\psi^2$ vs. r for a $1s$ orbital in Figure 1.11 is a radial distribution function. For comparison, plots of $r^2\psi^2$ vs. r for $1s$ and $2s$ orbitals are shown below:

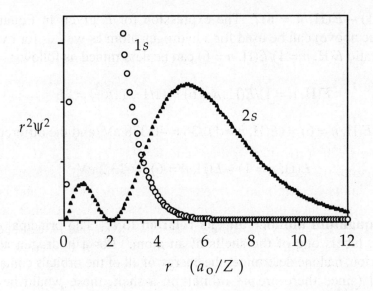

Whereas the radial distribution function for a $1s$ orbital has a single maximum, that for a $2s$ orbital has two maxima and a minimum (at $r = 2a_0/Z$ for hydrogenic $2s$ orbitals). The presence of the node at $r = 2a_0/Z$ for $\psi(2s)$ requires the presence of the two maxima and the minimum in the $2s$ radial distribution function. Using the same reasoning, the absence of a radial node for $\psi(2p)$ requires that the $2p$ radial distribution function has only a single maximum, as shown in Figure 1.12 and below:

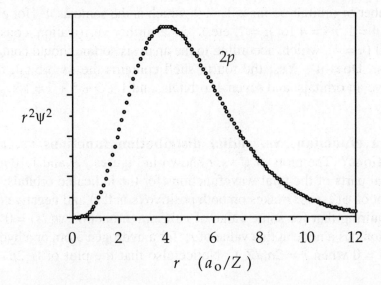

The angular wavefunctions are the familiar pictures that chemists draw to represent s, p, d, f, etc. orbitals, such as the ones in Figures 1.13–1.16. The familiar nodal plane for a $2p$ orbital is a property the orbital possesses because of the mathematical form of its angular wavefunction, not because of the mathematical form of its radial wavefunction.

1.9   *I(Ca) vs. I(Zn)?*   The first ionization energies of calcium and zinc are 6.11 and 9.39 eV, respectively (see Appendix 1). Both of these atoms have an electron configuration that ends with $4s^2$: Ca is [Ar]$4s^2$ and Zn is [Ar]$3d^{10}4s^2$. An atom of zinc has 30 protons in its nucleus and an atom of calcium has 20, so clearly zinc has a higher nuclear charge than calcium. Remember though, it is *effective* nuclear charge (Z_{eff}) that directly affects the ionization energy of an atom. Since I(Zn) > I(Ca), it would seem that Z_{eff}(Zn) > Z_{eff}(Ca). How can you demonstrate that this is as it should be? The actual nuclear charge can always be readily determined by looking at the periodic table and noting the atomic number of an atom. The effective nuclear charge cannot be directly determined, i.e., it requires some interpretation on your part. Read Section 1.6, *Penetration and shielding*, again. Study the trend for the period 2 p-block elements in Table 1.3. The pattern that emerges is that not only Z but also Z_{eff} rises from boron to neon. Each successive element has one additional proton in its nucleus and one additional electron to balance the charge. However, the additional electron never completely shields the other electrons in the atom. Therefore, Z_{eff} rises from B–Ne. Similarly, Z_{eff} rises through the d block from Sc–Zn, and that is why Z_{eff}(Zn) > Z_{eff}(Ca).

1.10   *I(Sr) vs. I(Ba) vs. I(Ra)?*   The first ionization energies of strontium, barium, and radium are 5.69, 5.21, and 5.28 eV, respectively (see Table 1.6). Normally, atomic radius increases and ionization energy decreases down a group in the periodic table. However, in this case I(Ba) < I(Ra). Study the periodic table, especially the alkaline earths. Notice that Ba is eighteen elements past Sr, but Ra is thirty-two elements past Ba. The difference between the two corresponds to the fourteen $4f$ elements between Ba and Lu. As explained above in the answer to Exercise 1.9, Z_{eff} rises with each successive element because of incomplete shielding. Therefore, even though radium would be expected to have a larger radius than barium, it has a higher first ionization energy because it has such a large Z_{eff}.

1.11   *I_2 of some period 4 elements?*   The second ionization energies of the elements calcium–manganese increase from left to right in the periodic table with the exception that $I_2(Cr) > I_2(Mn)$. The electron configurations of the elements are:

Ca	Sc	Ti	V	Cr	Mn
$[Ar]4s^2$	$[Ar]3d^14s^2$	$[Ar]3d^24s^2$	$[Ar]3d^34s^2$	$[Ar]3d^54s^1$	$[Ar]3d^54s^2$

Both the first and the second ionization processes remove electrons from the $4s$ orbital of these atoms, with the exception of Cr. In general, the $4s$ electrons are poorly shielded by the $3d$ electrons, so $Z_{eff}(4s)$ increases from left to right and I_2 also increases from left to right. While the I_1 process removes the sole $4s$ electron for Cr, the I_2 process must remove a $3d$ electron. The higher value of I_2 for Cr relative to Mn is a consequence of the special stability of half–filled subshell configurations.

1.12   **Ground-state electron configurations? (a) C?** Four elements past He. Helium ends period 1, therefore carbon is $[He]2s^22p^2$.

(b) F? Seven elements past He, therefore $[He]2s^22p^5$.

(c) Ca? Two elements past Ar, which ends period 3 leaving the $3d$ subshell empty, therefore $[Ar]4s^2$.

(d) Ga^{3+}? Thirteen elements, but only ten electrons, past Ar, and since it is a cation there is no doubt that $E(3d) < E(4s)$, therefore $[Ar]3d^{10}$.

(e) Bi? Twenty-nine elements past Xe, which ends period 5 leaving the $5d$ and the $4f$ subshells empty, therefore $[Xe]4f^{14}5d^{10}6s^26p^3$.

(f) Pb^{2+}? Twenty-eight elements, but only twenty-six electrons, past Xe, which ends period 5 leaving the $5d$ and the $4f$ subshells empty, therefore $[Xe]4f^{14}5d^{10}6s^2$.

1.13   **More ground-state electron configurations? (a) Sc?** Three elements past Ar, therefore $[Ar]3d^14s^2$.

(b) V^{3+}? Five elements, but only two electrons, past Ar, and since it is a cation there is no doubt that $E(3d) < E(4s)$, therefore $[Ar]3d^2$.

(c) Mn^{2+}? Seven elements, but only five electrons, past Ar, therefore $[Ar]3d^5$.

(d) Cr^{2+}? Six elements, but only four electrons, past Ar, therefore $[Ar]3d^4$.

(e) Co^{3+}? Nine elements, but only six electrons, past Ar, therefore $[Ar]3d^6$.

(f) Cr^{6+}? Six elements past Ar, but with a +6 charge it has the *same* electron configuration as Ar, which is written as [Ar]. Sometimes inorganic chemists will write the electron configuration as $[Ar]3d^0$ to emphasize that there are no d electrons for this d-block metal ion in its highest oxidation state.

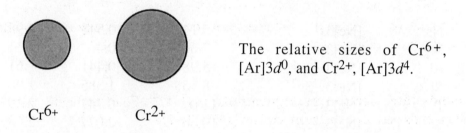

The relative sizes of Cr^{6+}, $[Ar]3d^0$, and Cr^{2+}, $[Ar]3d^4$.

Cr^{6+} Cr^{2+}

(g) Cu? Eleven elements past Ar, but its electron configuration is not $[Ar]3d^94s^2$. The special stability experienced by completely filled subshells causes the actual electron configuration of Cu to be $[Ar]3d^{10}4s^1$.

(h) Gd^{3+}? Ten elements, but only seven electrons, past Xe, which ends period 5 leaving the $5d$ and the $4f$ subshells empty, therefore $[Xe]4f^7$.

1.14 **More ground-state electron configurations? (a) W?** Twenty elements past Xe, fourteen of which are the $4f$ elements. If you assumed that the configuration would resemble that of chromium, you would write $[Xe]4f^{14}5d^56s^1$. It turns out that the actual configuration is $[Xe]4f^{14}5d^46s^2$. The configurations of the heavier d- and f-block elements show some exceptions to the trends for the lighter d-block elements.

(b) Rh^{3+}? Nine elements, but only six electrons, past Kr, therefore $[Kr]4d^6$.

(c) Eu^{3+}? Nine elements, but only six electrons, past Xe, which ends period 5 leaving the $5d$ and the $4f$ subshells empty, therefore $[Xe]4f^6$.

(d) Eu^{2+}? This will have one more electron than Eu^{3+}. Therefore, the ground-state electron configuration of Eu^{2+} is $[Xe]4f^7$.

(e) V^{5+}? Five elements past Ar, but with a 5+ charge it has the *same* electron configuration as Ar, which is written as [Ar] or [Ar]$3d^0$.

(f) Mo^{4+}? Six elements, but only two electrons, past Kr, therefore [Kr]$4d^2$.

1.15 I_1, A_e, and χ for period 3? The following values were taken from Tables 1.6, 1.7, and 1.8:

element	electron configuration	I_1 (eV)	A_e (eV)	χ
Na	[Ne]$3s^1$	5.14	0.548	0.93
Mg	[Ne]$3s^2$	7.64	–0.4	1.31
Al	[Ne]$3s^23p^1$	5.98	0.441	1.61
Si	[Ne]$3s^23p^2$	8.15	1.385	1.90
P	[Ne]$3s^23p^3$	11.0	0.747	2.19
S	[Ne]$3s^23p^4$	10.36	2.077	2.58
Cl	[Ne]$3s^23p^5$	13.10	3.617	3.16
Ar	[Ne]$3s^23p^6$	15.76	–1.0	

In general, I_1, A_e, and χ all increase from left to right across period 3 (or from top to bottom in the table above). All three quantities reflect how tightly an atom holds on to its electrons, or how tightly it holds on to additional electrons. The cause of the general increase across the period is the gradual increase in Z_{eff}, which itself is caused by the incomplete shielding of electrons of a given value of n by electrons with the same n. The exceptions are explained as follows: I_1(Mg) > I_1(Al) and A_e(Na) >A_e(Al): both of these are due to the greater stability of $3s$ electrons relative to $3p$ electrons; A_e(Mg) and A_e(Ar) < 0: filled subshells impart a special stability to an atom or ion (in these two cases the additional electron must be added to a higher energy subshell (for Mg) or shell (for Ar)); I_1(P) > I_1(S) and A_e(Si) > A_e(P): the loss of an electron from S and the gain of an additional electron by Si both result in an ion with a half-filled p subshell, which, like filled subshells, imparts a special stability to an atom or ion.

1.16 Metallic radii of Nb and Ta? If you look at the elements just before these two in Table 1.4, you will see that this is a general trend. Normally, the period 6 elements would be expected to have larger metallic radii than their period 5 vertical neighbors; only Cs and Ba follow this trend: Cs is larger than Rb and Ba is larger than Sr. Lutetium, Lu, is significantly smaller than yttrium, Y, and Hf is just barely the same size as Zr. After Nb and Ta, the "normal" expectation is observed. There are no intervening elements between Sr and Y, but there are fourteen intervening elements, the lanthanides, between Ba and Lu. A contraction of the radii of the elements starting with Lu is due to incomplete shielding by the 4*f* electrons. For a more detailed discussion of this concept, study Section 1.8(a), *Atomic and ionic radii.*

1.17 **Electronegativities across period 2?** Plots of electronegativity across period 2 and ionization energies across period 2 are superimposed on the figure below. The general trend is the same in both plots; both χ and I_1 increase from left to right across a period, and this is because the effective nuclear charge increases for the $n = 2$ orbitals across period 2. The two deviations in the upper plot result from different phenomena. For boron, the outermost electron occupies a 2*p* orbital, which has a higher energy than a beryllium atom's 2*s* orbital. The higher energy of the 2*p* electron offsets a boron atom's greater nuclear charge. For oxygen, two electrons are paired in one of the 2*p* orbitals, and the mutual repulsion they experience offsets an oxygen atom's greater nuclear charge relative to a nitrogen atom (see Example 1.5). Even though this exercise did not require the use of electron affinity values (Table 1.7), it is useful to think about the "connections" between the various atomic properties. Note that the electron affinities of beryllium and nitrogen are negative, and the explanation for these apparent anomalies is the same as the explanation given above for the departure of $I_1(\text{B})$ and $I_1(\text{O})$ from the general upward trend.

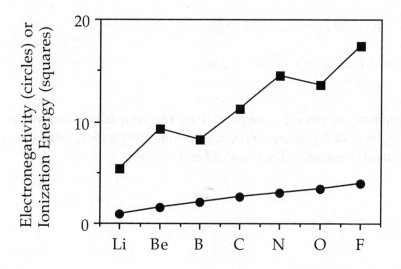

1.18   **Frontier orbitals of Be?** Recall from Section 1.8(c), *Electron affinity*, that the frontier orbitals are the highest occupied and the lowest unoccupied orbitals of a chemical species (atom, molecule, or ion). Since the ground-state electron configuration of a beryllium atom is $1s^2 2s^2$, the frontier orbitals are the $2s$ orbital (highest occupied) and the $2p$ orbitals (lowest unoccupied). Note that there can be more than two frontier orbitals if either the highest occupied and/or lowest unoccupied energy levels are degenerate.

1.19   **Broad trends in *I*, atomic radius, and electronegativity?** In general, as you move across a period from left to right, the ionization energy, *I*, increases, the atomic radius decreases, and the electronegativity increases. All three trends are the result of the increase in effective nuclear charge, Z_{eff}, from left to right.

Guide to Solutions Quiz

1   Fill in the blank in each of the following nuclear reactions, which have been used to synthesize some of the heaviest elements known:

(a) $^{253}_{99}\text{Es}$ + _____ $\rightarrow$ $^{256}_{101}\text{Md}$ + $^{1}_{0}\text{n}$

(b) $^{246}_{96}\text{Cm}$ + $^{12}_{6}\text{C}$ $\rightarrow$ _____ + $5\,^{1}_{0}\text{n}$

(c) $^{243}_{95}\text{Am}$ + _____ $\rightarrow$ $^{256}_{103}\text{Lr}$ + $5\,^{1}_{0}\text{n}$

2   Without consulting a periodic table, fill in the immediate neighbors of phosphorus in the drawing below. Practice learning the periodic table this way with other "central" elements. Try it with Cl and with Ag.

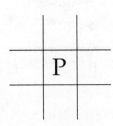

3   Calculate and compare the energy gaps, in electron volts, between the $n = 1$ and $n = 2$ levels of a hydrogen atom and the $n = 11$ and $n = 12$ levels. What seems to happen as n becomes very large?

4   What is the color of light that is twice as energetic as 800 nm wavelength near-infrared light?

5   How many angular nodes does an f orbital have? How many radial nodes does a $4f$ orbital have?

6   Refer to Appendix 1, *Electronic properties of the elements*. The figure below is a plot of first ionization energy (y axis) vs. atomic number (x axis) for the Group III/13 elements B–Tl. Explain the alternation in ionization energies starting with aluminum. Draw a similar plot for the alkali metals. Explain any exception(s) to the normal trend.

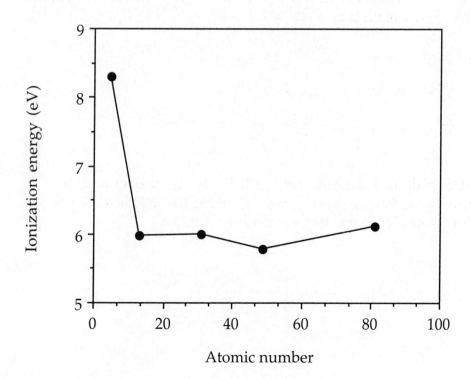

7   Refer again to Appendix 1. Excluding the alkali metals, which metal has the
        highest second ionization energy? What bearing do you think this has on this
        element's common oxidation state(s)?

8   Refer once again to Appendix 1. List all ground-state atoms between hydrogen
        and krypton that have exactly two unpaired electrons. List all that are
        diamagnetic.

9   Table 1.7, *Electron affinities of the main-group elements*, lists two groups of
        elements that have negative electron affinities. What other group in the periodic
        table do you think behaves similarly? Explain.

10   Write ground-state electron configurations of: (a) In^+ and In^{3+}; (b) Fe^{2+} and
        Fe^{3+}; (c) Au^+ and Au^{3+}. In each case, which ion has the higher effective
        nuclear charge? Explain.

2 The structures of simple solids

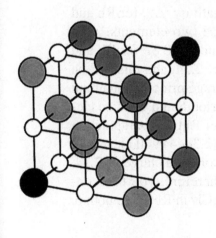

The rock-salt structure of NaCl is an example of an ionic solid built up from a close-packed array of anions. In the drawing at the left, the small open circles represent Na^+ ions. The larger circles represent Cl^- ions, which are stacked together in a cubic close-packed (i.e. ABCABC...) array. The direction perpendicular to the plane of close-packed Cl^- ions is along a body diagonal of the unit cell cube: the black Cl^- ions are in A positions, the lightly shaded ones are in B positions, and the heavily shaded ones are in C positions. The Na^+ ions are in *all* of the octahedral holes formed by the Cl^- ion array.

S2.1 **The size of a tetrahedral hole?** Note that line $S–T = r + r_h$, by definition. Therefore, you must express S–T in terms of r. Note also that S–T is the hypoteneuse of the right triangle STM, with sides S⋯M and M–T. Point M is at the midpoint of line S⋯S, and since S⋯S = 2r, S⋯M = r. The angle θ is 54.74°, one-half of the tetra-hedral angle S–T–S (109.48°). Therefore, $\sin 54.74° = r/(r + r_h)$, and $r_h = 0.225r$. This is the same as $r_h = ((3/2)^{1/2} – 1)r$.

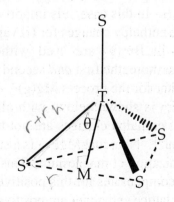

S2.2   **How many ions are in the CsCl unit cell?**   The unit cell is shown in Figure 2.11. An ion that is completely inside of the cell counts as 1 ion in that cell, and the single Cs^+ ion in the center of the unit cell fits this description. An ion that is at a corner of the unit cell is shared by eight different unit cells and counts as 1/8 of an ion in each of the eight cells. There are eight Cl^- ions in this category. Therefore, the total number of ions is $1(1) + 8(1/8) = 2$. As in the Example, the two ions in the unit cell are one cation and one anion, since the CsCl structure is only found for 1:1 salts.

S2.3   **The coordination number of Ti in rutile?**   The rutile structure is shown in Figure 2.16. There is a Ti^{4+} ion in the center of the structure, and it is connected to six O^{2-} ions to form an octahedral TiO_6 coordination unit. Therefore, the coordination number of Ti in rutile is 6.

S2.4   **Predict the coordination environment of RbCl?**   To put RbCl on the map, you need to determine the difference in electronegativity, $\Delta\chi$, for Rb and Cl and the average principal quantum number for these two elements. The Pauling electronegativities can be found in Table 1.8, and are 0.82 for Rb and 3.16 for Cl. Therefore, the difference $\Delta\chi = 3.16 - 0.82 = 2.34$, and this will be the x coordinate for RbCl on the structure map. The average principal quantum number for Rb and Cl is 4, since Rb is an element in period 5 and Cl is an element in period 3. Therefore, 4 will be the y coordinate for RbCl. Extrapolation of the line in Figure 2.22 shows that the (x, y)coordinate (2.34, 4) will clearly lie in the C.N. = 6 region of the map. In fact, with a value of $\Delta\chi$ as large as 2.34, almost any 1:1 salt will lie in the C.N. = 6 region. Therefore, you can confidently predict that the coordination numbers of Rb^+ and Cl^- in RbCl are both 6.

S2.5   **Calculate $\Delta_L H$ for MgBr$_2$?**   You should proceed as in the example, calculating the total enthalpy change for the Born–Haber cycle and setting it equal to $\Delta_L H$. In this case, it is important to recognize that two Br^- ions are required, so the enthalpy changes for (i) vaporization of $Br_2(l)$ and (ii) breaking the Br–Br bond in $Br_2(g)$ are used without dividing by 2, as was done for KCl. Furthermore, the first *and* second ionization enthalpies for Mg(g) must be added together for the process $Mg(g) \rightarrow Mg^{2+}(g) + 2e^-$. The Born–Haber cycle for MgBr$_2$ is shown below, with all of the enthalpy changes given in kJ mol^{-1}. These enthalpy changes are not to scale. The lattice enthalpy is equal to 2421 kJ mol^{-1}. Note that MgBr$_2$ is a stable compound despite the enormous enthalpy of ionization of magnesium. This is because the very large lattice enthalpy more than compensates for this positive enthalpy term. Note the standard convention used; lattice enthalpies are positive enthalpy changes.

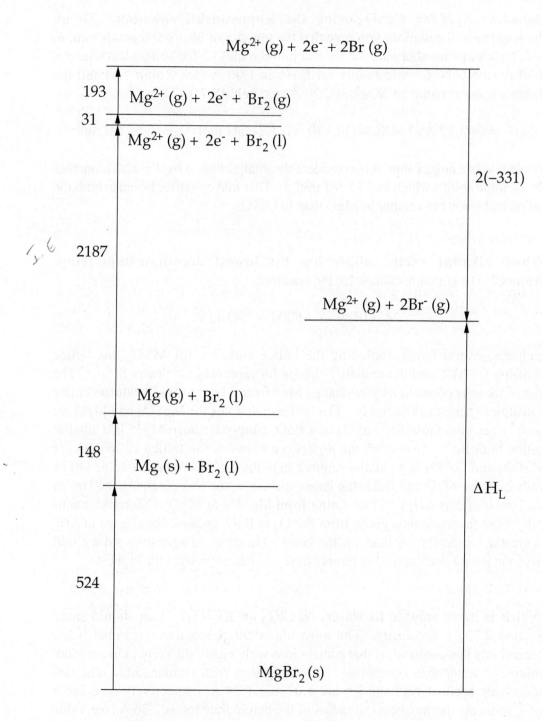

Mg^{2+} (g) + 2e⁻ + 2Br (g)

193   Mg^{2+} (g) + 2e⁻ + Br_2 (g)

31

Mg^{2+} (g) + 2e⁻ + Br_2 (l)

2(−331)

2187

Mg^{2+} (g) + 2Br⁻ (g)

Mg (g) + Br_2 (l)

148

Mg (s) + Br_2 (l)

ΔH_L

524

$MgBr_2$ (s)

Born–Haber cycle for $MgBr_2$ (Self-Test S2.5.)

S2.6   **Calculate $\Delta_L H$ for $CaSO_4$ using the Kapustinskii equation?** To use the Kapustinskii equation, you note that the number of ions per formula unit, n, is 2, the charge numbers are +2 for calcium ion and –2 for sulfate ion, and $d =$ 3.30 Å (the C.N. 6 ionic radius for Ca^{2+} is 1.00 Å (see Table 1.5) and the thermochemical radius of SO_4^{2-} is 2.30 Å (see Table 2.7)). Therefore:

$$\Delta_L H = (8/3.30 \text{ Å})(1 - (0.345)/(3.30 \text{ Å}))(1.21 \text{ MJ mol}^{-1}) = 2.63 \text{ MJ mol}^{-1}$$

While this is a large value, it is considerably smaller than ΔH_L for MgO (another 2+/2– ionic solid), which is 3.85 MJ mol^{-1}. This makes sense, because both the cation and anion are smaller in MgO than in $CaSO_4$.

S2.7   **Which alkaline earth sulfate has the lowest decomposition temperature?** The enthalpy change for the reaction:

$$MSO_4(s) \rightarrow MO(s) + SO_3(g)$$

includes several terms, including the lattice enthalpy for MSO_4, the lattice enthalpy for MO, and the enthalpy change for removing O^{2-} from SO_4^{2-}. The last of these is constant as you change M^{2+} from Mg^{2+} to Ba^{2+}, but the lattice enthalpies change considerably. The lattice enthalpies for $MgSO_4$ and MgO are both larger than those for $BaSO_4$ and BaO, simply because Mg^{2+} is a smaller cation than Ba^{2+}. However, the *difference* between the lattice enthalpies for $MgSO_4$ and $BaSO_4$ is a smaller number than the difference between the lattice enthalpies for MgO and BaO (the larger the anion, the less changing the size of the cation affects ΔH_L). Thus, going from $MgSO_4$ to MgO is thermodynamically more favorable than going from $BaSO_4$ to BaO, because the *change* in ΔH_L is greater for the former than for the latter. Therefore, magnesium sulfate will have the lowest decomposition temperature and barium sulfate the highest.

S2.8   **Which is more soluble in water, $NaClO_4$ or $KClO_4$?** You should study Section 2.12(c), *Solubility*. The most important concept to remember is the general rule that compounds that contain ions with widely different radii are more soluble in water than compounds containing ions with similar radii. The six-coordinate radii of Na^+ and K^+ are 1.02 and 1.38 Å, respectively (see Table 1.5), while the thermochemical radius of the perchlorate ion is 2.36 Å (see Table 2.7). Therefore, since the radii of Na^+ and ClO_4^- differ more than the radii of K^+ and ClO_4^-, the salt $NaClO_4$ should be more soluble in water than $KClO_4$.

2.1 **Which of the following are close-packed?** **(a) ABCABC. . .?** Any ordering scheme of planes is close-packed if no two adjacent planes have the same position (i.e. if no two planes are *in register*). When two planes are in register, the packing looks like the figure below and to the right, whereas the packing in a close-packed structure allows the atoms of one plane to fit more efficiently into the spaces between the atoms in an adjacent plane, like the figure below and to the left. Notice that the empty spaces between the atoms in the figure to the left are much smaller than in the figure to the right. The efficient packing exhibited by close-packed structures is why, for a given type of atom, close-packed structures are more dense than any other possible structure. In the case of an ABCABC. . . structure, no two adjacent planes are in register, so the ordering scheme is close-packed.

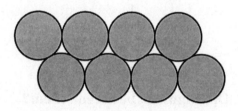 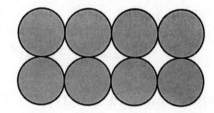

(b) ABAC. . .? Once again, no two adjacent planes are in register, so the ordering scheme is close-packed.

(c) ABBA. . .? The packing of planes using this sequence will put two B planes next to each other as well as two A planes next to each other, so the ordering scheme is not close-packed.

(d) ABCBC. . .? No two adjacent planes are in register, so the ordering scheme is close-packed.

(e) ABABC. . .? No two adjacent planes are in register, so the ordering scheme is close-packed.

(f) ABCCB. . .? The packing of planes using this sequence will put two C planes next to each other, so the ordering scheme is not close-packed.

2.2 **Mark B and C positions in a layer of atoms at A positions?** In the drawing below, the spheres represent atoms in A positions, the ⊗ symbols are at B positions, and the O symbols are at C positions.

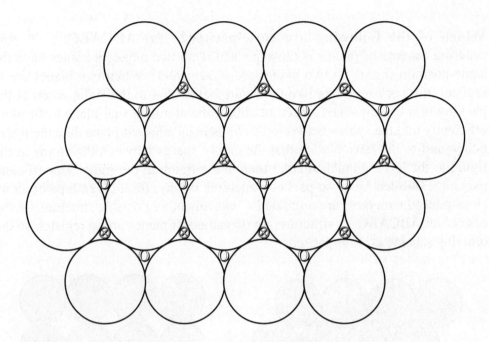

2.3   **Polymorph and polytype?** (a) **Distinguish between them?** A polymorph is a particular crystalline form of a substance, which is stable under certain conditions of temperature and pressure. On the other hand, a polytype is a solid structure that differs from another polytype in the third dimension only. Therefore, two polytypes must have the same structure in two dimensions.

(b) **Give an example of each?** Examples of polymorphs are (1) diamond and graphite, (2) α-Fe and γ-Fe, (3) wurtzite and sphalerite, and (4) gray tin and white tin, among others. The most straightforward example of polytypes is the pair of close-packed lattices hcp and ccp.

2.4   **Describe the structure of WC?** A solid-state structure is generally described, when appropriate, as a close-packed array of one type of atom or ion with other types of atoms or ions in the interstitial holes between the packing spheres. Tungsten atoms are considerably larger than carbon atoms, so in this case it is best to envision WC as a close-packed lattice of W atoms with C atoms in interstitial holes. Since the compound has the rock-salt structure, the structure is described as a *cubic* close-packed array of W atoms with C atoms in *all* of the *octahedral* holes.

2.5   **Rubidium chloride?** (a) **Coordination numbers?** The rock-salt polymorph of RbCl is based on a ccp array of Cl⁻ ions in which the Rb⁺ ions occupy all of the octahedral holes. An octahedron has six vertices, so the Rb^+ ions are six-coordinate. Since RbCl is a 1:1 salt, the Cl^- ions must be six-coordinate as well. The cesium-chloride polymorph is based on a cubic array of Cl^- ions with Rb^+ ions at the unit cell centers. A cube has eight vertices, so the Rb^+ ions are eight-coordinate, and therefore the Cl^- ions are also eight-coordinate.

(b) Larger Rb^+ radius? If more anions are packed around a given cation, the hole that the cation sits in will be larger. You saw an example of this when the radii of tetrahedral ($0.225r$) and octahedral holes ($0.414r$) were compared (r is the radius of the anion). Therefore, the cubic hole in the cesium-chloride structure is larger than the octahedral hole in the rock-salt structure. A larger hole means a longer distance between the cation and anion, and hence a larger apparent radius of the cation. Therefore, the apparent radius of rubidium is larger when RbCl has the cesium-chloride structure and smaller when RbCl has the rock-salt structure.

2.6 **The ReO_3 structure? (a) Coordination numbers?** A drawing of the unit cell of ReO_3 is shown at the right. The Re^{6+} ions are the small shaded spheres at the cube corners and the O^{2-} ions are the open spheres on the cube edges. Each O^{2-} ion is close to only two rhenium cations, so the coordination number of oxygen is 2. Since the stoichiometry of this compound is 1:3, the coordination number of rhenium must be 6.

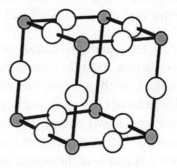

(b) New structure type with a cation at the center? If a cation, M^{n+}, were located at the center of the cell shown above, it would be equidistant from the twelve O^{2-} ions. Therefore it would be 12-coordinate. The stoichiometry M:Re:O would be 1:1:3 and the formula unit would be $MReO_3$. The new structure would be identical to the structure of perovskite, shown in Figure 2.17.

2.7 **Which elements form ionic compounds?** The ions that form solids that conform well to the ionic model are those that are small, hard, not polarizable, and not highly charged (i.e. they are +1, +2, −1, or −2). In the drawing below, note that Be has been omitted, although all of the other +2 alkaline earths form solids that are well described by the ionic model. This is because Be^{2+} is so small that it is too polarizing, and compounds like BeF_2 and BeO have considerable covalent character.

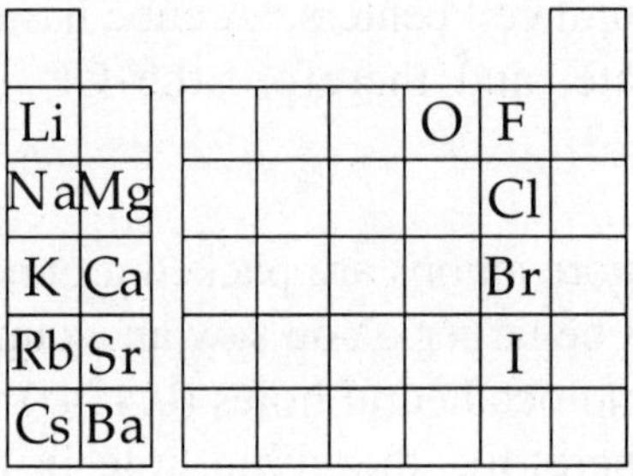

Li					O	F	
Na	Mg					Cl	
K	Ca					Br	
Rb	Sr					I	
Cs	Ba						

2.8 **The rock-salt structure of NaCl?** **(a) Coordination number?** The structure of NaCl is shown below. The Na^+ ions are the small open spheres while the Cl^- ions are the larger shaded spheres. Each type of ion has six nearest neighbors which are the opposite type of ion.

(b) Second-nearest neighbors? Each Na^+ ion has twelve second-nearest neighbors, and these are Na^+ ions (one along each of the twelve edges of the cubic unit cell shown below).

(c) Close-packed plane of Cl^- ions? The six more heavily shaded Cl^- ions are in a single "close-packed" plane (the Cl^- ions cannot really be close-packed; if they were, the Na^+ ions would not fit into the octahedral holes). These have been redrawn in the plane of the page to the right of the unit cell. If you look down one of the four C_3 axes that the unit cell cube possesses, the x marks the spot where this axis intersects the close-packed plane. Note that the Na^+ ion that is at the center of the unit cell cube lies on the C_3 axis. This ion lies at position x but not in the plane of the page. Instead, it lies midway between this close-packed plane and an adjacent one.

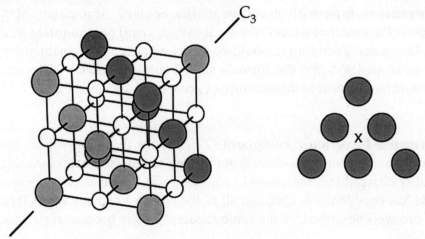

2.9   **The cesium-chloride structure?**   **(a) Coordination number?**   The unit cell for this structure is shown in Figure 2.11. Each type of ion has eight nearest neighbors which are the opposite type of ion, so the coordination number for each type of ion is 8.

(b) Second-nearest neighbors?   Each unit cell is surrounded by six equivalent unit cells, each of which shares a face with its six neighbors. Since each of these unit cells contains a Cs^+ ion at its center, each Cs^+ ion has six second-nearest neighbors that are Cs^+ ions, one in the center of each of the six neighbor unit cells.

2.10   **How many ions are in?**   **(a) The CsCl unit cell?**   The Cs^+ ion in the center of the unit cell, shown in Figure 2.11, is contained entirely within that cell. So, each unit cell contains one Cs^+ ion. The 1:1 stoichiometry of this compound demands that there must be exactly one Cl^- contained in the cell. The Cl^- ions at the corners of a unit cell contribute equally to eight different unit cells, so each cell does indeed contain one Cl^- ion $(8(1/8) = 1)$.

(b) The sphalerite (ZnS) unit cell?   See the answer to Example 2.2. There are four cations and four anions in the sphalerite unit cell.

2.11   **The stoichiometry of rutile?**   The structure of a unit cell of the rutile form of TiO_2 is shown in Figure 2.16. The cell contains one Ti^{4+} ion at the center, which belongs completely to the cell. It also contains 1/8 of the eight Ti^{4+} ions at the corners of the unit cell. Thus, the unit cell contains two Ti^{2+} ions. Four of the O^{2-} ions surrounding the central Ti^{4+} ion are on faces of the unit cell (although they are not at the centers of the faces). Two faces have two O^{2-} ions each while the other four faces have none. One-half of these four oxide ions belong to the unit cell shown in the figure, while the other halves belong to adjacent cells. Since the other two O^{2-} ions surrounding the central Ti^{4+} ion are completely within the unit cell, the cell contains four net O^{2-} ions $((4/2) + 2 = 4)$. Therefore, while the unit cell contains two Ti^{4+} ions, it contains four O^{2-} ions, and the 1:2 stoichiometry of rutile is consistent with the structure.

2.12   **The stoichiometry of perovskite?**   The structure of a unit cell of perovskite is shown in Figure 2.17. The cell contains one Ca^{2+} ion at its center. It also contains 1/8 of the eight Ti^{4+} ions at the corners of the cell, for a total of 1 Ti^{4+} ion and a Ca:Ti ratio of 1:1. Finally, the unit cell contains 1/4 of the twelve O^{2-} ions that are centered on the twelve edges of the cubic unit cell, for a total of 3

O^{2-} ions and a Ca:Ti:O ratio of 1:1:3. Therefore, the stoichiometry of perovskite, $CaTiO_3$, is consistent with the structure.

2.13   An MX_2 structure based on the bcc CsCl structure? Several unit cells of the CsCl structure are shown at the right. The shaded circles and the black circle are Cl^- ions while the slightly smaller open circles are Cs^+ ions. Two of the four Cs^+ ions have been removed, following the instructions. If you focus your attention on the black Cl^- ion, you will see that it will have four nearest neighbor Cs^+ ions, the two shown and

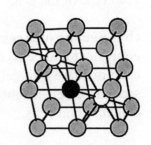

two more in a layer of unit cells that is adjacent to these four and out of the plane of the page. Those two Cs^+ ions will be contained in the lower left and upper right unit cells in the layer not shown, and the coordination geometry around the black Cl^- ion (as well as *all* Cl^- ions) will be tetrahedral. The structure you have formed is the fluorite structure, exhibited by CaF_2 and other ionic solids. The fluorite structure is shown in Figure 2.13.

2.14   Unit cell dimensions? Refer to the drawing accompanying the answer to Exercise 2.8. Note that the unit cell edge length for the rock-salt structure is equal to the hypotenuse of an isosceles right triangle. If the Se^{2-} ions are just in contact in MgSe, then the legs of the isosceles right triangle are equal to twice the radius of a Se^{2-} ion. Since the hypotenuse (edge length) in MgSe is 5.45 Å, the ionic radius of Se^{2-} is (5.45 Å)/(2√2) = 1.93 Å. For the ionic solids MgSe, CaSe, SrSe, and BaSe, the hypotenuses (edge lengths) are 5.45, 5.91, 6.23, and 6.62 Å, respectively, and are equal to twice the cation radius plus 5.45 Å, which is twice the radius of Se^{2-}. Thus, the ionic radii of Mg^{2+}, Ca^{2+}, Sr^{2+}, and Ba^{2+} are, respectively, (5.45 Å – 2(1.93) Å)/2 = 0.80 Å, (5.91 Å – 2(1.93) Å)/2 = 1.03 Å, (6.23 Å – 2(1.93) Å)/2 = 1.19 Å, and (6.62 Å – 2(1.93) Å)/2 = 1.38 Å. This assumption of ion contact in certain limiting cases is one of the traditional ways of apportioning cation and anion radii in ionic compounds.

2.15   Which are well described by an ionic model? **(a) LiF?** Both elements in this compound are in period 2, so the average principal quantum number, n, is 2. The difference in Pauling electronegativities, $\Delta\chi$, is 3.98 – 0.98 = 3.00 (see Table 1.8). Therefore, LiF is way off the structure map in Figure 2.22, far to the right, and is certainly in the C.N. 6 region of the map. A very reasonable prediction is that LiF has the rock-salt structure. Since both ions are hard (nonpolarizable) and not too highly charged, an ionic model is a good one for this compound.

(b) RbBr? The average n is 4.5, and $\Delta\chi$ is $2.96 - 0.82 = 2.14$. This puts RbBr in the C.N. 6 region of the map, so a reasonable prediction is that this compound has the rock-salt structure. An ionic model is still a good one for RbBr, but not as good as it was for LiF (compare CsBr and LiF in Table 2.6).

(c) SrS? The average n is 4, and $\Delta\chi$ is $2.58 - 0.95 = 1.63$. This puts SrS in the C.N. 6 region of the map, so a reasonable prediction is that this compound also has the rock-salt structure. Due to the polarizability of the S^{2-} ion, an ionic model is probably only approximately valid for this compound. The lattice enthalpy calculated using an ionic model will probably be lower than the actual lattice enthalpy.

(d) BeO? The average n is 2, and $\Delta\chi$ is $3.44 - 1.57 = 1.87$. This compound is right on the borderline between the C.N. 4 and C.N. 6 regions of the structure map. If you consider that *covalent* compounds containing a period 2 central atom are generally limited to four atoms around the small central atom, you predict that the two types of ions in BeO will exhibit C.N. 4. This is in fact the case: beryllium oxide has the wurtzite structure. An ionic model is not a good one for this compound.

2.16   **Calculate $\Delta_f H$ for KF_2?** You can use a Born–Haber approach to this exercise (see Figure 2.24). The unknown will be the enthalpy of formation, which is the bottommost arrow on the left side of the diagram. You will need to calculate or estimate ΔH_L for the hypothetical compound KF_2, which you can do using the Born–Mayer equation once you have estimated the radius of K^{2+}. To do this, you consult Table 1.5, and notice that $r(K^+) = 1.51$ Å, $r(Ca^{2+}) = 1.12$ Å, and $r(Sr^{2+}) = 1.25$ Å (these are C.N. 8 radii, since you were instructed to assume that KF_2 would have the fluorite structure, in which the cations are 8-coordinate). You can guess that $r(K^{2+}) > r(Ca^{2+})$, since $Z_{eff}(K^{2+}) < Z_{eff}(Ca^{2+})$. However, you can also guess that $r(K^{2+}) < r(Sr^{2+})$, since strontium has a higher principal quantum number than potassium. Thus, a reasonable estimate for $r(K^{2+})$ is 1.2 Å. Since the C.N. 4 radius of F^- is 1.31 Å, the interionic distance in KF_2 is estimated to be 1.2 Å + 1.31 Å = 2.5 Å. Therefore:

$$\Delta H_L(KF_2) = (1.39 \text{ MJ mol}^{-1})(2/2.5 \text{ Å})(1 \text{ Å} - (0.345 \text{ Å})/(2.5 \text{ Å}))(2.519)$$

$$\Delta H_L(KF_2) = 2.4 \text{ MJ mol}^{-1}$$

(the Madelung constant for the CaF_2 structure is the last term in the expression for ΔH_L, 2.519). This will be a favorable (negative) term as far as the formation of $KF_2(s)$ from $K(s)$ and $F_2(g)$ is concerned. The only other favorable term is the electron affinity enthalpy change for two F atoms, $2(328 \text{ kJ mol}^{-1}) = 656 \text{ kJ mol}^{-1}$ (see Table 1.7). The positive enthalpy terms will be the sublimation enthalpy of potassium, 89 kJ mol^{-1} (see Figure 2.24), the first *and second* ionization enthalpies of potassium, 419 kJ mol^{-1} and 3069 kJ mol^{-1} (see Table 1.6), respectively, and the bond enthalpy of F_2, 155 kJ mol^{-1}. Summing these together (see below) yields the estimated enthalpy of formation of KF_2, which is 676 kJ mol^{-1}, a rather large positive number. The predicted instability of KF_2 can be seen to be the result of the enormous second ionization enthalpy of potassium.

Positive enthalpy terms in the formation of hypothetical KF_2		Negative enthalpy terms in the formation of hypothetical KF_2	
Subl. of K	89 kJ mol^{-1}	Lattice enth.	−2400 kJ mol^{-1}
Ionization of K	3488 kJ mol^{-1}	e$^-$ affin. of 2F	−656 kJ mol^{-1}
1/2 F_2 bond enth.	155 kJ mol^{-1}		
Totals	3732 kJ mol^{-1}		−3056 kJ mol^{-1}

$$\Delta H_f(KF_2) = 3732 \text{ kJ mol}^{-1} + (-3056 \text{ kJ mol}^{-1}) = 676 \text{ kJ mol}^{-1}$$

2.17   Estimate the order of lattice energies for MgO, NaCl, and LiF? Since the coulombic attraction accounts for the bulk of the lattice enthalpy, the compound with the largest coulombic potential energy, V, will have the largest lattice enthalpy (see Section 2.11(b), *Coulombic contributions to lattice enthalpies*). As long as all three compounds have the same structure (in this case they all have the rock-salt structure), $V \propto z_A z_B/d$. You can see that the factor $z_A z_B$ is 4 for MgO but only 1 for NaCl and LiF. Therefore, the lattice enthalpy for MgO is the largest, because d will certainly not vary by a factor of 4 for the compounds under consideration. Between NaCl and LiF, the latter will have the larger lattice enthalpy, because d will be smaller; both Li$^+$ and F$^-$ are smaller than their counterparts in NaCl. Therefore, the order of lattice enthalpies for these three simple ionic solids is NaCl < LiF < MgO.

2.18   Which is more stable to thermal decomposition?   (a) MgCO$_3$ or CaCO$_3$? As discussed in the answer to S2.7, the enthalpy changes that are not constant for the two reactions

$$MgCO_3 \rightarrow MgO + CO_2 \qquad \text{and} \qquad CaCO_3 \rightarrow CaO + CO_2$$

are the four lattice enthalpies, for $MgCO_3$, MgO, $CaCO_3$, and CaO. Since Mg^{2+} is smaller than Ca^{2+}, the enthalpy change $(\Delta H_L(MgCO_3) - \Delta H_L(MgO))$ is a *larger* negative number (and hence makes for a more favorable reaction) than the corresponding enthalpy change $(\Delta H_L(CaCO_3) - \Delta H_L(CaO))$. Therefore, since the decomposition is more favorable for magnesium carbonate, calcium carbonate is more stable to thermal decomposition.

(b) CsI_3 or $N(CH_3)_4I_3$? The same reasoning may be applied here as above; the compound with the smaller cation will decompose more readily to a covalent compound and an ionic solid with a smaller anion (as large an anion as I^- is, it is smaller than I_3^-). Thus, you need to compare the ionic radii of Cs^+ and $N(CH_3)_4^+$. The radius for Cs^+, taken from Table 1.5, is 1.67 Å (for C.N. 6). The radius for $N(CH_3)_4^+$ is not found in any table in Chapters 1 or 2, but you can make a reasonable guess that it is larger than the radius of Cs^+. Therefore, since the decomposition is more favorable for cesium triiodide, tetramethyl-ammonium triiodide is more stable to thermal decomposition.

2.19 **Which is more soluble?** **(a) $MgSO_4$ or $SrSO_4$?** In general, difference in size of the ions favors solubility in water. The thermochemical radius of sulfate ion is 2.30 Å while the 6-coordinate radii of Mg^{2+} and Sr^{2+} are 0.72 and 1.16 Å, respectively. Since the difference in size between Mg^{2+} and SO_4^{2-} is greater than the difference in size between Sr^{2+} and SO_4^{2-}, $MgSO_4$ is predicted to be more soluble in water than $SrSO_4$, and this is in fact the case: $K_{sp}(MgSO_4) > K_{sp}(SrSO_4)$.

(b) NaF or $NaBF_4$? This exercise can be answered without refering to tables in the text. The ions Na^+ and F^- are isoelectronic, so clearly Na^+ is smaller than F^-. It should also be obvious that the radius of BF_4^- is larger than the radius of F^-. Therefore, the difference in size between Na^+ and BF_4^- is greater than the difference in size between Na^+ and F^-; $NaBF_4$ is more soluble in water than NaF.

Guide to Solutions Quiz

1 The structure of metallic cesium at ordinary conditions is body-centered cubic. At high pressures, however, its structure changes to a close-packed polymorph. Explain why this is reasonable.

2   The last element in Group VI/16 is polonium, Po. Its structure is simple cubic, edge length 3.37 Å. (a) How many nearest-neighbor atoms surround each Po atom? (b) How many polonium atoms occupy each unit cell? (c) Calculate the density of ^{209}Po in g cm^{-3}.

3   Consider a cubic close-packed array of spheres. (a) How many octahedral holes surround each packing sphere? (b) What shape (geometry) do these octahedral holes form around each packing sphere? Hint: consider the structure of NaCl.

4   Use the radius ratio to predict the coordination number of the calcium ions in CaF_2, $CaCl_2$, and $CaBr_2$.

5   (a) List as many similarities and differences as you can for the sphalerite and wurtzite structures. (b) Table 2.5 shows that their Madelung constants are not exactly the same. Suggest a reason.

6   The lattice enthalpies of LiF, CsF, LiI, and CsI are 1034, 744, 718, and 584 kJ mol^{-1}, respectively. Explain why the *difference* between LiF and CsF is greater than the *difference* between LiI and CsI.

7   Using the data in Question 6, above, estimate the lattice enthalpy of CsCl. Compare your estimate with the value determined using the Born–Haber cycle (the enthalpy of sublimation of cesium is 76 kJ mol^{-1}).

8   Calculate the lattice enthalpy for CsCl using the Born–Mayer equation and compare the result to your answer to Question 7. The interionic distance in CsCl is 3.56 Å.

9   The enthalpy of formation of H$^-$(g), +145 kJ mol^{-1}, is large and positive. Despite this, the salt Na$^+$H$^-$ is thermodynamically stable ($\Delta_f H° \sim -50$ kJ mol^{-1}). The enthalpy of formation of He$^-$(g), +48 kJ mol^{-1}, is much smaller than for H$^-$(g). Nevertheless, the salt Na$^+$He$^-$ is unknown and probably has a positive $\Delta_f H°$. Using enthalpy change concepts you learned in this chapter, suggest a reason why Na$^+$He$^-$ is thermodynamically unstable.

10   The thermochemical radius of the NH$_4^+$ ion is 1.51 Å. Using Table 2.7 and Figure 2.22, determine which is more soluble in water, NH$_4$F or NH$_4$I.

3 Molecular structure and bonding

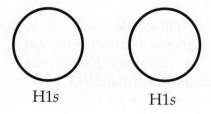

H1s   H1s

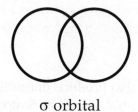

σ orbital

One of the most powerful concepts in chemistry is that molecular orbitals result from the overlap of atomic orbitals. In this figure, two hydrogen 1s atomic orbitals overlap to form a σ molecular orbital.

S3.1 Lewis structure for PCl_3? The four atoms supply $5 + (3 \times 7) = 26$ valence electrons. Since P is less electronegative than Cl, it is likely to be the central atom, so the 13 pairs of electrons are distributed as shown at the right. In this case, each atom obeys the octet rule. Whenever it is possible to follow the octet rule without violating other electron counting rules, you should do so.

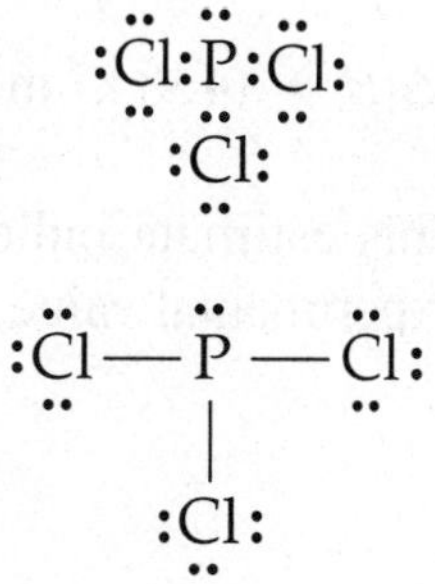

S3.2   **Resonance structures for NO_2^-?**   First, draw a Lewis structure for the ion, trying to satisfy the octet rule.   Then, permute any multiple bonds among equivalent atoms (i.e. those of the same chemical and structural type).   In this case, the octet rule can be satisfied by drawing one N–O single bond and one N=O double bond.   Two resonance structures, shown at the right, are necessary to depict the observation that both N–O bonds in the nitrite ion are equal in length and strength.

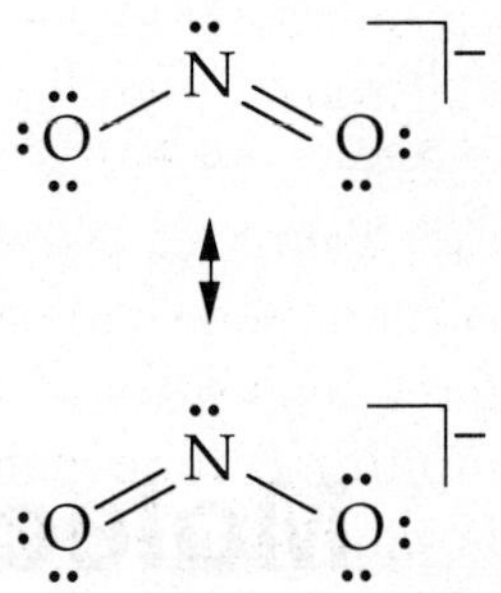

S3.3   **Oxidation numbers?**   **(a) O_2^+?**   The charge on the oxygenyl ion is +1, and that charge is shared by two oxygen atoms.   Therefore, O.N.(O) = +1/2.   This is an unusual oxidation number for oxygen.

(b) Phosphorus in PO_4^{3-}?   The charge on the phosphate ion is –3, so O.N.(P) + 4 × O.N.(O) = –3.   Oxygen is normally given an oxidation number of –2.   Therefore, O.N.(P) = –3 – (4)(–2) = +5.   The central phosphorus atom in the phosphate ion has the maximum oxidation number for Group V/15.

S3.4   **Estimate $\Delta_f H$ for H_2S?**   You can "form" this molecule by considering the following reaction:

$$1/8 \ S_8(g) \ + \ H_2(g) \ \rightarrow \ H_2S(g)$$

On the left side, you must break one H–H bond and also produce one sulfur atom from cyclic S_8.   Since there are eight S–S bonds holding eight S atoms together, you must supply the mean S–S bond enthalpy *per S atom*.   On the right side, you form two H–S bonds.   From the values given in Table 3.5, you can estimate:

$$\Delta_f H \approx (436 \ kJ \ mol^{-1}) \ + \ (264 \ kJ \ mol^{-1}) \ - \ 2(338 \ kJ \ mol^{-1}) = 24 \ kJ \ mol^{-1}$$

This estimate indicates a slightly endothermic enthalpy of formation, but the experimental value, –21 kJ mol^{-1}, is slightly exothermic.

S3.5   **Predict the shape of XeF_2?**   The Lewis structure is shown below, accommodating an octet for the 4 F atoms and an expanded valence shell of 10 electrons for the Xe atom with the $8 + (2 \times 7) = 22$ valence electrons provided by the three atoms. The five electron pairs around the central Xe atom will arrange themselves at the corners of a trigonal bipyramid (like in PF_5). The three lone pairs will be in the equatorial plane, to minimize lone pair/lone pair repulsions. The resulting shape of the molecule, shown at the right, is linear (i.e. the F–Xe–F bond angle is 180°).

$$:\ddot{X}e \diagdown \begin{array}{c} \ddot{F}: \\[4pt] \ddot{F}: \end{array}$$

$$F \textemdash Xe \textemdash F$$

S3.6   **Electron configurations for S_2^{2-} and Cl_2^-?**   The first of these two anions has the same Lewis structure as peroxide, O_2^{2-}. It also has a similar electron configuration to that of peroxide, except for the use of sulfur atom valence $3s$ and $3p$ atomic orbitals instead of oxygen atom $2s$ and $2p$ orbitals. There is no need to use sulfur atom $3d$ atomic orbitals, which are higher in energy than the $3s$ and $3p$ orbitals, since the $2 \cdot 6 + 2 = 14$ valence electrons of S_2^{2-} will not completely fill the stack of molecular orbitals constructed from sulfur atom $3s$ and $3p$ atomic orbitals. Thus, the electron configuration of S_2^{2-} is $1\sigma_g^2 2\sigma_u^2 3\sigma_g^2 1\pi_u^4 2\pi_g^4$. The Cl_2^- anion contains one more electron than S_2^{2-}, so its electron configuration is $1\sigma_g^2 2\sigma_u^2 3\sigma_g^2 1\pi_u^4 2\pi_g^4 4\sigma_u^1$.

S3.7   **Molecular orbitals of ClO^-?**   Study Section 3.9, *Heteronuclear diatomic molecules*, again. After that, you should conclude that the bonding molecular orbitals of ClO^- are predominantly oxygen in character, since oxygen is more electronegative than chlorine. Similarly, the antibonding molecular orbitals are predominantly chlorine in character. A chlorine atom uses its $3s$ and $3p$ valence-shell orbitals for bonding. An oxygen atom uses its $2s$ and $2p$ orbitals. The MO diagram for ClO^- will look very similar to Figure 3.24 except that Cl, with its $3s$ and $3p$ orbitals, will replace iodine on the left-hand side of the figure and O, with its $2s$ and $2p$ orbitals, will replace chlorine on the right-hand side of the figure.

S3.8   **Is any XH_2 molecule linear?**   According to Figure 3.38, a XH_2 molecule is expected to be linear if it contains four or fewer electrons. This is because the bottom two orbitals, which can contain up to four electrons, are lowest in energy when the H–X–H bond angle is 180°. Based on this analysis, both NaH_2 and MgH_2 should be linear, because they contain three and four valence electrons, respectively. The molecule AlH_2 contains five electrons and so is not expected to be linear.

S3.9   **The conductance of Ge at 370 K?** A plot of the data given in Example 3.9 is shown below (the squares are the data points). As explained in the answer in the text, a plot of ln G vs. $1/T$ is linear. A linear least-squares fit to the data yields the equation ln $G = 11.1 - (4.24 \times 10^3)(1/T)$. At 370 K, you can interpolate ln G from the graph (the circle) or calculate ln G using the fitted equation. At 370 K, ln G is found to be –0.348, so $G = 0.706$ S.

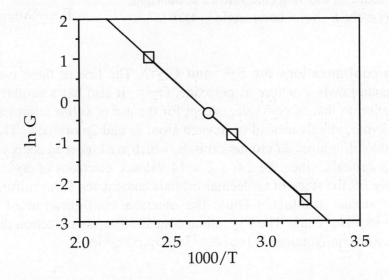

3.1   **Lewis structures and VSEPR theory? (a) GeCl$_3^-$?** The Lewis structure is shown below. Only one resonance structure is important (each atom has an octet).

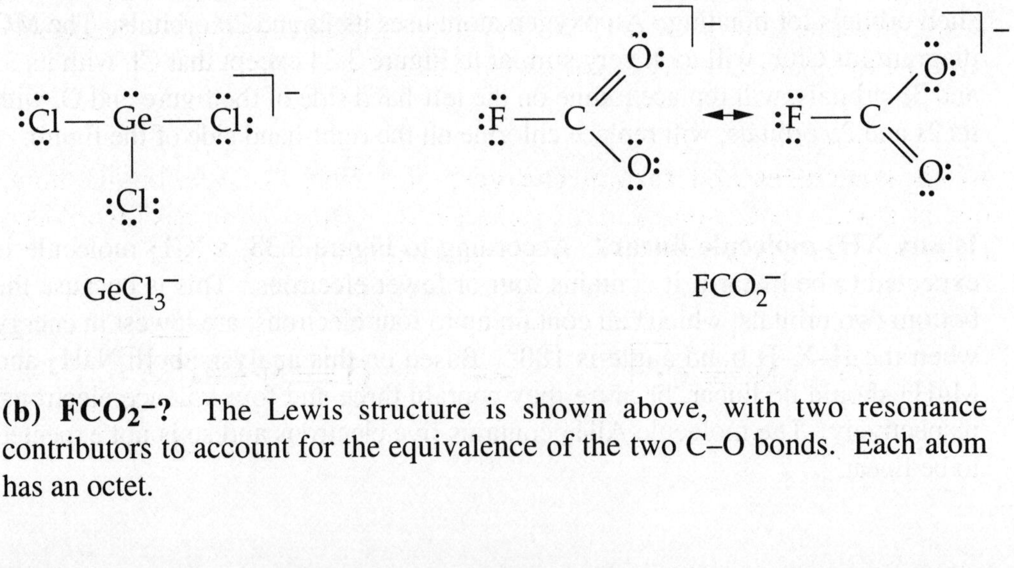

$GeCl_3^-$ FCO_2^-

(b) FCO$_2^-$? The Lewis structure is shown above, with two resonance contributors to account for the equivalence of the two C–O bonds. Each atom has an octet.

(c) CO_3^{2-}? Three resonance structures are necessary to account for the fact that the three C–O bonds in the carbonate anion are equivalent, and these are shown below. Each atom has an octet in all three resonance structures.

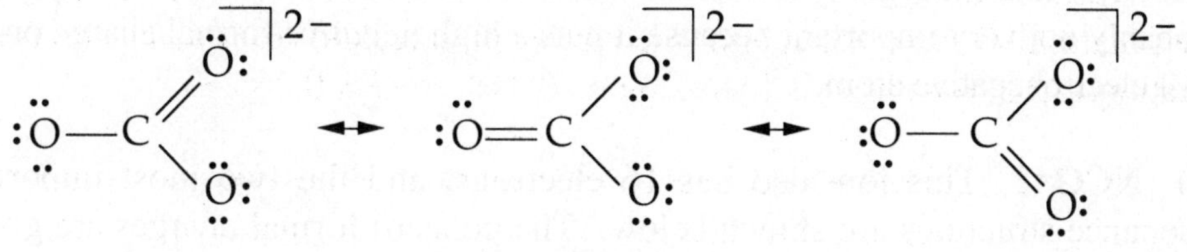

$$CO_3^{2-}$$

(d) $AlCl_4^-$? Only one resonance structure is necessary to achieve an octet around each atom and to account for the equivalence of the four Al–Cl bonds, and it is shown below.

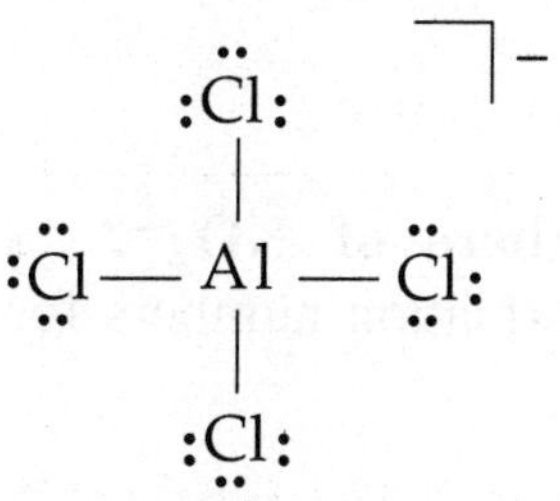

$$AlCl_4^-$$ 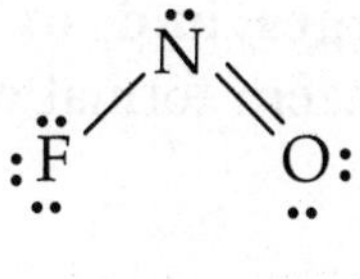 FNO

(e) FNO? The least electronegative atom is likely to be the central atom. The Lewis structure of this molecule is shown above, requiring only one resonance structure to achieve an octet around all three atoms.

3.2 **Lewis structures and formal charges?** **(a) ONC^-?** With one O atom, one N atom, one C atom, and a –1 charge, the ONC^- anion has 16 valence electrons. You can arrange them in two resonance forms as follows:

$\quad$ –1 $\quad$ +1 $\quad$ –1 $\qquad\qquad\qquad\qquad$ +1 $\quad$ –2

The nonzero formal charges for the atoms are given. These were calculated using the formula in Section 3.1(b), *Formal charges*. The number of lone pair electrons (not the number of lone pairs) and half the number of shared electrons are subtracted from the number of valence electrons on the parent atom. The resonance structure on the left is likely to be the dominant one, since it contains smaller formal charges. In addition, the resonance structure on the right is probably not very important because it puts a high *negative* formal charge on the least electronegative atom.

(b) NCO⁻? This ion also has 16 electrons, and the two most important resonance structures are shown below. The nonzero formal charges are given. The resonance structure on the left is likely to be the dominant one, since it puts the negative formal charge on O, the most electronegative atom.

$$:N \equiv C - \ddot{O}: \quad \rule{0pt}{0pt}^{-} \quad \longleftrightarrow \quad \ddot{N} = C = \ddot{O} \quad \rule{0pt}{0pt}^{-}$$

$$\qquad\qquad -1 \qquad\qquad\qquad\qquad\qquad -1$$

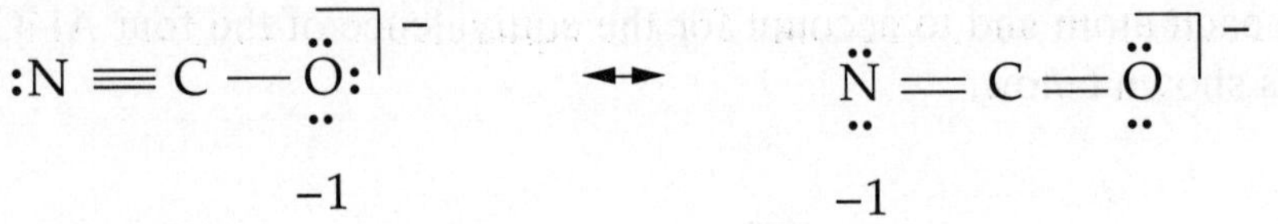

3.3   Formal charges and oxidation numbers of NO_2^-? The resonance structures, nonzero formal charges, and oxidation numbers for nitrite ion are shown below:

Formal Charges	−1					−1
Oxidation No.	−2	+3	−2		−2	+3	−2

The two structures shown are similar in that they both contain a N–O single bond and a N=O double bond. Therefore, they each contribute equally. The nitrogen atom in nitrite ion can be oxidized or reduced, since it is in the +3 oxidation state and since nitrogen can have oxidation numbers that range from −3 to +5. Both the oxidation number (+3) and the formal charge (0) on the nitrogen atom fail to give an accurate picture of its actual charge, which is slightly negative (that is, the negative charge is shared by all three atoms).

3.4   **More Lewis structures?**
(a) **XeF$_4$?** There is a straightforward way to draw Lewis
structures with a central atom
and some number of atoms.

$$:\ddot{Xe}: \; + \; 4 \cdot \ddot{F}: \quad \longrightarrow$$

In this case, start with the Lewis structure of a xenon atom. Then consider how
many fluorine atoms there are in the molecule, and use one of xenon's electrons
for each Xe–F bond The complete Lewis structure of XeF$_4$ is shown below.

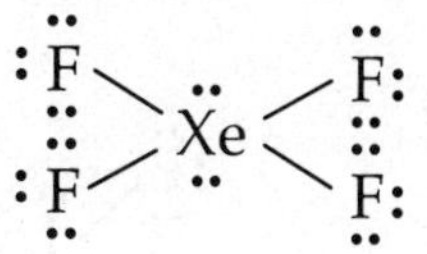

XeF$_4$ 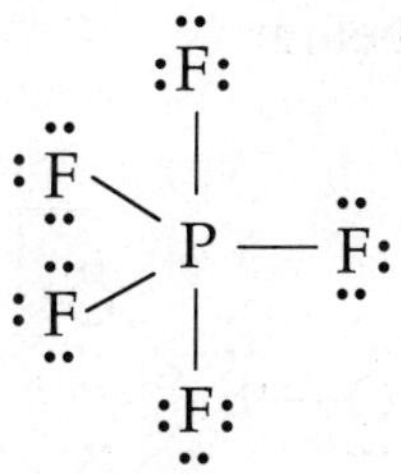   PF$_5$ 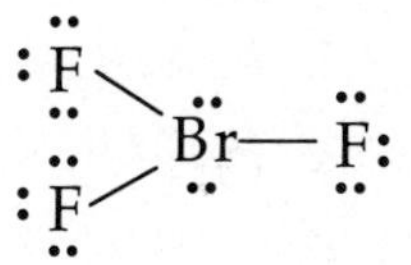   BrF$_3$

(b) **PF$_5$?** The Lewis structure, above, shows that the central P atom is
surrounded by five bonding electron pairs.

(c) **BrF$_3$?** The Lewis structure for this molecule is also shown above. Like
the P atom in PF$_5$, the Br atom in BrF$_3$ is surrounded by five electron pairs,
three bonding pairs and two lone pairs.

(d) **TeCl$_4$?** The Lewis structure for this molecule is shown below.

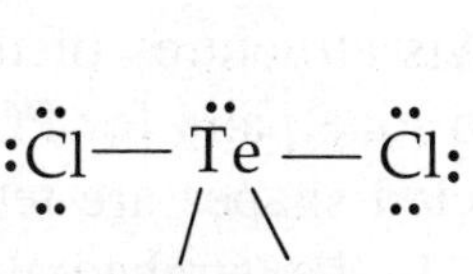

TeCl$_4$ 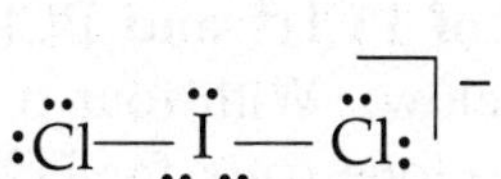   ICl$_2^-$

(e) **ICl$_2^-$?** The Lewis structure is shown above. The central I atom has five
electron pairs.

3.5   **What shape would you expect for: (a) SO$_3$?** The Lewis structure of
sulfur trioxide is shown below. With three σ bonds and no lone pairs, you

should expect a trigonal-planar geometry (like BF$_3$). The shape of SO$_3$ is also shown below.

(b) SO$_3{}^{2-}$? The Lewis structure of sulfite ion is shown below. With three σ bonds and one lone pair, you should expect a trigonal-pyramidal geometry (like NH$_3$). The shape of SO$_3{}^{2-}$ is also shown below.

(c) IF$_5$? The Lewis structure of iodine pentafluoride is shown below. With five σ bonds and one lone pair, you should expect a square-pyramidal geometry. The shape of IF$_5$ is also shown below.

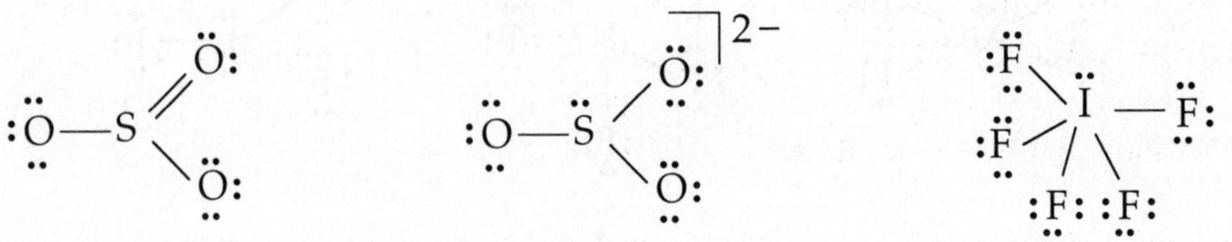

SO$_3$ SO$_3{}^{2-}$ IF$_5$

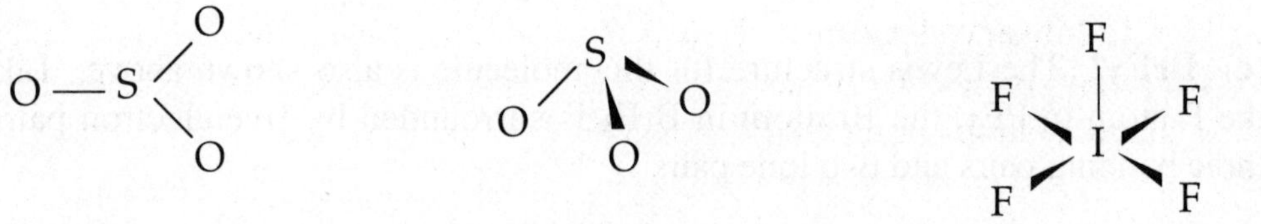

3.6   **The shapes of PCl$_4{}^+$ and PCl$_6{}^-$?** The Lewis structures of these two ions are shown below. With four σ bonds and no lone pairs for PCl$_4{}^+$ and six σ bonds and no lone pairs for PCl$_6{}^-$, the expected shapes are tetrahedral (like CCl$_4$) and octahedral (like SF$_6$), respectively. In the tetrahedral PCl$_4{}^+$ ion, all P–Cl bonds are the same length and all Cl–P–Cl bond angles are 109.5°. In the octahedral PCl$_6{}^-$ ion, all P–Cl bonds are the same length and all Cl–P–Cl bond angles are either 90° or 180°. The P–Cl bond distances in the two ions would not necessarily be the same length.

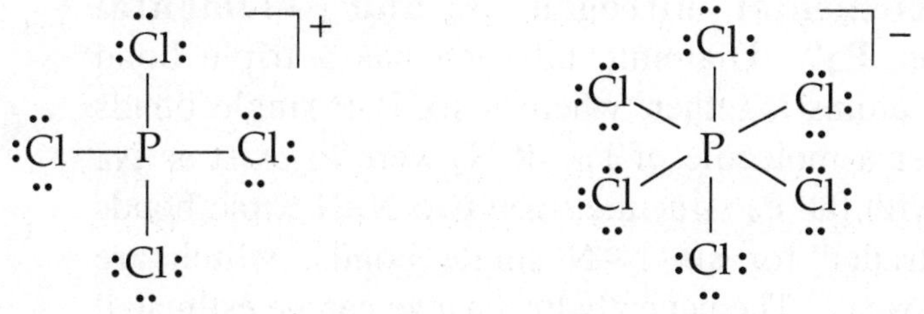

$$:\overset{\displaystyle ..}{\underset{\displaystyle ..}{Cl}}: \quad \Big]^+$$

$$:\overset{..}{\underset{..}{Cl}} - \overset{|}{\underset{|}{P}} - \overset{..}{\underset{..}{Cl}}:$$

$$:\overset{..}{\underset{..}{Cl}}:$$

$$PCl_4^+$$

$$:\overset{..}{\underset{..}{Cl}}: \quad \Big]^-$$

$$:\overset{..}{\underset{..}{Cl}} \diagdown \overset{|}{P} \diagup \overset{..}{\underset{..}{Cl}}:$$

$$:\overset{..}{\underset{..}{Cl}} \diagup \quad \diagdown \overset{..}{\underset{..}{Cl}}:$$

$$:\overset{..}{\underset{..}{Cl}}:$$

$$PCl_6^-$$

3.7 **Calculate bond lengths: (a) CCl_4 (observed value = 1.77 Å)?** From the covalent radii values given in Table 3.4, 0.77 Å for C and 0.99 Å for Cl, the C–Cl bond length in CCl_4 is predicted to be 0.77 Å + 0.99 Å = 1.76 Å. The agreement with the experimentally observed value is excellent.

(b) $SiCl_4$ (observed value = 2.01 Å)? The covalent radius for Si is 1.18 Å. Therefore, the Si–Cl bond length in $SiCl_4$ is predicted to be 1.18 Å + 0.99 Å = 2.17 Å. This is 8% longer than the observed bond length, so the agreement is not as good in this case.

(c) $GeCl_4$ (observed value = 2.10 Å)? The covalent radius for Ge is 1.22 Å. Therefore, the Ge–Cl bond length in $GeCl_4$ is predicted to be 2.21 Å. This is 5% longer than the observed bond length.

3.8 **Si=O or Si–O in silicon–oxygen compounds?** This question is similar to the one addressed in Example 3.4. You need to consider the enthalpy difference between one mole of Si=O double bonds and two moles of Si–O single bonds. The difference is:

$$2(\text{Si–O}) - (\text{Si=O}) = 2(466 \text{ kJ}) - (640 \text{ kJ}) = 292 \text{ kJ}$$

Therefore, the two single bonds will always be better enthalpically than one double bond. If silicon atoms only have single bonds to oxygen atoms in silicon–oxygen compounds, the structure around each silicon atom will be tetrahedral: each silicon will have four single bonds to four different oxygen atoms.

3.9   **Why is elemental nitrogen N_2 and elemental phosphorus P_4?**   Diatomic nitrogen has a triple bond holding the atoms together, whereas six P–P single bonds hold together a molecule of P_4. If N_2 were to exist as N_4 molecules with the P_4 structure, then two N≡N triple bonds would be traded for six N–N single bonds, which are intrinsically weak. The net enthalpy change can be estimated from the data in Table 3.5 to be 2(945 kJ) – 6(163 kJ) = 912 kJ, which indicates that the tetramerization of nitrogen is *very* unfavorable. On the other hand, multiple bonds between period 3 and larger atoms are not as strong as two times the analogous single bond, so P_2 molecules, each with a P≡P triple bond, would not be as stable as P_4 molecules, containing only P–P single bonds. In this case, the net enthalpy change for $2P_2 \rightarrow P_4$ can be estimated to be 2(481 kJ) – 6(201 kJ) = –244 kJ.

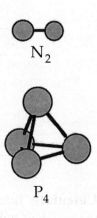

3.10   **Calculate ΔH from mean bond enthalpies?**   For the reaction:

$$2\,H_2(g) + O_2(g) \rightarrow 2\,H_2O(g)$$

Since you must break two moles of H–H bonds and one mole of O=O bonds on the left-hand side of the equation and form four moles of O–H bonds on the right-hand side, the enthalpy change for the reaction can be estimated as:

$$\Delta H \approx 2(436\text{ kJ}) + 497\text{ kJ} - 4(463\text{ kJ}) = -483\text{ kJ}$$

The experimental value is –484 kJ, which is in closer agreement with the estimated value than ordinarily expected. Since Table 3.5 contains average bond enthalpies, there is frequently a small error when comparing estimates to a specific reaction.

3.11   **Predict standard enthalpies?**   Consider the first reaction:

$$S_2^{2-}(g) + 1/4\,S_8(g) \rightarrow S_4^{2-}(g)$$

Hypothetically, two S–S single bonds (of S_8) are broken to produce two S atoms, which combine with S_2^{2-} to form two new S–S single bonds in the product S_4^{2-}. Since two S–S single bonds are broken and two are made, the net enthalpy change is zero. Now consider the second reaction:

$$O_2^{2-}(g) + O_2(g) \rightarrow O_4^{2-}(g)$$

Here there is a difference. A mole of O=O double bond of O_2 is broken, and two moles of O–O single bonds are made. The overall enthalpy change, based on the mean bond enthalpies in Table 3.5, is:

$$O=O - 2(O–O) = 497 \text{ kJ} - 2(146 \text{ kJ}) = 205 \text{ kJ}$$

The large positive value indicates that this is not a favorable process.

3.12   How many unpaired electrons? (a) O_2^-? You must write the electron configurations for each species, using Figure 3.14, and then apply the Pauli exclusion principle to determine the situation for incompletely filled degenerate orbitals. In this case the electron configuration is $1\sigma_g^2 2\sigma_u^2 3\sigma_g^2 1\pi_u^4 2\pi_g^3$. With three electrons in the pair of $2\pi_g$ molecular orbitals, one electron must be unpaired. Thus, the superoxide anion has a single unpaired electron.

(b) O_2^+? The configuration is $1\sigma_g^2 2\sigma_u^2 3\sigma_g^2 1\pi_u^4 2\pi_g^1$, so the oxygenyl cation also has a single unpaired electron.

(c) BN? You can assume that the energy of the $3\sigma_g$ molecular orbital is *higher* than the energy of the $1\pi_u$ orbitals, since that is the case for CO (see Figure 3.22). Therefore, the configuration is $1\sigma_g^2 2\sigma_u^2 1\pi_u^4$, and, as observed, this diatomic molecule has no unpaired electrons. If the configuration were $1\sigma_g^2 2\sigma_u^2 3\sigma_g^2 1\pi_u^2$, the molecule would have two unpaired electrons since each of the $1\pi_u$ orbitals would contain an unpaired electron, in accordance with the Pauli exclusion principle.

(d) NO^-? The exact ordering of the $3\sigma_g$ and $1\pi_u$ energy levels is not clear in this case, but it is not relevant either as far as the number of unpaired electrons is concerned. The configuration is either $1\sigma_g^2 2\sigma_u^2 1\pi_u^4 3\sigma_g^2 2\pi_g^2$ or it is $1\sigma_g^2 2\sigma_u^2 3\sigma_g^2 1\pi_u^4 2\pi_g^2$. In either case, this anion has two unpaired electrons, and these electrons occupy the set of antibonding $2\pi_g$ molecular orbitals.

3.13   Writing electron configurations? (a) Be_2? Only four valence electrons for two Be atoms gives the electron configuration $1\sigma_g^2 2\sigma_u^2$.

(b) B_2? The electron configuration is $1\sigma_g^2 2\sigma_u^2 1\pi_u^2$.

(c) C_2^-? The electron configuration is $1\sigma_g^2 2\sigma_u^2 1\pi_u^4 3\sigma_g^1$.

(d) F_2^+? The electron configuration is $1\sigma_g^2 2\sigma_u^2 3\sigma_g^2 1\pi_u^4 2\pi_g^3$.

3.14 **Determining bond orders?** The Lewis structures for the three species are shown below:

$$:\overset{..}{S}\!=\!\overset{..}{S}: \qquad :\overset{..}{\underset{..}{Cl}}\!-\!\overset{..}{\underset{..}{Cl}}: \qquad :\overset{..}{\underset{..}{N}}\!=\!\overset{..}{\underset{..}{O}}:\,\Big]^-$$

(a) S_2? The electron configuration of this diatomic molecule is $1\sigma_g^2 2\sigma_u^2 3\sigma_g^2 1\pi_u^4 2\pi_g^2$. The bonding molecular orbitals are $1\sigma_g$, $1\pi_u$, and $3\sigma_g$, while the antibonding molecular orbitals are $2\sigma_u$, and $2\pi_g$. Therefore, the bond order is $(1/2)((2 + 4 + 2) - (2 + 2)) = 2$, which is consistent with the double bond between the S atoms suggested by the Lewis structure.

(b) Cl_2? The electron configuration is $1\sigma_g^2 2\sigma_u^2 3\sigma_g^2 1\pi_u^4 2\pi_g^4$. The bonding and antibonding orbitals are the same as for S_2, above. Therefore, the bond order is $(1/2)((2 + 4 + 2) - (2 + 4)) = 1$, which is in harmony with the single bond between the Cl atoms indicated by the Lewis structure.

(c) NO^-? The electron configuration of NO^-, $1\sigma_g^2 2\sigma_u^2 1\pi_u^4 3\sigma_g^2 2\pi_g^2$, is the same as the configuration for S_2, shown above. Thus, the bond order for NO^- is 2, as for S_2, once again in harmony with the conclusion based on the Lewis structure.

3.15 **Changes in bond order and bond distance?** **(a) $O_2 \rightarrow O_2^+ + e^-$?** The molecular orbital electron configuration of O_2 is $1\sigma_g^2 2\sigma_u^2 3\sigma_g^2 1\pi_u^4 2\pi_g^2$. The two $2\pi_g$ orbitals are π-antibonding orbitals, so when one of the $2\pi_g$ electrons is removed, the oxygen–oxygen bond order increases from 2 to 2.5. Since the bond in O_2^+ becomes stronger, it should become shorter as well.

(b) $N_2 + e^- \rightarrow N_2^-$? The molecular orbital electron configuration of N_2 is $1\sigma_g^2 2\sigma_u^2 1\pi_u^4 3\sigma_g^2 2\pi_g^4$. The next electron must go into the $4\sigma_u$ orbital, which is σ-antibonding (refer to Figures 3.13 and 3.14). This will decrease the nitrogen–nitrogen bond order from 3 to 2.5. Therefore, N_2^- has a weaker and longer bond than N_2.

(c) NO → NO⁺ + e⁻? The configuration of the NO molecule is either $1\sigma_g^2 2\sigma_u^2 1\pi_u^4 3\sigma_g^2 2\pi_g^4 4\sigma_u^1$ or $1\sigma_g^2 2\sigma_u^2 3\sigma_g^2 1\pi_u^4 2\pi_g^4 4\sigma_u^1$. Removal of the $4\sigma_u$ antibonding electron will increase the bond order from 2.5 to 3. Therefore, NO⁺ has a stronger and shorter bond than NO. Notice that NO⁺ and N_2 are isoelectronic.

3.16 **Linear H_4 MOs?** Four atomic orbitals can yield four independent linear combinations. The four relevant ones in this case, for a hypothetical linear H_4 molecule, are shown at the right in order of increasing energy. The most stable orbital has the fewest nodes (i.e. the electrons in this orbital are not excluded from the internuclear regions), the next orbital in energy has only one node, and so on to the fourth and highest energy orbital, with three nodes (a node between each of the four H atoms).

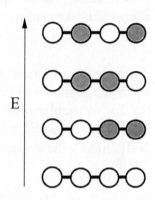

3.17 **Molecular orbitals of linear [HHeH]²⁺?** By analogy to linear H_3, the three atoms of [HHeH]²⁺ will form a set of three molecular orbitals, one bonding, one nonbonding, and one antibonding. They are shown below. The

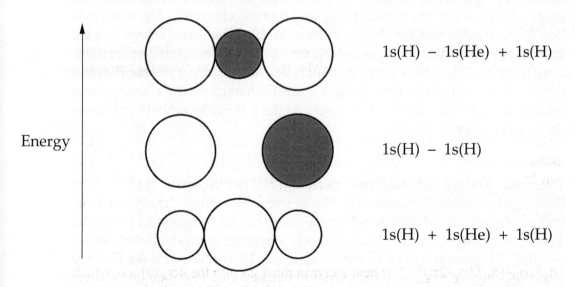

1s(H) − 1s(He) + 1s(H)

1s(H) − 1s(H)

1s(H) + 1s(He) + 1s(H)

forms of the wavefunctions are also shown, without normalizing coefficients. You should conclude that He is more electronegative than H because the ionization energy of He is nearly twice that of H. Therefore, the bonding MO has a larger coefficient (larger sphere) for He than for H, and the antibonding

MO has a larger coefficient for H than for He. The bonding MO is shown at lowest energy, since it has no nodes. The nonbonding and antibonding orbitals follow at higher energies, since they have one and two nodes, respectively. Since $[HHeH]^{2+}$ has four electrons, only the bonding and nonbonding orbitals are filled. However, the species is probably not stable in isolation because of +/+ repulsions. In solution it would be unstable with respect to proton transfer to another chemical species that can act as a base, such as the solvent or counterion. *Any* substance is more basic than helium.

3.18   **Average bond order in NH₃?**   The molecular orbital energy diagram for ammonia is shown in Figure 3.33. The interpretation given in the text was that the $2a_1$ molecular orbital is almost nonbonding, so the electron configuration $1a_1^2 1e^4 2a_1^2$ results in only three bonds $((2 + 4)/2 = 3)$. Since there are three N–H "links", the average N–H bond order is 1 $(3/3 = 1)$.

3.19   **Describe the character of the HOMOs and LUMOs of SF₆?**   The molecular orbital energy diagram for sulfur hexafluoride is shown in Figure 3.34. The nonbonding *e* HOMOs are pure F atom symmetry-adapted orbitals, and they do not have any S atom character whatsoever. They could only have S atom character if they were bonding or antibonding orbitals composed of atomic orbitals of both types of atoms in the molecule. On the other hand, the antibonding *t* orbitals have both sulfur and fluorine character. Since sulfur is less electronegative than fluorine, its valence orbitals lie at higher energy than the valence orbitals of fluorine (from which the *t* symmetry-adapted combinations were formed). Thus, the *t* bonding orbitals lie closer in energy to the F atom *t* combinations and hence they contain more F character; the *t* antibonding orbitals, the LUMOs, lie closer in energy to the S atom $3p$ orbitals and hence they contain more S character.

3.20   **Electron precise or electron deficient?   (a) Square H₄²⁺?**   The drawing below shows a square array of four hydrogen atoms. Clearly, each line connecting any two of the atoms is not a (2c,2e) bond, because this molecular ion has only two electrons. Instead, this is a hypothetical example of (4c,2e) bonding. We cannot write a Lewis structure for this species. It is not likely to exist; it should be unstable with respect to two separate H_2^+ diatomic species with (2c,1e) bonds.

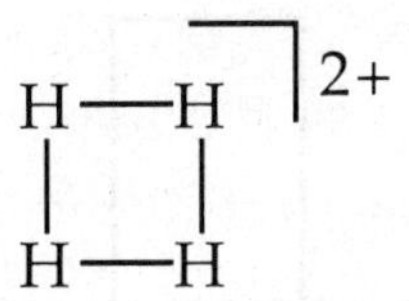

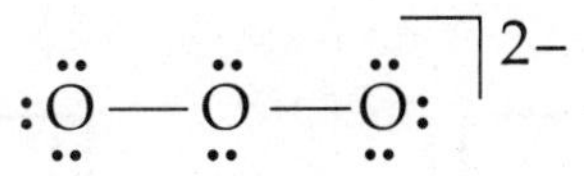

(b) Bent O_3^{2-}? A proper Lewis structure for this 20-electron ion is shown above. Therefore, it is electron precise. It could very well exist.

3.21 **Find isolobal fragments?** **(a) CH_3^-?** The fragment CH_3^- has a pyramidal structure with a single lone pair of electrons in the carbon atom. Ammonia, NH_3, has exactly the same molecular and electronic structure (see below), and hence is isolobal with CH_3^-.

(b) O atom? An O atom has six valence electrons, and in the ground state four of them are paired and two of them are unpaired in separate p orbitals. An excited state of O has three pairs of electrons in three orbitals along with an empty orbital. The fragment BH_3 is isolobal with this state of an O atom.

(c) $[Mn(CO)_5]^-$? The fragment $[Mn(CO)_5]^-$ has many more valence electrons than any possible NH_n fragment. Nevertheless, this metal carbonyl fragment has a single lone pair of electrons on the Mn atom, and hence is isolobal with NH_3 (see below).

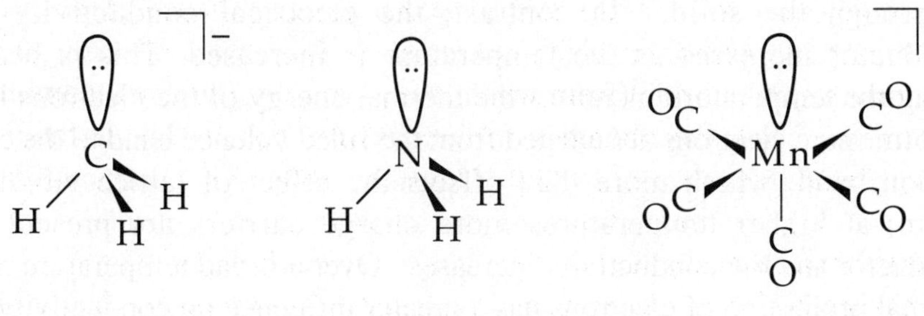

3.22 **Distinguish between a metal and a semiconductor?** **(a) Simple band picture?** The band pictures for a metal and for two types of semiconductors are shown below:

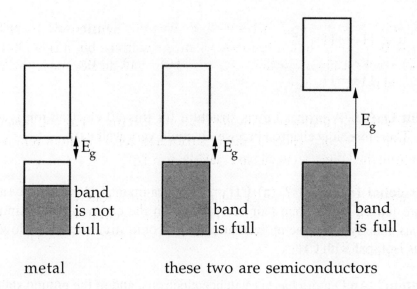

metal   these two are semiconductors

The band gap, E_g, for semiconductors is an important property. For the one on the left, E_g is about the same size as RT_{room}, and its conductivity will be much larger than for the one on the right, for which $E_g \gg RT_{room}$.

(b) Temperature dependence? The electrical conductivity of a metal decreases as the temperature is increased. This is because increasing the temperature increases lattice vibrations, which reduce the freedom of electrons to move through the solid. In contrast, the electrical conductivity of a semiconductor increases as the temperature is increased. This is because increasing the temperature increases the thermal energy of the electrons in the solid. Thus, more electrons are excited from the filled valence band to the empty conduction band, which more than offsets the effect of lattice vibrations. Therefore, at higher temperatures more charge carriers are present in a semiconductor and the conductivity increases. Over a broad temperature range, this thermal promotion of electrons has a greater influence on conductivity than the vibrations which decrease conductivity.

(c) Insulator vs. semiconductor? The conductivity of an insulator is very low but it does increase as the temperature is increased. Therefore, it is not possible to distinguish between a semiconductor and an insulator by the temperature dependence of their conductivities. In fact, the two types of materials are both thought of as semiconductors. They correspond to the two different situations shown in the band pictures above (the insulator has $E_g \gg RT_{room}$).

3.23 **Which are n-type and which are p-type semiconductors? (a) As-doped Ge?** Germanium has four valence electrons, but arsenic has five. Therefore, each As atom substituted for a Ge atom adds an electron to the lattice, so the doped material is n-type.

(b) Ga-doped Ge? Gallium, with three valence electrons, has one fewer than germanium. Therefore, each Ga atom substituted for a Ge atom removes an electron from the lattice, so the doped material is p-type.

(c) Si-doped Ge? Silicon and germanium both have four valence electrons, so substitution of Si for Ge does not change the total electron count of the lattice. Therefore, the doped material is neither n-type nor p-type. The dopant should have little effect on Ge.

3.24 **Calculate E_g for TiO_2?** The band gap is equal to a change in energy, ΔE, that is equivalent to a photon of light with a wavelength of 350 nm or less. Therefore:

$$E_g = \Delta E = h\nu = hc/\lambda = (6.626 \times 10^{-34} \text{ J s})(3.00 \times 10^8 \text{ m s}^{-1})/(350 \times 10^{-9} \text{ m})$$

$$E_g = 5.68 \times 10^{-19} \text{ J} = 3.55 \text{ eV} \qquad (1 \text{ eV} = 1.60 \times 10^{-19} \text{ J})$$

3.25 **Doping of TiO_2 with Ti(III)?** Titanium(IV) has no valence electrons, but Ti(III) has one. Therefore, each Ti^{3+} ion substituted for a Ti^{4+} ion adds an electron to the lattice, so the doped material is n-type. You can see, in general, that doping a material with a lower oxidation state results in n-doping while doping with a higher oxidation state results in p-doping.

3.26 **Is GaAs-doped with Se n-doped or p-doped?** Selenium is electronegative, like arsenic, and so will substitute for arsenic in the compound GaAs. Since Se, with six valence electrons, has one more than As, which only has five, each Se atom substituted for a As atom adds an electron to the lattice, resulting in an n-type doped material.

3.27 **Onset of absorption for photoconduction in CdS?** As in Exercise 3.24, the band gap, in this case 2.4 eV, is equal to a change in energy that is equivalent to a photon of light with a wavelength of l or less. Therefore:

$$E_g = \Delta E = h\nu = hc/\lambda$$

$$\lambda = hc/E_g = (6.626 \times 10^{-34} \text{ J s})(3.00 \times 10^8 \text{ m s}^{-1})/(2.4 \text{ eV})(1.60 \times 10^{-19} \text{ J eV}^{-1})$$

$$\lambda = 5.18 \times 10^{-7} \text{ m} = 518 \text{ nm}$$

Thus, light that has a wavelength of 518 nm *or less* will excite an electron from the valence band to the conduction band. Light that has a wavelength of 518 nm has just enough energy to bridge the band gap, and light that has a shorter wavelength has more than enough energy.

3.28   **Calculate $\sigma(373 \text{ K})/\sigma(273 \text{ K})$ for Si?** From Section 3.15, *Semiconduction*, you learned that the conductivity of a semiconductor shows an Arrhenius-like temperature dependence:

$$\sigma = \sigma_0 e^{-E_a/kT} \qquad \text{where } E_a \approx (1/2)E_g$$

Therefore:

$$\sigma(373 \text{ K})/\sigma(273 \text{ K}) = e^{-(E_a/k)((1/373) - (1/273))} =$$

$$\exp(-((1.12 \text{ eV})/2)(1.60 \times 10^{-19} \text{ J eV}^{-1})(-9.82 \times 10^{-4} \text{ K}^{-1})/(1.38 \times 10^{-23} \text{ J K}^{-1})))$$

$$\sigma(373 \text{ K})/\sigma(273 \text{ K}) = e^{6.38} = 590$$

So, the conductivity of silicon increases by a factor of 590 between 0°C and 100°C. The band pictures below represent this behavior in general for an intrinsic semiconductor (the separations between shaded and light areas should be smooth curves, not straight lines). At temperature T_1, a few charge carriers have been promoted from the filled valence band into the empty conduction band, but at the higher temperature T_2, more charge carriers have been promoted and the conductivity is higher.

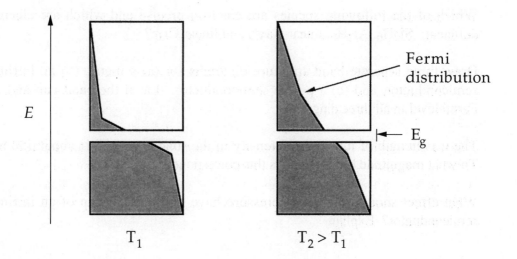

Guide to Solutions Quiz

1   Draw Lewis dot structures for PF_2^-, PF_2^+, PF_4^-, and PF_4^+.

2   (a) Draw resonance structures for the thiosulfate ion, $S_2O_3^{2-}$, which has a central sulfur atom. (b) Decide which resonance structures are more important and which are less important.

3   (a) Draw a Lewis dot structure for ClNO (central N atom). (b) What are the oxidation numbers and formal charges on the three atoms?

4   Compare and contrast the enthalpy changes for the reaction of ethylene, C_2H_4, with each of the halogens F_2, Cl_2, Br_2, and I_2. Explain any trends you that you observe.

5   Use a molecular orbital diagram to decide how many unpaired electrons each of the following species contains and which one has the strongest bond: CN, CN^+, CN^-.

6   (a) Use a molecular orbital diagram to find the average sulfur–fluorine bond order in SF_6. (b) What happens to the bond order when SF_6 is oxidized or reduced to produce SF_6^+ or SF_6^-, respectively?

7   Which of the following species are electron precise and which are electron deficient: Si_2H_6, Al_2H_6, linear Na_3^+, and linear Cl_3^-?

8   Draw three separate band structure diagrams for (a) a metal, (b) an intrinsic semiconductor, and (c) a p-type semiconductor. Label the band gap and the Fermi level in all three diagrams.

9   The wavelength of maximum intensity in the solar spectrum is about 550 nm. To what magnitude band gap does this correspond?

10   What effect should increasing pressure have on the band gap of an intrinsic semiconductor? Explain.

4   Molecular symmetry

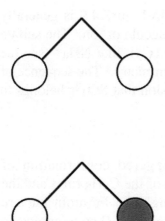

The H atom $1s$ orbitals in H_2O form the two linear combinations shown. The unshaded lobes are + throughout and the shaded lobe is − throughout. These symmetry-adapted orbitals are labelled a_1 (top) and b_2 (bottom). The molecular orbitals of H_2O are formed by making suitable linear combinations of these symmetry-adapted orbitals with O atomic orbitals of a_1 and b_2 symmetry.

S4.1   **Identify the C_3 axes of an NH_4^+ ion?**   Each of the N–H bond vectors corresponds to a C_3 axis for this ion, so there are four such axes. Since for a given C_3 axis the other three H atoms are related to each other by rotations of 120° or 240°, all four H atoms in the ammonium ion are symmetry related.

S4.2   **(a) BF_3 point group?**   All molecules possess the identity, E. Since BF_3 is trigonal planar, it is obvious that it possesses a 3-fold rotation axis (C_3) and a mirror plane of symmetry that coincides with the molecular plane (σ_h, since it is perpendicular to the C_3 axis, which is the "major" axis). There are also three 2-fold axes (C_2) that coincide with the three B–F bond vectors, and three more mirror planes, each of which contains a B–F bond and is perpendicular to σ_h (these are called σ_v since they are parallel to the major axis). The set of elements (E, C_3, $3C_2$, σ_h, $3\sigma_v$) corresponds to the group D_{3h}. Refer to Table 4.2 and note that the D_{3h} point group also contains an S_3 improper rotation axis. However, in many cases it is not necessary to find the complete set of symmetry elements to uniquely determine the point group of a molecule or ion. In fact, if you use the decision tree shown in Figure 4.9 to determine the point group of BF_3, you will find that the smaller set of elements (C_3, $3C_2$, σ_h) uniquely corresponds to D_{3h}.

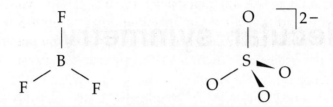

(b) SO_4^{2-} point group? Using the decision tree in Figure 4.9 is generally the easiest way to determine the point group of a molecule or ion. The sulfate ion (i) is non-linear, (ii) possesses four 3-fold axes (C_3), like NH_4^+ (see the answer to S4.1), and (iii) does not have a center of symmetry. The sequence of "no, yes, no" on the decision tree leads to the conclusion that SO_4^{2-} belongs to the T_d point group.

S4.3   **Is ferrocene polar?** Like ruthenocene, the staggered conformation of ferrocene has a C_5 axis passing through the centroids of the C_5H_5 rings and the Fe atom. It also has five C_2 axes that pass through the Fe atom but are perpendicular to the major C_5 axis, so it belongs to one of the D point groups, in this case D_{5d} (it lacks the σ_h plane of symmetry that a D_{5h} molecule like ruthenocene possesses). Since the D_{5d} conformation of ferrocene has a C_5 axis *and* perpendicular C_2 axes, it is not polar (see Section 4.3). You may find it difficult to find the n C_2 axes for a D_{nd} structure. However, if you draw the mirror planes, the C_2 axes lie between them. In this case, one C_2 axis interchanges the front vertex of the top ring with one of the two front vertices of the bottom ring, while a second C_2 axis, rotated exactly 36° from the first one, interchanges the same vertex on top with the other front bottom one.

S4.4   **Is the skew form of H_2O_2 chiral?** Except for the identity, E, the only element of symmetry that this conformation of hydrogen peroxide possesses is a C_2 axis that passes through the midpoint of the O–O bond and bisects the two O–O–H planes (these are *not* mirror planes of symmetry). Hence this form of H_2O_2 belongs to the C_2 point group, and it is chiral since this group does not contain any S_n axes. In general, any structure that belongs to a C_n or D_n point group is chiral, as are molecules that are asymmetric (C_1 symmetry).

S4.5 **What is the maximum possible degeneracy for an O_h molecule?** This question, although asked about SF_6, could have been asked about any molecule with rigorous octahedral symmetry, i.e. a molecule that belongs to the O_h point group. If you refer to the character table for this group, which is given in Appendix 3, you find that there are characters of 1, 2, *and* 3 in the column headed by the identity element, E. Therefore, the *maximum* possible degree of degeneracy of the orbitals in SF_6 is 3 (although non-degenerate and two-fold degenerate orbitals are allowed). As an example, the sulfur atom valence p orbitals, and any molecular orbitals formed using them, are triply degenerate in SF_6.

S4.6 **Orbital symmetry for a square-planar array of H atoms?** You must adopt some conventions to answer this one. First, you assume that the combination of H atom $1s$ orbitals given looks like the figure shown below. This array of H atoms has D_{4h} symmetry. Inspection of the character table for this group, which is given in Appendix 3, reveals that there are three different types of C_2 rotation axes, i.e. there are three columns labelled C_2, C_2', and C_2''. The first of these is the C_2 axis that is coincident with the C_4 axis; the second type, C_2', represents two axes in the H_4 plane that do not pass through any H atoms; the third type, C_2'', represents two axes in the H_4 plane that pass through pairs of opposite H atoms. Now, instead 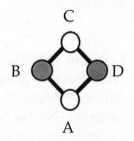 of applying operations from all ten columns to this array, to see if it changes into itself (i.e. the +/– signs of the lobes stay the same) or if it changes sign, you can make use of a shortcut. Notice that the array changes into itself under the inversion operation through the center of symmetry. Thus the character for this operation, i, is 1. This means that the symmetry label for this array is one of the first four in the character table, A_{1g}, A_{2g}, B_{1g}, or B_{2g}. Notice also that for these four, the symmetry type is uniquely determined by the characters for the first five columns of operations, which are:

E	C_4	C_2	C_2'	C_2''
1	−1	1	−1	1

These match the characters of the B_{2g} symmetry label.

S4.7   **Orbital symmetry for a tetrahedral array of H atoms?** The molecule CH_4 has T_d symmetry, and the combination of H atom $1s$ orbitals given also has T_d symmetry. The group theory jargon that is used in this case is to say that the combination has the *total* symmetry of the molecule. This is true because each time the H atom array of orbitals is subjected to an operation in the T_d point group, the array changes into itself. In each case, the character is 1. Each point group has one symmetry label (one row) for which all the characters are one, and for the T_d point group it is called A_1 (see Appendix 3). Thus, the symmetry label of the given combination of H atom $1s$ orbitals is A_1.

S4.8   **Can the bending mode of N_2O be Raman active?** The Lewis structure of nitrous oxide is shown at the right. Based on this structure, you can predict that its geometry should be linear. However, unlike CO_2, with which it is isoelectronic, N_2O does not have a center of symmetry. Therefore, the exclusion rule does not apply, and a band that is IR active *can* be Raman active as well.

$$\overset{..}{N}\!=\!N\!=\!\overset{..}{\underset{..}{O}}$$

S4.9   **Confirm that the symmetric mode is A_g?** All of the operations of this group leave the displacement vectors unchanged. Therefore, all of the operations have a character of +1. This corresponds to the first row in the D_{2h} character table, which is the A_g symmetry type.

S4.10   **Is the IR band for SF_6 A_{1g} or T_{1u}?** Consult the O_h character table in Appendix 3. The A_{1g} symmetry type has the same symmetry as the function $x^2 + y^2 + z^2$. Therefore, a vibration with this symmetry type would be Raman active, not IR active. On the other hand, the T_{1u} symmetry type has the same symmetry as the functions x, y, or z, and so a T_{1u} vibration would be IR active. Therefore, the single IR absorption band for SF_6 arises from the T_{1u} mode, not the A_{1g} mode.

4.1   **Draw sketches of C_n axes and σ planes? (a) NH_3?** In the drawings below, the circle represents the nitrogen atom of ammonia and the diamonds represent the hydrogen atoms. The mirror plane drawn is in the plane of the page, and it contains the nitrogen atom and the hydrogen atom on the left. The other two hydrogen atoms are out of this plane, one behind the page and one in front.

C_3

C_4

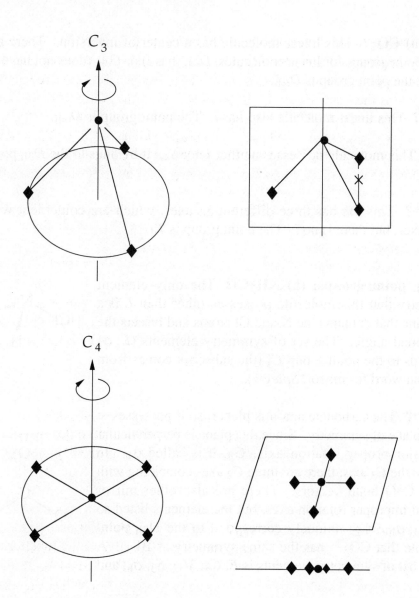

(b) The $PtCl_4^{2-}$ ion? In the drawings above, the circle represents the platinum atom of the tetrachloroplatinate anion and the diamonds represent the chlorine atoms. The mirror plane drawn is in the plane of the page, and it contains all five atoms. This plane is also drawn on its side, so that all five atoms seem to lie on a single line. In addition, there are four more mirror planes perpendicular to the first plane.

4.2 S_4 or i: (a) CO_2? This linear molecule has a center of inversion. There are only two point groups for linear molecules, $D_{\infty h}$ (has i) or $C_{\infty v}$ (does not have i). Therefore, the point group is $D_{\infty h}$.

(b) C_2H_2? This linear molecule also has i. The point group is $D_{\infty h}$.

(c) BF_3? This molecule possesses neither i nor S_4. It belongs to the D_{3h} point group.

(d) $SO_4{}^{2-}$? This ion has three different S_4 axes, which are coincident with three C_2 axes, but there is no i. The point group is T_d.

4.3   **Assigning point groups: (a) NH_2Cl?** The only element of symmetry that this molecule possesses other than E is a mirror plane that contains the N and Cl atoms and bisects the H–N–H bond angle. The set of symmetry elements (E, σ) corresponds to the point group C_s (the subscript comes from the German word for mirror, *Spiegel*).

(b) $CO_3{}^{2-}$? The carbonate anion is planar, so it possesses at least one plane of symmetry. Since this plane is perpendicular to the major proper rotation axis, C_3, it is called σ_h. In addition to the C_3 axis, there are three C_2 axes coinciding with the three C–O bond vectors. There are also other mirror planes and improper rotation axes, but the elements listed so far $(E, C_3, \sigma_h, 3C_2)$ uniquely correspond to the D_{3h} point group (note that $CO_3{}^{2-}$ has the same symmetry as BF_3). A complete list of symmetry elements is $E, C_3, 3C_2, S_3, \sigma_h,$ and $3\sigma_v$.

(c) SiF_4? This molecule has four C_3 axes, one coinciding with each of the four Si–F bonds. In addition, there are six mirror planes of symmetry (any pair of F atoms and the central Si atom define a mirror plane, and there are always six ways to choose a pair of objects out of a set of four). Furthermore, there is no center of symmetry. Thus, the set $(E, 4C_3, 6\sigma,$ no $i)$ describes this molecule and corresponds to the T_d point group. A complete list of symmetry elements is $E, 4C_3, 3C_2, 3S_4,$ and $6\sigma_d$.

(d) HCN? Hydrogen cyanide is linear, so it belongs to either the $D_{\infty h}$ or the $C_{\infty v}$ point group. Since it does not possess a center of symmetry, which is a requirement for the $D_{\infty h}$ point group, it belongs to the $C_{\infty v}$ point group.

(e) SiFClBrI? This molecule does not possess any element of symmetry other than the identity element, E. Thus, it is asymmetric and belongs to the C_1 point group, the simplest possible point group.

(f) BrF$_4$⁻? This anion is square planar. It has a C_4 axis and four perpendicular C_2 axes. It also has a σ_h mirror plane. These symmetry elements uniquely correspond to the D_{4h} point group. A complete list of symmetry elements is E, C_4, a parallel C_2, four perpendicular C_2, S_4, i, σ_h, $2\sigma_v$, and $2\sigma_d$.

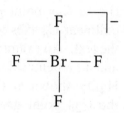

4.4   **The symmetry elements of orbitals? (a) An s orbital?** An s orbital, which has the shape of a sphere, possesses an infinite number of C_n axes where n can be any number from 1 to ∞, plus an infinite number of mirror planes of symmetry. It also has a center of inversion, i. A sphere has the highest possible symmetry.

(b) A p orbital? The + and – lobes of a p orbital are not equivalent and therefore cannot be interchanged by potential elements of symmetry. Thus, a p orbital does not possess a mirror plane of symmetry perpendicular to the long axis of the orbital. It does, however, possess an infinite number of mirror planes that pass through both lobes and include the long axis of the orbital. In addition, the long axis is a C_n axis, where n can be any number from 1 to ∞ (in group theory this is referred to as a C_∞ axis).

(c) A d_{xy} orbital? The two pairs of + and – lobes of a d_{xy} orbital are interchanged by the center of symmetry that this orbital possesses. It also possesses three mutually perpendicular C_2 axes, each one coincident with one of the three Cartesian coordinate axes. Furthermore, it possesses three mutually perpendicular mirror planes of symmetry, which are coincident with the the xy plane and the two planes that are rotated by 45° about the z axis from the xz plane and the yz plane.

(d) A d_{z^2} orbital? Unlike a p_z orbital, a d_{z^2} orbital has two large + lobes along its long axis, and a – torus (or doughnut) around the middle. In addition to the symmetry elements possessed by a p orbital (see above), the infinite number of mirror planes that pass through both lobes and include the long axis of the orbital as well as the C_∞ axis, a d_{z^2} orbital also possesses (i) a center of symmetry, (ii) a mirror plane that is perpendicular to the C_∞ axis, (iii) an infinite number of C_2 axes that pass through the center of the orbital and are perpendicular to the C_∞ axis, and (iv) an S_∞ axis.

4.5 **Which species are polar?** **(a) The criteria?** The sets of symmetry elements that independently require that a molecule is *nonpolar* are (i) a C_n axis and a perpendicular C_2 axis (i.e. a D point group), (ii) a C_n axis and a σ_h plane (i.e., a C_{nh} point group), (iii) a center of inversion, i, or (iv) multiple non-collinear C_n axes with n > 2. The first two sets, (i) and (ii), were discussed in the text. An example of a species that has a center of symmetry but not a C_n axis and a perpendicular C_2 axis is the planar conformation of H_2O_2 shown at the right. This molecule belongs to the C_{2h} point group and is nonpolar. The fourth set (iv) includes the tetrahedron, the octahedron, and the icosahedron. For these symmetries, there is no distinction between the x, y, and z directions (i.e. x, y, and z are triply degenerate).

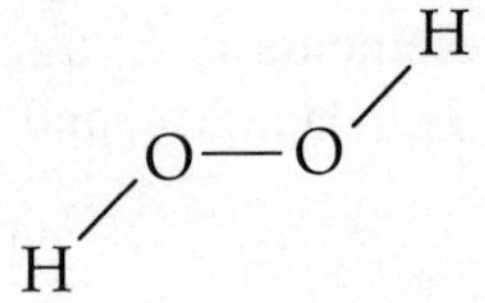

(b) Examples: NH$_2$Cl? This molecule does not meet any of the four criteria stated above, and so it is polar.

CO$_3^{2-}$? The point group is D_{3h}, so on the basis of criterion (i) it is nonpolar.

SiF$_4$? The point group is T_d, so on the basis of criterion (iv) it is nonpolar.

HCN? This molecule does not meet any of the four criteria stated above, and so it is polar.

SiFClBrI? This molecule too does not meet any of the four criteria stated above, and so it is polar.

BrF$_4^-$? This ion belongs to the D_{4h} point group, so on the basis of criterion (i) it is nonpolar.

4.6 **Which species are chiral?** **(a) The criteria.** The symmetry criterion for chirality is the absence of an S_n element of symmetry. Recall that $S_1 = \sigma$ and $S_2 = i$.

(b) Examples: NH$_2$Cl? This molecule does possess a mirror plane of symmetry, so it is not chiral.

CO_3^{2-}? The carbonate anion possesses four different planes of symmetry, so it is not chiral.

SiF_4? This molecule, belonging to the T_d point group, possesses six different planes of symmetry, so it is not chiral. You should note that a tetrahedron also possesses three S_4 improper rotation axes that coincide with its C_2 axes.

HCN? Since this molecule possesses an infinite number of mirror planes of symmetry, it is not chiral.

SiFClBrI? This molecule does meet the criterion stated above for chirality since it does not possess any S_n element of symmetry (since it is C_1, or asymmetric, it does not possess *any* element of symmetry other than E).

BrF_4^-? This D_{4h} ion possesses *many* S_n elements of symmetry, including σ_h, i, $2\sigma_v$, $2\sigma_d$, and S_4. Therefore, it is not chiral.

4.7 **Point group and degenerate MOs of the SO_3^{2-} ion?** **(a) Point group?** Using the decision tree shown in Figure 4.9, you will find the point group of this anion to be C_{3v} (it is nonlinear, it only has one proper rotation axis, a C_3 axis, and it has three σ_v mirror planes of symmetry).

(b) Degenerate MOs? Inspection of the C_{3v} character table (Appendix 3) shows that the characters under the column headed by the identity element, E, are 1 and 2. Therefore, the maximum degeneracy possible for molecular orbitals of this anion is 2.

(c) Which p orbitals have the maximum degeneracy? According to the character table, the S atom $3s$ and $3p_z$ orbitals are each singly degenerate (and belong to the A_1 symmetry type), but the $3p_x$ and $3p_y$ orbitals are doubly degenerate (and belong to the E symmetry type). Thus, the $3p_x$ and $3p_y$ atomic orbitals on sulfur can contribute to molecular orbitals that are two-fold degenerate.

4.8 **Point group and degenerate MOs of PF$_5$: (a) Point group?** As above, you use the decision tree to assign the point group, in this case concluding that PF$_5$ has D_{3h} symmetry (it has a trigonal bipyramidal structure, by analogy with PCl$_5$ (see Table 3.7); it is nonlinear, has only one high-order proper rotation axis, a C_3 axis, it has three C_2 axes that are perpendicular to the C_3 axis, and it has a σ_h mirror plane of symmetry).

(b) Degenerate MOs? Inspection of the D_{3h} character table (Appendix 3) reveals that the characters under the E column are 1 and 2, so the maximum degeneracy possible for a molecule with this symmetry is 2.

(c) Which p orbitals have the maximum degeneracy? The P atom $3p_x$ and $3p_y$ atomic orbitals, which are doubly degenerate and are of the E' symmetry type (i.e. they *have* E' symmetry), can contribute to molecular orbitals that are two-fold degenerate. In fact, if they contribute to molecular orbitals at all, they *must* contribute to two-fold degenerate ones.

4.9   **Orbitals for a square-planar H_4 array:**   **(a) The symmetry-adapted orbitals?** Since our basis set of H atom orbitals consists of four H $1s$ orbitals (one per atom), you can construct exactly four symmetry-adapted combinations, which are shown below (the lines between the H atoms have no special meaning):

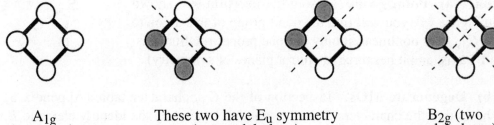

| A_{1g} | These two have E_u symmetry | B_{2g} (two |
| (no nodes) | (one nodal plane) | nodal planes) |

(b) Point group and symmetry labels? The MH_4L_2 complex has D_{4h} symmetry (a single C_4 axis, four C_2 axes perpendicular to it, and a σ_h mirror plane of symmetry). For a review of how to assign the symmetry label to a given orbital, see Self-test S4.7. The nodal planes in the drawings above are indicated by the dashed lines. Note that the symmetry labels are written using capital letters. However, if you refer to the orbital with A_{1g} symmetry, you call it an a_{1g} orbital (this is also called the totally symmetric orbital). Note also that the nodal planes for the e_u orbitals are perpendicular to each other: this is a necessary requirement for independent functions. The choice of x and y axes that led to the B_{2g} assignment for the fourth orbital is to have them coincide with the two nodes. If you choose x and y to bisect the nodes, then the orbital will have B_{1g} symmetry.

(c) **Which metal *d* orbitals have the the correct symmetry?** Inspection of the D_{4h} character table (Appendix 3) shows that d_{z2} has A_{1g} symmetry, so it can form molecular orbitals with the first of the H atom symmetry-adapted orbitals. There are no *d* orbitals with E_u symmetry, so no MOs can be formed using the H atom e_u orbitals and metal *d* orbitals. Finally, since d_{xy} has B_{2g} symmetry, it can form MOs with the fourth H atom symmetry-adapted orbital.

4.10 Vibrational modes of SO_3? (a) In the plane of the nuclei? The trigonal planar structure of SO_3 is shown in the answer to Exercise 3.5(a). If you consider the C_3 axis to be the z axis, then each of the four atoms has two independent displacements in the xy plane, namely along the x and along the y axis. The product (4 atoms)(2 displacement modes/atom) gives 8 displacement modes, not all of which are vibrations. There are two translation modes in the plane of the nuclei, one each along the x and y axes. There is also one rotational mode around the z axis. Therefore, if you subtract these 3 non-vibrational displacement modes from the total of 8 displacement modes, you arrive at a total of 5 vibrational modes in the plane of the nuclei.

(b) Perpendicular to the molecular plane? You can use your answer to part (a), above, to answer this question. Since there are four atoms in the molecule, there are $3(4) - 6 = 6$ vibrational modes. You discovered that there are 5 vibrational modes in the plane of the nuclei for SO_3, so there must be only 1 vibrational mode perpendicular to the molecular plane.

4.11 Vibrations that are IR and Raman active? (a) SF_6? Since sulfur hexafluoride has a center of symmetry, the mutual exclusion rule applies. Therefore, none of the vibrations of this molecule can be *both* IR and Raman active. A quick glance at the O_h character table in Appendix 3 confirms that the functions x, y, and z (required for IR activity) have the T_{1u} symmetry type and that all of the binary product functions such as x^2, xy, etc. (required for Raman activity) have different symmetry types.

(b) BF_3? Boron trifluoride does not have a center of symmetry. Therefore, it is possible that some vibrations are both IR and Raman active. You should consult the D_{3h} character table in Appendix 3. Notice that the pairs of functions (x, y) and $(x^2 - y^2, xy)$ have the E' symmetry type. Therefore, any E' symmetry vibration will be observed as a band in both IR and Raman spectra.

4.12   Vibrations of a C_{6v} molecule that are neither IR nor Raman active?
You will need to consult the C_{6v} character table in Appendix 3. If a vibration is neither IR nor Raman active, then the symmetry type of the vibration must be different than the symmetry types for the functions x, y, and z (required for IR activity) and all of the binary product functions such as x^2, xy, etc. (required for Raman activity). The symmetry types A_2, B_1, and B_2 satisfy this criterion. Therefore, any A_2, B_1, or B_2 vibrations of a C_{6v} molecule will not be observed in either the IR spectrum or the Raman spectrum.

Guide to Solutions Quiz

1   Determine the point group of PF_2^-, PF_2^+, PF_4^-, PF_4^+, $S_2O_3^{2-}$ (central sulfur atom), and ClNO (central N atom).

2   Determine the point group of o-dichlorobenzene, m-dichlorobenzene, p-dichlorobenzene, 1,2,3-trichlorobenzene, and 1,2,4-trichlorobenzene.

3 (a) Determine the point group of a tennis ball, including the seam. (b) Determine the formula and structure of a hydrocarbon containing only three carbon atoms that has the same symmetry.

4 Generally, but not always, molecules with degenerate molecular orbitals possess at least one C_3 axis or higher C_n axis. Inspect Appendix 3 and find a point group with a doubly degenerate symmetry type but without a C_3 or higher order C_n axis. Based on your discovery, formulate a more rigorous "rule" to replace the first sentence of this question.

5 How many unique mirror planes does a tetrahedron possess? How many does an octahedron possess? Hint: using three-dimensional models can really make a difference when trying to find symmetry elements.

6 Find all of the symmetry elements for BF_3, PF_3, and ClF_3.

7 Use symmetry criteria to determine whether the following molecules are polar: BF_3, PF_3, ClF_3, SO_2Cl_2, *cis*-$C_2H_2Cl_2$, *trans*-$C_2H_2Cl_2$.

8 (a) What are the five mathemetical abbreviations for the d orbitals? Draw all five of them, labelling the appropriate axes. (b) Without referring to Appendix 3, determine the D_2 point group symmetry type for each orbital. The D_2 character table is shown below.

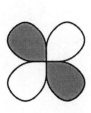 

D_2	E	$C_2(z)$	$C_2(y)$	$C_2(x)$
A	1	1	1	1
B_1	1	1	−1	−1
B_2	1	−1	1	−1
B_3	1	−1	−1	1

9   The XeF_5^- ion was recently prepared and was found to have a pentagonal-planar geometry (this is the first time this geometry has been observed). (a) How many vibrational modes does this ion have in the plane of the nuclei? (b) How many vibrational modes does it have perpendicular to the molecular plane?

10   What are the symmetries of the vibrational modes of the XeF_5^- ion that are (a) IR active only, (b) Raman active only, (c) both IR and Raman active?

5 Acids and bases

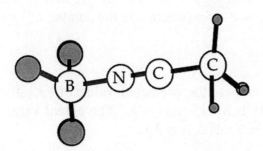

A simple Lewis acid–Lewis base complex, formed by mixing the Lewis acid BF_3 with the Lewis base CH_3CN. In this complex, the structure of the Lewis acid has changed from planar to pyramidal, while the structure of the Lewis base has hardly changed at all.

S5.1 Identifying acids and bases? (a) $HNO_3 + H_2O \rightleftharpoons H_3O^+ + NO_3^-$? The compound HNO_3 transfers a proton *to* water, so it is an acid. The nitrate ion is its conjugate base. In this reaction, H_2O accepts a proton, so it is a base. The hydronium ion, H_3O^+, is its conjugate acid.

(b) $CO_3^{2-} + H_2O \rightleftharpoons HCO_3^- + OH^-$? Carbonate ion accepts a proton from water, so it is a base. The hydrogen carbonate, or bicarbonate, ion is its conjugate acid. In this reaction, H_2O donates a proton, so it is an acid. Hydroxide ion is its conjugate base.

(c) $NH_3 + H_2S \rightleftharpoons NH_4^+ + HS^-$? Ammonia accepts a proton from hydrogen sulfide, so it is a base. The ammonium ion, NH_4^+, is its conjugate acid. Since hydrogen sulfide donated a proton, it is an acid, while HS^- is its conjugate base.

S5.2 **Arrange in order of increasing acidity?** $[Na(H_2O)_6]^+$, $[Sc(H_2O)_6]^{3+}$, $[Mn(H_2O)_6]^{2+}$, and $[Ni(H_2O)_6]^{2+}$? Since the acid strength of aqua acids increases as the electrostatic parameter, $\xi = z^2/(r + d)$, the strongest acid will have the highest charge, at least for a group of acids for which $r + d$ does not vary too much. Thus $[Na(H_2O)_6]^+$, with the lowest charge, will be the weakest of the four aqua acids, and $[Sc(H_2O)_6]^{3+}$, with the highest charge, will be the strongest. The remaining two aqua acids have the same charge, and so the one with the smaller ionic radius, r, will have the smaller value of $r + d$ and hence the greater acidity. Since Ni^{2+} has a greater Z_{eff} than Mn^{2+}, it has a smaller radius, and so $[Ni(H_2O)_6]^{2+}$ is more acidic than $[Mn(H_2O)_6]^{2+}$. The order of increasing acidity is $[Na(H_2O)_6]^+ <$ $[Mn(H_2O)_6]^{2+} < [Ni(H_2O)_6]^{2+} < [Sc(H_2O)_6]^{3+}$.

S5.3 **Predict pK_a values?** **(a) H_3PO_4?** Pauling's first rule for predicting the pK_a of a mononuclear oxoacid is $pK_a \approx 8 - 5p$ (where p is the number of oxo groups attached to the central element).

Since $p = 1$, the predicted value of pK_a for H_3PO_4 is $8 - (5 \times 1) = 3$. The actual value, given in Table 5.2, is 2.1.

(b) $H_2PO_4^-$? Pauling's second rule for predicting the pK_a of a mononuclear oxoacid is that successive pK_a values for polyprotic acids increase by five units for each successive proton transfer. Since $pK_a(1)$ for H_3PO_4 was predicted to be 3 (see above), the predicted value of pK_a for $H_2PO_4^-$, which is $pK_a(2)$ for H_3PO_4, is $3 + 5 = 8$. The actual value, given in Table 5.2, is 7.4.

(c) HPO_4^{2-}? The pK_a for HPO_4^{2-} is the same as $pK_a(3)$ for H_3PO_4, so the predicted value is $3 + (2 \times 5) = 13$. The actual value, given in Table 5.2, is 12.7.

S5.4 **What happens to Ti(IV) in aqueous solution as the pH is raised?** According to Figure 5.5, Ti(IV) is amphoteric. Treatment of an aqueous solution containing Ti(IV) ions with ammonia causes the precipitation of TiO_2, but further treatment with NaOH causes the TiO_2 to redissolve.

S5.5   **Identify the acids and bases?** A general rule that works in many, but not all, instances is that negatively charged ions are Lewis bases and positively charged ions are Lewis acids. This rule certainly works in the three parts of this exercise. However, be alert for the possibility that a charged species may be neither acidic nor basic, just as an electrically neutral species may be neither acidic nor basic.

(a) $FeCl_3 + Cl^- \rightarrow [FeCl_4]^-$? The acid $FeCl_3$ forms a complex, $[FeCl_4]^-$, with the base Cl^-.

(b) $I^- + I_2 \rightarrow I_3^-$? The acid I_2 forms a complex, I_3^-, with the base I^-.

(c) $[SnCl_3]^- + (CO)_5MnCl \rightarrow (CO)_5Mn{-}SnCl_3 + Cl^-$? The acid, $Mn(CO)_5^+$, is displaced from its complex with the base Cl^- by the base $[:SnCl_3]^-$, and the new complex $(CO)_5Mn{-}SnCl_3$ is formed.

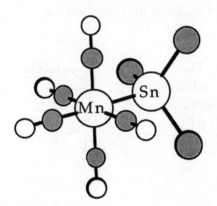

A drawing of the $(CO)_5Mn{-}SnCl_3$ molecule, a complex of $[Mn(CO)_5]^+$ and $[SnCl_3]^-$. The carbon atoms of the CO ligands are bound to the Mn atom. The O atoms are unshaded.

In part (c), both the Lewis bases are species that are stable whether or not they are part of the acid–base complex. This is true for many, but not all, Lewis acids. In this case, the Lewis acid is not a stable species that has an independent existence. That is, the cation $[Mn(CO)_5]^+$ cannot be isolated as a simple salt, only as complexes. A more familiar example of a Lewis acid that does not have an independent existence in solution or in the solid state is the proton, or hydrogen ion, H^+. Recall that the free proton is always bound to a solvent molecule or another basic species.

S5.6   **The difference in structure between $(H_3Si)_3N$ and $(H_3C)_3N$?** If the N atom lone pair of $(H_3Si)_3N$ is delocalized onto the three Si atoms, it cannot exert its normal steric influence as predicted by VSEPR rules. Therefore, the N atom of $(H_3Si)_3N$ is trigonal planar, whereas the N atom of $(H_3C)_3N$ is trigonal pyramidal.

The structures of $(H_3Si)_3N$ and $(H_3C)_3N$, excluding the hydrogen atoms.

(structure diagrams of $(H_3Si)_3N$ and $(H_3C)_3N$)

S5.7 Aluminosilicate or sulfide minerals? Cd, Rb, Cr, Pb, Sr, and Pd? The bases in aluminosilicate minerals are hard silicate oxo anions, while the base in sulfide minerals is the soft base S^{2-}. Given these two choices, hard metals will be found in aluminosilicate minerals and soft metals will be found in sulfides. Of the list of metals given, Rb, Cr, and Sr are hard, and are found in aluminosilicates. The metals Rb and Sr are alkali metals, all of which are hard. Chromium is an early *d*-block metal, and all of these are hard too. The metals Cd, Pb, and Pd are soft, and are found in sulfides. Lead is a heavy *p*-block metal, all of which are soft. The metals Cd and Pd are 2nd row late *d*-block metals. Along with the period 6 *d*-block metals, all of these are soft too.

S5.8 Draw the structure of $BF_3 \cdot OEt_2$? The ether oxygen atom will form a dative bond with the boron atom of BF_3. The structure around the boron atom will go from trigonal-planar in BF_3 to tetrahedral in $F_3B–OEt_2$. The structure of the complex is shown at the right.

(structure diagram of $BF_3 \cdot OEt_2$)

5.1 Sketch an outline of the *s* and *p* blocks of the periodic table, showing the elements that form acidic, basic, and amphoteric oxides? See the diagram below. If you cannot write out the *s*- and *p*-blocks from memory, you should spend some time learning that part of the periodic table. This knowledge will permit you to integrate many chemical facts into a logical pattern of trends.

Li	Be		B	C	N		
Na	Mg		Al	Si	P	S	Cl
K	Ca		Ga	Ge	As	Se	Br
Rb	Sr		In	Sn	Sb	Te	I
Cs	Ba		Tl	Pb	Bi		

The elements that form basic oxides in plain type, those forming acidic oxides in outline type, and those forming amphoteric oxides in boldface type. Note the diagonal region from upper left to lower right that includes the elements forming amphoteric oxides. The elements Ge, Sn, Pb, As, Sb, and Bi form amphoteric oxides only in their lower oxidation states (II for Ge, Sn, and Pb; III for As, Sb, and Bi). They form acidic oxides in their higher oxidation state (IV for Ge, Sn, and Pb; V for As, Sb, and Bi).

5.2 **Identify the conjugate bases of the following acids:** **(a)** $[Co(NH_3)_5(OH_2)]^{3+}$? A conjugate base is a species with one fewer proton than the parent acid. Therefore, the conjugate base in this case is $[Co(NH_3)_5(OH)]^{2+}$, shown below (L = NH_3).

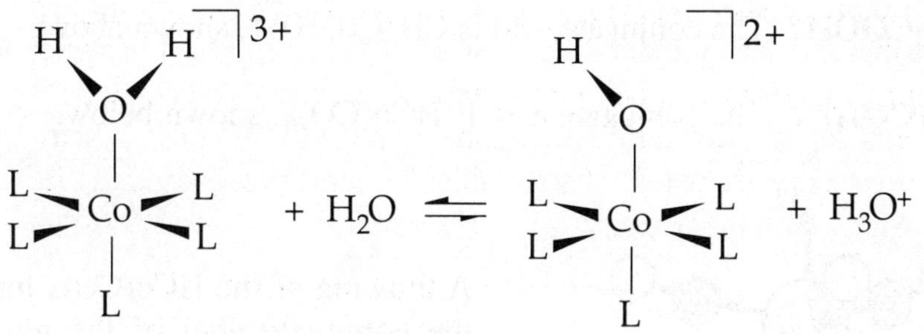

(b) HSO_4^-? The conjugate base is SO_4^{2-}.

(c) CH_3OH? The conjugate base is CH_3O^-.

(d) $H_2PO_4^-$? The conjugate base is HPO_4^{2-}.

(e) $Si(OH)_4$? The conjugate base is $SiO(OH)_3^-$.

(f) HS^-? The conjugate base is S^{2-}.

5.3 **Identify the conjugate acids of the following bases?** **(a)** C_5H_5N **(pyridine)?** A conjugate acid is a species with one more proton than the parent base. Therefore, the conjugate acid in this case is the pyridinium ion, $C_5H_6N^+$, shown below.

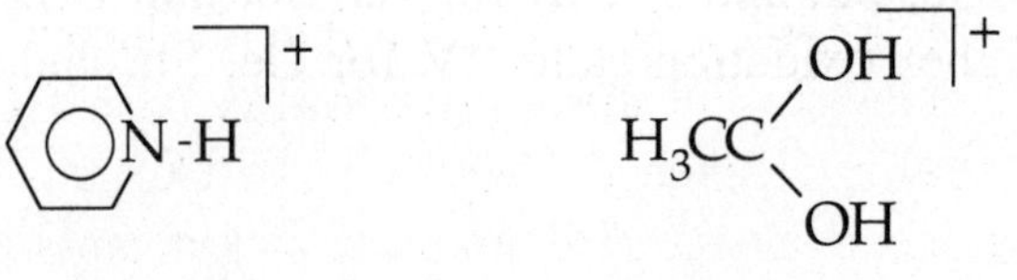

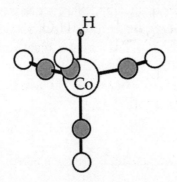

$C_5H_6N^+$ $CH_3C(OH)_2^+$

(b) HPO_4^{2-}? The conjugate acid is $H_2PO_4^-$.

(c) O^{2-}? The conjugate acid is OH^-.

(d) CH_3COOH? The conjugate acid is $CH_3C(OH)_2^+$, shown above.

(e) $[Co(CO)_4]^-$? The conjugate acid is $HCo(CO)_4$, shown below.

A drawing of the $HCo(CO)_4$ molecule, the conjugate acid of the tetrahedral $Co(CO)_4^-$ anion. The C atoms of the CO ligands are bound to the Co atom. The O atoms are unshaded.

(f) CN^-? The conjugate acid is HCN.

5.4   **List the bases HS^-, F^-, I^-, and NH_2^- in order of increasing proton affinity?** You should make use of Table 5.1 to answer this question. The species with the greatest proton affinity will be the strongest base, and its conjugate acid will be the weakest acid. The weakest acid will have the smallest value of K_a (or the most positive value of pK_a). Since Table 5.1 shows that HI is a stronger acid than HF which is a stronger acid than H_2S, a partial order of proton affinity is $I^- < F^- < HS^-$. Since NH_3 is a very weak acid, NH_2^- must be a very strong base. Therefore, our final list, in order of increasing proton affinity, is $I^- < F^- < HS^- < NH_2^-$.

5.5   **Which bases are too strong or too weak to be studied experimentally? (a) CO_3^{2-} O^{2-}, ClO_4^-, and NO_3^- in water?** You can interpret the term "studied experimentally" to mean that the base in question exists in water (i.e. it is not completely protonated to its conjugate acid) *and* that the base in question can be partially protonated (i.e. it is not so weak that the strongest acid possible in water, H_3O^+, will fail to produce a measurable amount of the conjugate acid). Using these criteria, the base CO_3^{2-} is of directly measurable base strength, since the equilibrium $CO_3^{2-} + H_2O \rightleftharpoons HCO_3^- + OH^-$ produces measurable amounts of reactants and products. The base O^{2-}, on the other hand, is completely protonated in water to produce OH^-, so the oxide ion is too strong to be studied experimentally in water. The bases ClO_4^- and NO_3^- are conjugate bases of very strong acids, which are completely deprotonated in water. Therefore, since it is not possible to protonate either perchlorate or nitrate ion in water, they are too weak to be studied experimentally.

(b) HSO_4^-, NO_3^-, and ClO_4^-, in H_2SO_4? The hydrogen sulfate ion, HSO_4^-, is the strongest base possible in liquid sulfuric acid. However, since acids can protonate it, it is not too strong to be studied experimentally. Nitrate ion is a weaker base than HSO_4^-, a consequence of the fact that its conjugate acid, HNO_3, is a stronger acid than H_2SO_4. However, nitrate is not so weak that it cannot be protonated in sulfuric acid, so NO_3^- is of directly measurable base strength in liquid H_2SO_4. On the other hand, ClO_4^-, the conjugate base of one of the strongest known acids, is so weak that it cannot be protonated in sulfuric acid, and hence cannot be studied in sulfuric acid.

5.6   Is the –CN group electron donating or withdrawing? A comparison of the aqueous pK_a values is necessary to answer this question. These are:

HOCN, 4	H_2NCN, 10.5	CH_3CN, 20
H_2O, 14	NH_3, very large	CH_4, very large

You know that the values of pK_a are very large for ammonia and methane because these compounds are not normally thought of as acids (this implies that they are extremely weak acids). Now, in all three cases, the cyano-containing compound has a lower pK_a (a *higher* acidity) than the parent compound. In the case of H_2O and HOCN, the latter compound is 10 orders of magnitude more acidic than water. The deprotonation equilibrium involves the formation of an anion, the conjugate base of the acid in question. For example:

$$HOCN + H_2O \rightleftharpoons OCN^- + H_3O^+$$

$$H_2O + H_2O \rightleftharpoons OH^- + H_3O^+$$

Since a lower pK_a means a larger K_a, this suggests that the anion OCN^- is better stabilized than OH^-. This occurs because the $-CN$ group is more electron withdrawing than the $-H$ substituent.

5.7 **Is the pK_a for $HAsO_4^{2-}$ consistent with Pauling's rules?** Pauling's first rule for predicting the pK_a of a mononuclear oxoacid is $pK_a \approx 8 - 5p$ (where p represents the number of oxo groups attached to the central element).

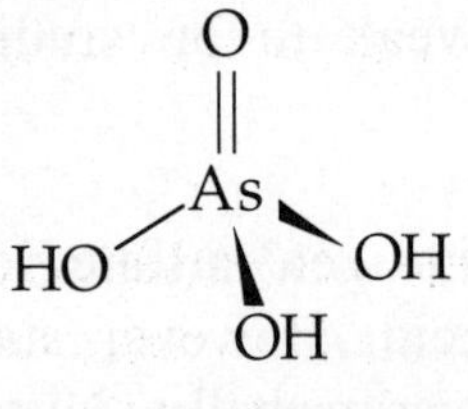

Since $p = 1$, the predicted value of $pK_a(1)$ for H_3AsO_4 is $8 - (5 \times 1) = 3$.

Pauling's second rule for predicting the pK_a of a mononuclear oxoacid is that successive pK_a values for polyprotic acids increase by five units for each successive proton transfer. Since $pK_a(1)$ for H_3AsO_4 was predicted to be 3, the predicted value of pK_a for $HAsO_4^{2-}$, which is $pK_a(3)$ for H_3AsO_4, is $3 + (2 \times 5) = 13$. The actual value, which differs by 1.5 pK_a units, is 11.5. This illustrates that Pauling's rules are only approximate.

5.8 **Account for the trends in the pK_a values of the conjugate acids of SiO_4^{4-}, PO_4^{3-}, SO_4^{2-}, and ClO_4^-?** The structures of these four anions, which can be determined to be tetrahedral using VSEPR, are shown below. As can be seen, the charge on the anions decreases from -4 for the silicon-containing species to -1 for the chlorine-containing species. The charge differences alone would make SiO_4^{4-} the most basic species. Hence $HSiO_4^{3-}$ is the least acidic conjugate acid. The acidity of the four conjugate acids increases in the order $HSiO_4^{3-} < HPO_4^{2-} < HSO_4^- < HClO_4$.

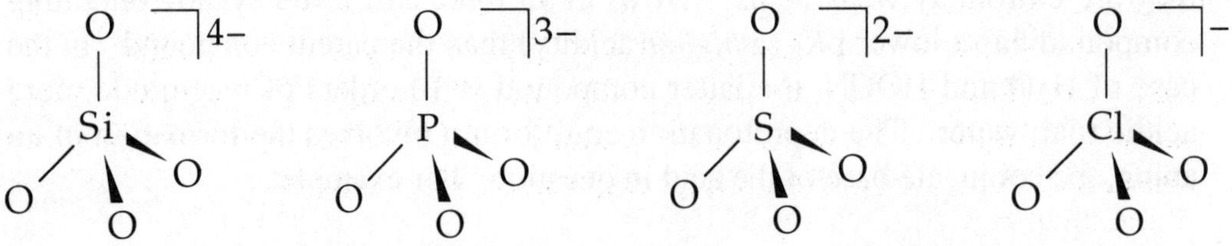

5.9   **Which of the following is the stronger acid: (a) $[Fe(OH_2)_6]^{3+}$ or $[Fe(OH_2)_6]^{2+}$?**   The Fe(III) complex, $[Fe(OH_2)_6]^{3+}$, is the stronger acid by virtue of the higher charge. The electrostatic parameter, $\xi = z^2/(r + d)$, will be considerably higher for $z = 3$ than for $z = 2$. The minor decrease in $r + d$ on going from the Fe(II) to the Fe(III) species will enhance the differences in ξ for the two species.

(b) $[Al(OH_2)_6]^{3+}$ or $[Ga(OH_2)_6]^{3+}$?   In this case, z is the same but $r + d$ is different. Since the ionic radius, r, is smaller for period 3 Al^{3+} than for period 4 Ga^{3+}, $r + d$ for $[Al(OH_2)_6]^{3+}$ is smaller than $r + d$ for $[Ga(OH_2)_6]^{3+}$ and the aluminum–containing species is more acidic.

(c) $Si(OH)_4$ or $Ge(OH)_4$?   As in part (b), above, z is the same but the $r + d$ parameter is different for these two compounds. The comparison here is also between species containing period 3 and period 4 central atoms in the same group, and the species containing the smaller central atom, $Si(OH)_4$, is more acidic.

(d) $HClO_3$ or $HClO_4$?   These two acids are shown below. According to Pauling's rule 1 for mononuclear oxoacids, the species with more oxo groups has the lower pK_a and is the stronger acid. Thus, $HClO_4$ is a stronger acid than $HClO_3$. Note that the oxidation state of the central chlorine atom in the stronger acid (+7) is higher than in the weaker acid (+5).

$$HClO_3 \qquad\qquad HClO_4$$

(e) H_2CrO_4 or $HMnO_4$?   As in part (d), above, the oxidation states of these two acids are different, VI for the chromium atom in H_2CrO_4 and VII for the manganese atom in $HMnO_4$. The species with the higher central-atom oxidation state, $HMnO_4$, is the stronger acid. Note that this acid has more oxo groups, three, than H_2CrO_4, which has two.

(f) H_3PO_4 or H_2SO_4?   The oxidation state of sulfur in H_2SO_4 is VI while the oxidation state of phosphorus in H_3PO_4 is only V. Furthermore, sulfuric acid has two oxo groups attached to the central sulfur atom while phosphoric acid has only one oxo group attached to the central phosphorus atom. Therefore, on both counts (which by now you can see are really manifestations of the same thing) H_2SO_4 is a stronger acid than H_3PO_4.

5.10   **Arrange the following oxides in order of increasing basicity? Al_2O_3, B_2O_3, BaO, CO_2, Cl_2O_7, and SO_3?**   First you pick out the intrinsically acidic oxides, since these will be the *least* basic. The compounds B_2O_3, CO_2, Cl_2O_7, and SO_3 are acidic, since the central element for each of them is found in the acidic region of the periodic table (see the *s* and *p* block diagram in the answer to Exercise 5.1 as well as Figure 5.4). The most acidic compound, Cl_2O_7, has the highest central-atom oxidation state, +7, while the least acidic, B_2O_3, has the lowest, +3. Of the remaining compounds, Al_2O_3 is amphoteric, which puts it on the borderline between acidic and basic oxides, and BaO is basic. Therefore, a list of these compounds in order of increasing basicity is $Cl_2O_7 < SO_3 < CO_2 < B_2O_3 < Al_2O_3 < BaO$.

5.11   **Arrange the following in order of increasing acidity? HSO_4^-, H_3O^+, H_4SiO_4, CH_3GeH_3, NH_3, and HSO_3F?**   The weakest acids, CH_3GeH_3 and NH_3, are easy to pick out of this group since they do not contain any –OH bonds. Ammonia is the weaker acid of the two, since it has a lower central-atom oxidation state, III, than that for the germanium atom in CH_3GeH_3, which is IV. Of the remaining species, note that HSO_3F is very similar to H_2SO_4 as far as structure and sulfur oxidation state (VI) are concerned, so it is reasonable to suppose that HSO_3F is a very strong acid, which it is. The anion HSO_4^- is a considerably weaker acid than HSO_3F, for the same reason that it is a considerably weaker acid than H_2SO_4, namely Pauling's rule 2 for mononuclear oxoacids. Since HSO_4^- is not completely deprotonated in water, it is a weaker acid than H_3O^+, which is the strongest possible acidic species in water. Finally, it is difficult to place exactly $Si(OH)_4$ in this group. It is certainly more acidic than NH_3 and CH_3GeH_3, and it turns out to be *less* acidic than HSO_4^-, despite the negative charge of the latter species. Therefore, a list of these species in order of increasing acidity is $NH_3 < CH_3GeH_3 < H_4SiO_4 < HSO_4^- < H_3O^+ < HSO_3F$.

The structures of H_2SO_4 and HSO_3F.

5.12 **Which aqua ion is the stronger acid, Na^+ or Ag^+?** Even though these two ions have about the same ionic radius, Ag^+–OH_2 bonds are much more covalent than Na^+–OH_2 bonds, a common feature of the chemistry of d-block vs. s-block metal ions. The greater covalence of the Ag^+–OH_2 bonds has the effect of delocalizing the positive charge of the cation over the whole aqua complex. As a consequence, the departing proton is repelled more by the positive charge of Ag^+(aq) than by the positive charge of Na^+(aq), and the former ion is the stronger.

5.13 **Which of the following elements form oxide polyanions or polycations: Al, As, Cu, Mo, Si, B, Ti?** As discussed in Sections 5.7 and 5.8, the aqua ions of metals that have amphoteric oxides generally undergo polymerization to polycations. The elements Al, Cu, and Ti fall into this category. On the other hand, polyoxoanions (oxide polyanions) are important for some of the early d-block metals, especially for V, Mo, and W in high oxidation states. Furthermore, many of the p-block elements form polyoxoanions, including As, B, and Si.

5.14 **The change in charge upon aqua ion polymerization?** One example of aqua ion polymerization is:

$$2\,[Al(OH_2)_6]^{3+} + H_2O \rightarrow [(H_2O)_5Al–O–Al(OH_2)_5]^{4+} + 2H_3O^+$$

The charge per aluminum atom is +3 for the mononuclear species on the left-hand side of the equation but only +2 for the dinuclear species on the right-hand side. Thus, poly*cation* formation reduces the average positive charge per central M atom by +1 per M.

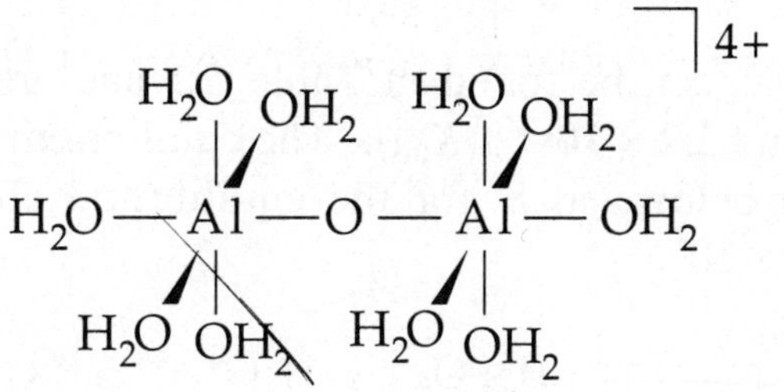

5.15   **Write balanced equations for the formation of $P_4O_{12}^{4-}$ from PO_4^{3-} and for the formation of $[(H_2O)_4Fe(OH)_2Fe(OH_2)_4]^{4+}$ from $[Fe(OH_2)_6]^{3+}$?** The two balanced equations are shown below. Note that the condensation reactions involve a neutralization of charge, either by adding H^+ to a highly charged anion or by removing H^+ from a highly charged cation. The structure of $P_4O_{12}^{4-}$, which is called *cyclo*tetrametaphosphate, is also shown below.

$$4\,PO_4^{3-} + 8\,H_3O^+ \rightarrow P_4O_{12}^{4-} + 12\,H_2O$$

$$2\,[Fe(OH_2)_6]^{3+} \rightarrow [(H_2O)_4Fe(OH)_2Fe(OH_2)_4]^{4+} + 2\,H_3O^+$$

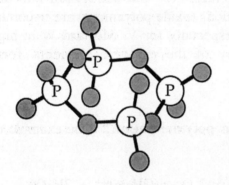

The structure of the $[P_4O_{12}]^{4-}$ ion in the salt $[NH_4]_4[P_4O_{12}]$.

5.16   **More balanced equations: (a) H_3PO_4 and Na_2HPO_4?** You can use the successive K_a values for phosphoric acid to estimate the equilibrium constant for the equilibrium below:

$$H_3PO_4 + HPO_4^{2-} \rightleftharpoons 2H_2PO_4^- \qquad\qquad K = ?$$

The three K_a values can be found in Table 5.1 and are 7.5×10^{-3} (K_{a1}), 6.2×10^{-8} (K_{a2}), and 2.2×10^{-13} (K_{a3}). The equilibrium above is the sum of the two equilibria below, so K for the equilibrium above is the product $(K_{a1})(1/K_{a2}) = 1.2 \times 10^5$.

$$H_3PO_4 + H_2O \rightleftharpoons H_2PO_4^- + H_3O^+ \qquad K_{a1} = 7.5 \times 10^{-3}$$

$$HPO_4^{2-} + H_3O^+ \rightleftharpoons H_2PO_4^- + H_2O \quad 1/K_{a2} = 1.6 \times 10^7$$

(b) CO_2 and $CaCO_3$? Successive K_a values can also be used to show that the equilibrium below lies to the right.

$$CO_2 + CaCO_3 + H_2O \rightleftharpoons Ca^{2+} + 2HCO_3^-$$

5.17   Identifying elements that form Lewis acids? All of the p-block elements except nitrogen, oxygen, fluorine, and the lighter noble gases form Lewis acids in one of their oxidation states. Examples are as follows: BF_3, B_2O_3, and B_2H_6 are Lewis acids; CO_2, organic ketones, and carbonium ions are Lewis acids; Ga, In, and Tl all form +1 cations, which are Lewis acids, and $GaCl_3$, $InCl_3$, and $TlCl_3$ are Lewis acids; the dichlorides and tetrachlorides of Ge, Sn, and Pb are Lewis acids; the trifluorides and pentafluorides of P, As, Sb, and Bi are Lewis acids (PF_3 and AsF_3 are also Lewis bases towards d-block metals, so these compounds are amphoteric); the dioxides of S, Se, and Te are Lewis acids, as are the tetrafluorides and hexafluorides of Se and Te; the trifluorides and pentafluorides of Cl, Br, and I are Lewis acids, as is the heptafluoride IF_7; the tetrafluoride and hexafluoride of xenon, XeF_4 and XeF_6, are also Lewis acids.

5.18   Identifying acids and bases: (a) $SO_3 + H_2O \rightarrow HSO_4^- + H^+$? The acids in this reaction are the Lewis acids SO_3 and H^+ and the base is the Lewis base OH^-. The complex (or adduct) HSO_4^- is formed by the displacement of the proton from the hydroxide ion by the stronger acid SO_3. In this way, the water molecule is thought of as an adduct of H^+ and OH^-. Since the proton must be bound to a solvent molecule, even though this fact is not explicitly shown in the reaction, the water molecule exhibits Brønsted acidity. Note that it is easy to tell that this is a displacement reaction instead of just a complex formation reaction because, while there is only one base in the reaction, there are *two* acids. A complex formation reaction only occurs with a single acid and a single base. A double displacement, or metathesis, reaction only occurs with two acids and two bases.

(b) $Me[B_{12}]^- + Hg^{2+} \rightarrow [B_{12}] + MeHg^+$? (Note: $[B_{12}]$ designates the Co center of the macrocyclic complex called coenzyme B_{12}). This is a displacement reaction. The Lewis acid Hg^{2+} displaces the Lewis acid $[B_{12}]$ from the Lewis base CH_3^-.

(c) KCl + SnCl$_2$ → K$^+$ + [SnCl$_3$]$^-$? This is also a displacement reaction. The Lewis acid SnCl$_2$ displaces the Lewis acid K$^+$ from the Lewis base Cl$^-$.

(d) AsF$_3$(g) + SbF$_5$(g) → [AsF$_2$][SbF$_6$]? Even though this reaction is the formation of an ionic substance, it is *not* simply a complex formation reaction. It is a displacement reaction. The very strong Lewis acid SbF$_5$ (one of the strongest known) displaces the Lewis acid [AsF$_2$]$^+$ from the Lewis base F$^-$.

(e) EtOH readily dissolves in pyridine? A Lewis acid/base complex formation reaction between EtOH (the acid) and py (the base) produces the adduct EtOH–py, which is held together by the kind of dative bond that you refer to as a hydrogen bond.

5.19   **Select the compound with the named characteristic? (a) Strongest Lewis acid: BF$_3$, BCl$_3$, or BBr$_3$?** The simple argument that more electronegative substituents lead to a stronger Lewis acid does not work in this case. Boron tribromide is observed to be the strongest Lewis acid of these three compounds. The shorter boron–halogen bond distances in BF$_3$ and BCl$_3$ than in BBr$_3$ are believed to lead to stronger halogen-to-boron p–p π bonding (see Section 5.8). According to this explanation, the acceptor orbital (empty p orbital) on boron is involved to a greater extent in π bonding in BF$_3$ and BCl$_3$ than in BBr$_3$, the acidities of BF$_3$ and BCl$_3$ are diminished relative to BBr$_3$.

BeCl$_2$ or BCl$_3$? Boron trichloride is expected to be the stronger Lewis acid of the two for two reasons. The first reason, which is more obvious, is that the oxidation number of boron in BCl$_3$ is +3 while for the beryllium atom in BeCl$_2$ it is only +2. The second reason has to do with structure. The boron atom in BCl$_3$ is only three-coordinate, leaving a vacant site to which a Lewis base can coordinate. Since BeCl$_2$ is polymeric, each beryllium atom is four-coordinate, and some Be–Cl bonds must be broken before adduct formation can take place.

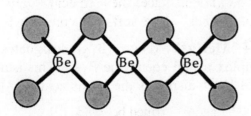
A piece of the infinite linear chain structure of BeCl$_2$. Each Be atom is four-coordinate, and each Cl atom is two-coordinate. The polymeric chains are formed by extending this piece to the right and to the left

B(n-Bu)$_3$ or B(t-Bu)$_3$? The Lewis acid with the unbranched substituents, B(n-Bu)$_3$, is the stronger of the two because, once the complex is formed, steric repulsions between the substituents and the Lewis base will be less than with the bulky, branched substituents in B(*t*-Bu)$_3$.

(b) More basic toward BMe$_3$: NMe$_3$ or NEt$_3$? These two bases have nearly equal basicities towards the proton in aqueous solution or in the gas phase. Steric repulsions between the substituents on the bases and the proton are negligible, since the proton is very small. However, steric repulsions between the substituents on the bases and *molecular* Lewis acids like BMe$_3$ are an important factor in complex stability, and so the smaller Lewis base NMe$_3$ is the stronger in this case.

2-Me-py or 4-Me-py? As above, steric factors favor complex formation with the smaller of two bases that have nearly equal Brønsted basicities. Therefore, 4-Me-py is the stronger base toward BMe$_3$, since the methyl substituent in this base cannot affect the strength of the B–N bond by steric repulsions with the methyl substituents on the Lewis acid.

5.20 **Which of the following reactions have $K_{eq} > 1$? (a) R$_3$P–BBr$_3$ + R$_3$N–BF$_3$ $\rightleftharpoons$ R$_3$P–BF$_3$ + R$_3$N–BBr$_3$?** From the discussion in Section 5.12, you know that phosphines are softer bases than amines. So, to determine the position of this equilibrium, you must decide which Lewis acid is softer, since the softer acid will preferentially form a complex with a soft base than with a hard base of equal strength. Boron tribromide is a softer Lewis acid than BF$_3$, a consequence of the relative hardness and softness of the respective halogen substituents. Therefore, the equilibrium position for this reaction will lie to the left, the side with the soft–soft and hard–hard complexes, so the equilibrium constant is less than 1. In general, it is found that soft substituents (or ligands) lead to a softer Lewis acid than for the same central element with harder substituents.

(b) SO$_2$ + Ph$_3$P–HOCMe$_3$ $\rightleftharpoons$ Ph$_3$P–SO$_2$ + HOCMe$_3$? In this reaction, the soft Lewis acid sulfur dioxide displaces the hard acid *t*-butyl alcohol from the soft base triphenylphosphine. The soft–soft complex is favored, so the equilibrium constant is greater than 1.

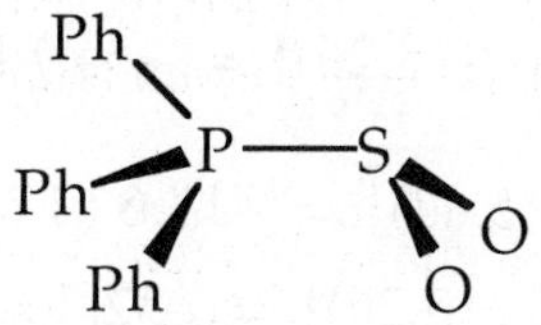

The adduct formed between triphenylphosphine and sulfur dioxide

(c) CH_3HgI + HCl $\rightleftharpoons$ CH_3HgCl + HI? Iodide is a softer base than chloride, an example of the general trend that elements later in a group are softer than their progenors. The soft acid CH_3Hg^+ will form a stronger complex with iodide than with chloride, while the hard acid H^+ will prefer chloride, the harder base. Thus, the equilibrium constant is less than 1.

(d) $[AgCl_2]^-(aq)$ + $2CN^-(aq)$ $\rightleftharpoons$ $[Ag(CN)_2]^-(aq)$ + $2Cl^-(aq)$? Cyanide is a softer and generally stronger base than chloride (see Table 5.3). Therefore, cyanide will displace the relatively harder base from the soft Lewis acid Ag^+. The equilibrium constant is greater than 1.

5.21   Choose between the two basic sites in Me_2NPF_2? The phosphorus atom in Me_2NPF_2 is the softer of the two basic sites, so it will bond more strongly with the softer Lewis acid BH_3 (see Table 5.3). The hard nitrogen atom will bond more strongly to the hard Lewis acid BF_3.

5.22   Why does Me_3N form a relatively weak complex with BMe_3? Since trimethylamine is the strongest Brønsted base in the gas phase, the reason that it does not form the most stable complex with trimethylboron can only be steric repulsions between the methyl substituents on the acid and those on the base.

5.23   Discuss relative basicities?   (a) Acetone and DMSO? Since both E_B and C_B are larger for DMSO than for acetone, DMSO is the stronger base regardless of how hard or how soft the Lewis acid is. The ambiguity for DMSO is that both the oxygen atom and sulfur atom are potential basic sites.

(b) Me_2S and DMSO? Dimethylsulfide has a C_B value that is two-and-a-half times larger than that for DMSO, while its E_B value is only one quarter that for DMSO. Thus, depending on the E_A and C_A values for the Lewis acid, either base could be stronger. For example, DMSO is the stronger base toward BF_3, while SMe_2 is the stronger base toward I_2. This can be predicted by calculating the ΔH of complex formation for all four combinations:

$DMSO–BF_3$, $\Delta H = -[(20.21)(2.76) + (3.31)(5.83)] = -75.1$ kJ mol^{-1}

$SMe_2–BF_3$, $\Delta H = -[(20.21)(0.702) + (3.31)(15.26)] = -64.7$ kJ mol^{-1}

$DMSO–I_2$ $\Delta H = -[(2.05)(2.76) + (2.05)(5.83)] = -17.6$ kJ mol^{-1}

$SMe_2–I_2$ $\Delta H = -[(2.05)(0.702) + (2.05)(15.26)] = -32.7$ kJ mol^{-1}

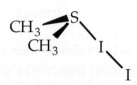

The stable complex of
DMSO and boron trifluoride

The stable complex of
dimethyl sulfide and I_2

5.24   **Write a balanced equation for the dissolution of SiO_2 by HF?**   The metathesis of two hard acids with two hard bases occurs as an equilibrium is established between solid, insoluble SiO_2 and soluble H_2SiF_6:

$$SiO_2 + 6\,HF \rightleftharpoons 2\,H_2O + H_2SiF_6 \quad\text{or}$$

$$SiO_2 + 4HF \rightleftharpoons 2H_2O + SiF_4$$

This is both a Brønsted acid-base reaction and a Lewis acid–base reaction. The Brønsted reaction involves the transfer of protons from HF molecules to O^{2-} ions, while the Lewis reaction involves complex formation between Si^{4+} ions and F^- ions.

5.25   **Write a balanced equation to explain the foul odor of damp Al_2S_3?**   The foul odor suggests a volatile compound is formed when Al_2S_3 comes in contact with water. The only volatile species that could be present, other than odorless water, is H_2S, which has the characteristic odor of rotten eggs. Thus, an equilibrium is established between two hard acids, Al(III) and H^+, and the bases O^{2-} and S^{2-}:

$$Al_2S_3 + 3\,H_2O \rightleftharpoons Al_2O_3 + 3\,H_2S$$

5.26   **Describe solvent properties?**   **(a) Favor displacement of Cl^- by I^- from an acid center?**   Since in this case you have no control over the hardness or softness of the acid center, you must do something else that will favor the acid–iodide complex over the acid–chloride complex. If you choose a solvent that decreases the activity of chloride relative to iodide, you can shift the following equilibrium to the right:

$$\text{acid–Cl}^- + \text{I}^- \rightleftharpoons \text{acid–I}^- + \text{Cl}^-$$

Such a solvent should interact more strongly with chloride (i.e. form an adduct with chloride) than with iodide. Thus, the ideal solvent properties in this case would be *weak, hard,* and *acidic.* It is important that the solvent be a weak acid, since otherwise the activity of both halides would be rendered negligible. An example of a suitable solvent is anhydrous HF. Another suitable solvent is H_2O.

(b) Favor basicity of R_3As over R_3N? In this case you wish to enhance the basicity of the soft base trialkylarsine relative to the hard base trialkyl-amine. You can decrease the activity of the amine if the solvent is a hard acid, since the solvent–amine complex would then be less prone to dissociate than the solvent–arsine complex. Alcohols such as methanol or ethanol would be suitable.

The Lewis acid–Lewis base complex of a trialkylamine (a hard base) and an alcohol (a hard acid).

(c) Favor acidity of Ag^+ over Al^{3+}? If you review the answers to parts (a) and (b) of this Exercise, a pattern will emerge. In both cases, a hard acid solvent was required to favor the reactivity of a soft base. In this part of the Exercise, you want to favor the acidity of a soft acid, so logically a solvent that is a hard base is suitable. Such a solvent will "tie up" (i.e. decrease the activity of) the hard acid Al^{3+} relative to the soft acid Ag^+. An example of a suitable solvent is diethyl ether. Another suitable solvent is H_2O.

(d) Promote the reaction $2\,FeCl_3 + ZnCl_2 \rightleftharpoons Zn^{2+} + 2\,[FeCl_4]^-$? Since Zn^{2+} is a softer acid than Fe^{3+}, a solvent that promotes this reaction will be a softer base than Cl^-. The solvent will then displace Cl^- from the Lewis acid Zn^{2+}, forming $[Zn(solv)_x]^{2+}$. The solvent must also have an appreciable dielectric constant, since ionic species are formed in this reaction. A suitable solvent is acetonitrile, MeCN.

5.27 **Why are acidic solvents useful for the preparation of reactive cations and basic solvents useful for the preparation of reactive anions?** The cationic species referred to, I_2^+ and Se_8^{2+}, are intrinsically acidic, since they bear a positive charge and are not coordinatively saturated. All but the weakest bases could potentially react with them to form complexes. Since it is the free cationic species that are frequently the targets of study, and not complexes of them, the base strength of any base present must be minimized. Therefore, to insure that the reaction mixture contains only the weakest of bases, you must use the most strongly acidic solvent possible. In the case of SbF_5/HSO_3F, the most basic species in solution would be HSO_3F! By the same token, anionic species such as those referred to, S_4^{2-} and Pb_9^{4-}, can only be studied in strongly basic solvents, because even the weakest of acids might form complexes with them.

5.28 **Propose a mechanism for the acylation of benzene?** The mechanism described in Section 5.8(b), *Aluminum halides*, involves the abstraction of Cl^- from CH_3COCl by $AlCl_3$ to form $AlCl_4^-$ and the Lewis acid CH_3CO^+ (an acylium cation). This cation then attacks benzene to form the acylated aromatic product. An alumina surface, such as the partially dehydroxylated one shown below, would also provide Lewis acidic sites that could abstract Cl^-:

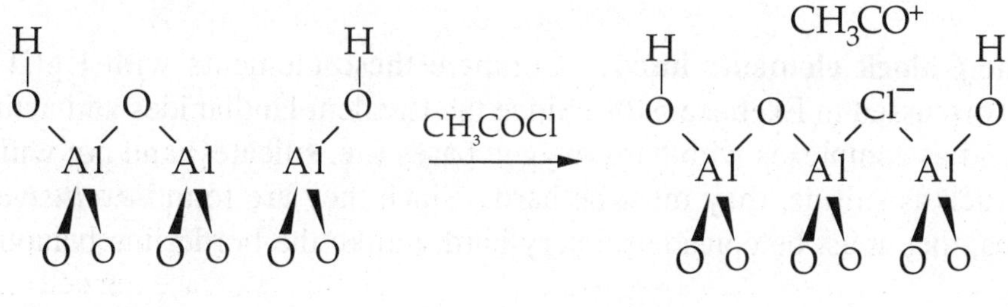

5.29 **Why does Hg(II) occur only as HgS?** Mercury(II) is a soft Lewis acid, and so is found in nature only combined with soft Lewis bases, the most common of which is S^{2-}. Sulfide can readily and permanently abstract Hg^{2+} from its complexes with harder bases in ore forming geological reaction mixtures. Zinc(II), which exhibits borderline behavior, is harder and forms stable compounds (i.e. complexes) with hard bases such as O^{2-}, CO_3^{2-}, and silicates as well as with S^{2-}. The particular ore that is formed with Zn^{2+} depends on factors including the relative concentrations of the competing bases.

5.30 **Write Brønsted acid–base reactions in liquid HF? (a) CH$_3$CH$_2$OH?**
Ethanol is a weaker acid than H$_2$O but a stronger base than H$_2$O due to the
electron donating property of the –C$_2$H$_5$ group. The balanced equation is:

$$CH_3CH_2OH + HF \rightleftharpoons CH_3CH_2OH_2^+ + F^-$$

(b) NH$_3$? The equation is NH$_3$ + HF $\rightleftharpoons$ NH$_4^+$ + F$^-$

(c) C$_6$H$_5$COOH? In this case, benzoic acid is a significantly stronger acid
than water. Therefore, it will protonate hydrogen fluoride:

$$C_6H_5COOH + HF \rightleftharpoons C_6H_5COO^- + H_2F^+$$

5.31 **The dissolution of silicates by HF?** This is both a Brønsted acid–base
reaction and a Lewis acid–base reaction. The reaction involves proton transfers
from HF to the silicate oxygen atoms (a Brønsted reaction) and the formation of
complexes such as SiF$_6^{2-}$ from the Lewis acid Si^{4+} and the Lewis base F$^-$ (a
Lewis reaction).

5.32 **Are the f-block elements hard?** Compare these elements with Hg(II) and
Zn(II), discussed in Exercise 5.29. Since the trivalent lanthanides and actinides
are found as complexes with hard oxygen bases (i.e. silicates) and not with soft
bases such as sulfide, they must be hard. Since they are found *exclusively* as
silicates, they must be considered very hard, unlike the borderline behavior of
Zn(II).

5.33 **Which is the stronger acid, SiO$_2$ or CO$_2$: CaCO$_3$(s) + SiO$_2$(s) →
[CaSiO$_3$]$_n$ + CO$_2$(g)?** Silicon dioxide is a stronger Lewis acid than CO$_2$,
since it abstracts an oxide ion, O^{2-}, from CO$_2$ in this important geochemical
reaction. The net reaction, which is a displacement reaction, is the following:

$$n\,CO_3^{2-} + n\,SiO_2 \rightarrow [SiO_3^{2-}]_n + n\,CO_2$$

5.34   **Identify the acids and bases in TiO$_2$ + Na$_2$S$_2$O$_7$ → Na$_2$SO$_4$ + TiO(SO$_4$)?** The net reaction is the displacement of the Lewis base SO$_4^{2-}$ from S$_2$O$_7^{2-}$, its adduct with the Lewis acid SO$_3$, by the strong Lewis base O^{2-}, as shown below:

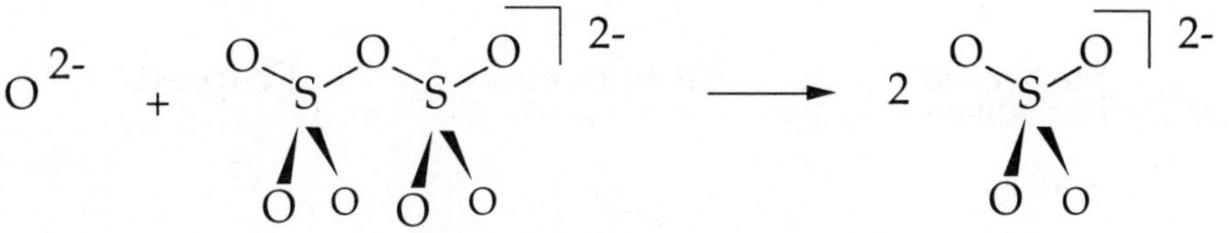

5.35   **AsF$_5$ and [AsF$_6$]$^-$?** An arsenic atom has five valence electrons. There are five As–F bonds in AsF$_5$ compound. Therefore, the central arsenic atom has 10 electrons in its valence shell in the compound (five single bonds), and the structure is a trigonal bipyramid. There are six As–F bonds in [AsF$_6$]$^-$. In this ion, the central arsenic atom has 12 electrons in its valence shell, and the structure is an octahedron.

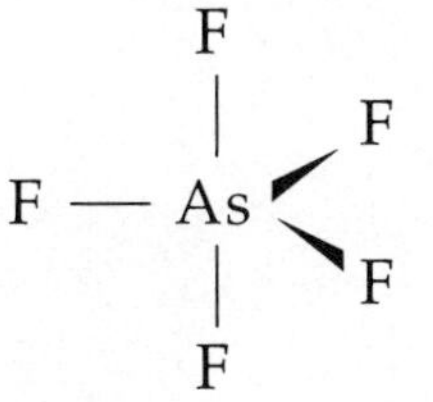

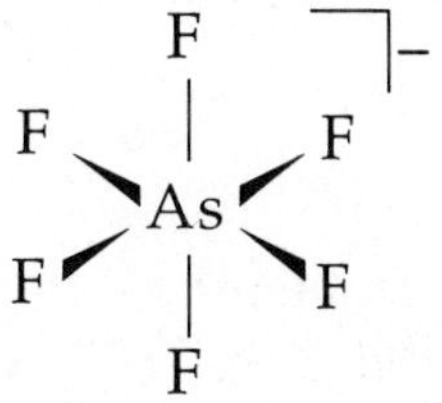

A trigonal bipyramid has a C_3 axis as its principal axis, three C_2 axes that are perpendicular to it, and two sets of mirror planes, a single horizontal plane of symmetry (σ_h) and three vertical planes of symmetry (σ_v). These symmetry elements define the D_{3h} point group. An octahedral ion such as [AsF$_6$]$^-$, with its three C_4 and four C_3 axes, has O_h symmetry.

X$_3$B–NH$_3$? The point group of this Lewis acid/base adduct depends on the conformation of the molecule. In either the eclipsed or staggered conformations, there are three vertical planes of symmetry besides the C_3 axis, and so the point group for these two conformers is C_{3v}. However, if the dihedral angle between the B–X and N–H bonds is anything other than 60° or 0°, the three mirror planes are lost and the point group is C_3. The projections below represent views down the B–N bond for the three types of conformers:

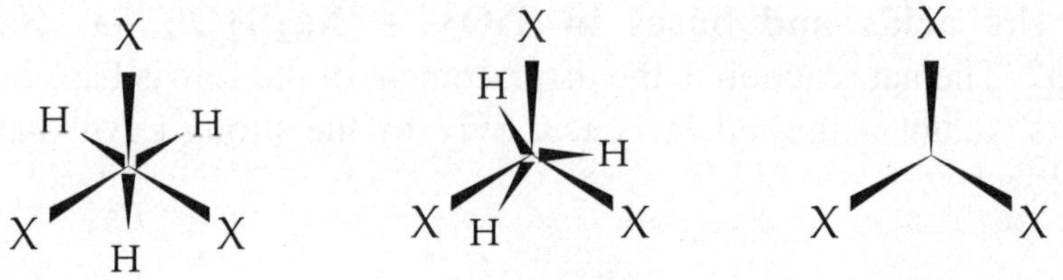

Staggered   In between   Eclipsed

C_{3v} C_3 C_{3v}

$Al_2Cl_6(g)$? This molecule has three mutually perpendicular C_2 axes, a center of symmetry, and three mutually perpendicular planes of symmetry. These symmetry elements define the D_{2h} point group.

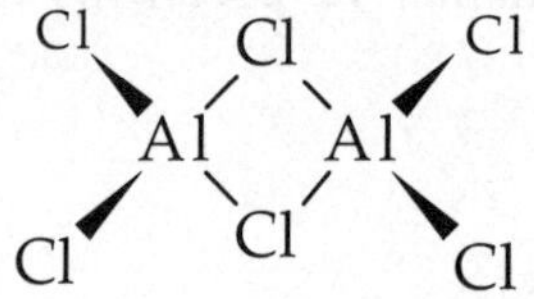

Guide to Solutions Quiz

1   List the following groups of Brønsted acids in order of increasing acidity in water: (a) $HClO_4$, $HClO$, $HClO_3$, $HClO_2$; (b) HF, HCl, HI, HCN.

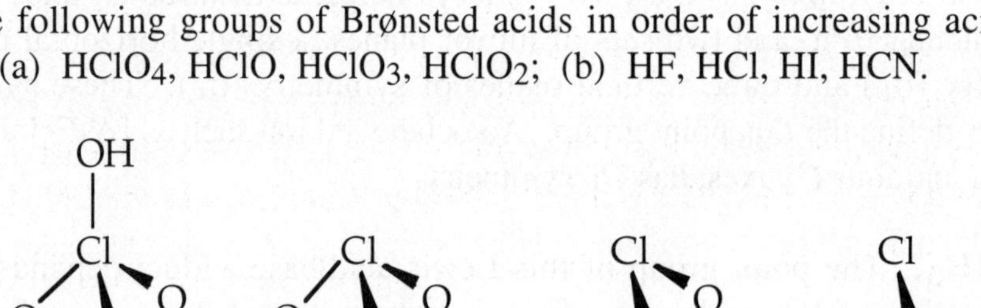

2   Pick the aqueous solution that contains the highest concentration of hydronium ion, $[H_3O]^+$: (a) 0.2 M HCl or 0.1 M H_2SO_4; (b) 0.01 M Fe^{3+} or 0.01 M Fe^{2+}; (c) 0.1 M NH_4^+ or 0.01 M $C_5H_5NH^+$ (pyridinium).

3   The compound Al_2O_3 is amphoteric. Write balanced equations for its reaction with a Brønsted acid and with a Brønsted base.

4   Identify the Lewis acid component and the Lewis base component in the following Lewis acid–base complexes: $SnCl_3^-$, I_3^-, HSO_4^-, CH_3CO^+, H_2F^+.

5   Sulfur trioxide, SO_3, is a stronger Lewis acid toward pyridine than is SO_2. In contrast, while SF_4 forms a complex with pyridine, SF_6 does not. Suggest an explanation.

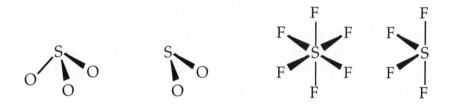

6   Using E and C parameters, and ignoring entropy contributions, calculate the equilibrium constant for the following displacement reaction:

$$(C_2H_5)_2O-B(CH_3)_3 + S(CH_3)_2 \rightleftharpoons (CH_3)_2S-B(CH_3)_3 + O(C_2H_5)_2$$

7   (a) The solid Lewis acid AlF_3 is not soluble in liquid HF. However, it dissolves if NaF is added to a mixture of AlF_3 and HF. Explain. (b) When the solution in part (a) is treated with excess BF_3, which is soluble in HF, AlF_3 precipitates. Explain.

8   The thiocyanate anion NCS^-, is ambidentate in that it can form complexes with Lewis acids through the nitrogen atom or the sulfur atom. (a) Decide which end will form a bond with Ca^{2+}, Cr^{2+}, and Hg^{2+}, and explain your reasoning. (b) How could you interpret infrared spectra of the complexes in the CN stretching region to confirm your predictions? Hint: use resonance structures.

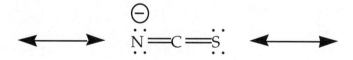

9   (a) BrF_3, a liquid molecular substance, has a small but measurable conductivity. Using balanced equations, show why this is reasonable. (b) Write balanced equations to show why the conductivity of BrF_3 is enhanced by adding either NaF or AlF_3.

10   Silicate minerals come in a wide variety of forms, many of which involve SiO_4 tetrahedra linked together to form polyoxoanions. Draw reasonable structures for the discrete anions $Si_2O_7^{2-}$ and $Si_4O_{12}^{8-}$ for the infinite single-chain anion $[SiO_3^{2-}]_n$, and for the infinite double-chain anion $[Si_2O_5^{2-}]_n$.

6   Oxidation and reduction

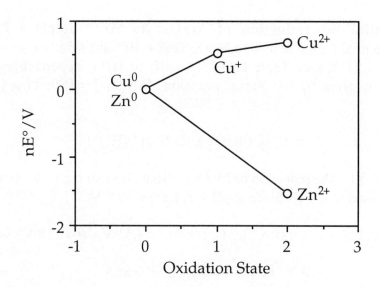

Frost diagrams, shown here for Cu and Zn in aqueous acid (pH 0), are useful representations of electrode potential data. These two graphs show that Cu^0 will not reduce $H^+(aq)$ to H_2, that Zn^0 will reduce $H^+(aq)$ to H_2 and will also reduce $Cu^{2+}(aq)$ to Cu^0, and that $Cu^+(aq)$ will disproportionate to Cu^0 and $Cu^{2+}(aq)$.

S6.1   The minimum temperature for reduction of MgO by carbon? At about 1800°C, the line for the reducing agent (C) dips below the MgO line, which means that the reactions $2Mg(l) + O_2(g) \rightarrow 2MgO(s)$ and $2C(s) + O_2(g) \rightarrow 2CO(g)$ have the same free energy change at that temperature. Thus, coupling the two reactions (i.e. subtracting the first from the second) yields the overall reaction $MgO(s) + C(s) \rightarrow Mg(l) + CO(g)$ with $\Delta G = 0$. At 1800°C or above, MgO can be conveniently reduced to Mg by carbon.

S6.2 **Can $Cr_2O_7^{2-}$ be used to oxidize Fe^{2+}, and would Cl^- oxidation be a problem?** The standard reduction potential for the $Cr_2O_7^{2-}/Cr^{3+}$ couple is 1.38 V, so it can oxidize any couple whose reduction potential is less than 1.38 V. In general terms, if the reduction potential for $Ox + e^- \rightarrow Red$ is X, then the overall potential E for the (unbalanced) reaction $Cr_2O_7^{2-} + Red \rightarrow Cr^{3+} + Ox$ is $(1.38 - X)$ V. Since $\Delta G = -NFE$, the reaction will have a negative free energy change as long as E is positive. Thus, as long as the *reduction* potential X is less than 1.38 V, E will be positive. Returning to the specific question at hand, since the reduction potential for the Fe^{3+}/Fe^{2+} couple is 0.77 V, Fe^{2+} will be oxidized to Fe^{3+} by dichromate. Since the reduction potential for the Cl_2/Cl^- couple is 1.36 V, oxidation of chloride ion is only slightly favored and in practice it is not observed to occur at an appreciable rate.

S6.3 **The potential for reduction of MnO_4^- to Mn^{2+} at pH = 7?** The standard potential for the reduction $MnO_4^-(aq) + 8H^+(aq) + 5e^- \rightarrow Mn^{2+}(aq) + 4H_2O(l)$ is 1.51 V (see Table 6.1). The pH, or $[H^+]$, dependence of the potential E is given by the Nernst equation ($[MnO_4^-] = [Mn^{2+}] = 1$ M) as follows:

$$E = E° - ((0.059 \text{ V})/5)(\log(1/[H^+]^8))$$

Note that $n = 5$ for the reduction of MnO_4^-. Note also that the factor 0.059 V/n can only be used at 25 °C. Since at pH = 7, $[H^+] = 10^{-7}$ M,

$$E = 1.51 \text{ V} - (0.0118 \text{ V})(\log(1/10^{-56})) = 1.51 \text{ V} - (0.0118 \text{ V})(56)$$

$$E = 1.51 \text{ V} - 0.66 \text{ V} = 0.85 \text{ V}$$

So, since the reduction potential is less positive, permanganate ion is a weaker oxidizing agent (i.e. it is less readily reduced) in neutral solution than at pH = 0.

S6.4 **What is the potential for the rapid oxidation of Mg by H^+ at 25 °C and pH = 7?** The balanced equation for the reaction in question is

$$Mg(s) + 2H^+(aq) \rightarrow Mg^{2+}(aq) + H_2(g)$$

The potential for this reaction, calculated using the Nernst equation, is

$$E = E° - ((0.059 \text{ V})/n)(\log Q)$$

where $n = 2$ and $Q = (P(H_2))[Mg^{2+}]/[H^+]^2$. The value of E° can be calculated by subtracting E° for the Mg^{2+}/Mg couple from E° for the H^+/H_2 couple. The former can be found in Appendix 2 (–2.356 V), and the latter, by definition, is 0, so $E^\circ = 0$ V – (–2.356 V) = 2.356 V. If you assume that $P(H_2) = 1$ bar and that $[Mg^{2+}] = 1$ M, then $Q = 1/[H^+]^2 = 10^{14}$ at pH 7, and $\log Q = 14$. Therefore,

$$E = 2.356 \text{ V} - (0.0295 \text{ V})(14) = 2.356 \text{ V} - 0.413 \text{ V} = 1.943 \text{ V}$$

This value is in excess of the 0.6 V overpotential needed for a rapid reaction.

S6.5   Can Fe^{2+} disproportionate under standard conditions? The disproportionation of Fe^{2+} involves the reduction of one equivalent of Fe^{2+} to Fe^0, a net gain of two equivalents of electrons, and the concomitant oxidation of two equivalents to Fe^{2+} to Fe^{3+}, a net loss of two equivalents of electrons:

$$3\,Fe^{2+}(aq) \rightarrow Fe^0(s) + 2\,Fe^{3+}(aq)$$

The value of E for this reaction can be calculated by subtracting E° for the Fe^{2+}/Fe couple (–0.44 V) from E° for the Fe^{3+}/Fe^{2+} couple (0.77 V),

$$E^\circ = -0.44 \text{ V} - 0.77 \text{ V} = -1.21 \text{ V}$$

This potential is large *and negative*, so the disproportionation will *not* occur.

S6.6   The fate of SO_2 emitted into clouds?   You are told that the standard reduction potential for the SO_4^{2-}/SO_2 couple is 0.17 V. Since the standard reduction potential for the O_2/H_2O couple is 1.23 V, the potential for the coupled reaction

$$2\,SO_2(aq) + O_2(g) + 2\,H_2O(l) \rightarrow 2\,SO_4^{2-}(aq) + 4\,H^+(aq)$$

is $E^\circ = 1.23$ V – 0.17 V = 1.06 V. Since this potential is large and positive, this reaction will be driven nearly to completion. The aqueous solution of SO_4^{2-} and H^+ ions precipitates as acid rain, which can have a pH as low as 2 (the pH of rain water that is not contaminated with sulfuric or nitric acid is ~5.6).

S6.7   Calculate E° for the ClO_3^-/ClO^- couple?   You should make use of the Latimer diagram for chlorine in acid solution (see Appendix 2) and the equation:

$$E_{13} = (n_1 E_{12} + n_2 E_{23})/(n_1 + n_2)$$

In this case E_{12} is the standard potential for the ClO_3^-/ClO_2^- couple:

$$ClO_3^-(aq) + 2H^+(aq) + 2e^- \rightarrow ClO_2^-(aq) + H_2O(l) \quad E_{12} = 1.18 \text{ V}, n_1 = 2$$

and E_{23} is the standard potential for the ClO_2^-/ClO^- couple:

$$ClO_2^-(aq) + 2H^+(aq) + 2e^- \rightarrow ClO^-(aq) + H_2O(l) \quad E_{23} = 1.67 \text{ V}, n_2 = 2$$

so $E_{13} = ((2)(1.18 \text{ V}) + (2)(1.67 \text{ V}))/4 = 1.43$ V. In this case, since $n_1 = n_2$, E_{13} is the simple average of E_{12} and E_{23}, whereas in the more general case $n_1 \neq n_2$, E_{13} is the *weighted* average of E_{12} and E_{23}.

S6.8   **Frost diagram for thallium in aqueous acid?** See the Figure below. This plot was made using the potentials given, $NE° = 0$ V for Tl^0 ($N = 0$), $NE° = -0.34$ V for Tl^+ ($N = 1$), and $NE° = 2.19$ V for Tl^{3+} ($n = 3$). Note that Tl^+ is stable with respect to disproportionation in aqueous acid. Note also that Tl^{3+} is a strong oxidant (i.e. it is very readily reduced), since the slope of the line connecting it with either lower oxidation state is large and positive.

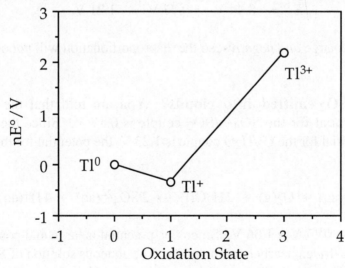

S6.9   **The oxidation number of manganese?** When permanganate, MnO_4^-, is used as an oxidant in aqueous solution, the manganese species that will remain is the most stable manganese species under the conditions of the reaction (i.e. acidic or basic). By "most stable" we mean most stable with respect to a reducing agent. Inspection of the Frost diagram for manganese, shown in

Figure 6.12, shows that $Mn^{2+}(aq)$ is the most stable species present, since it has the most negative $\Delta_f G$. Therefore, $Mn^{2+}(aq)$ will be the product of the redox reaction when MnO_4^- is used as an oxidizing agent in aqueous acid.

S6.10 Compare the strength of NO_3^- as an oxidizing agent in acidic and basic solution? As suggested in Section 6.9, *Frost diagrams*, the reduction of nitrate ion usually proceeds to NO, which is evolved from the solution, instead of proceeding to N_2O or all the way to N_2. If you compare the portion of the Frost diagram for nitrogen, below, containing the NO_3^-/NO couple in acidic and in basic solution, you see that the slope for the couple in acidic solution is positive while the slope for basic solution is negative. Therefore, nitrate is a stronger oxidizing agent (i.e. it is more readily reduced) in acidic solution than in basic solution. Even if the reduction of NO_3^- proceeded all the way to N_2, the slope of that line is still less than the slope of the line for the NO_3^-/NO couple in acid solution.

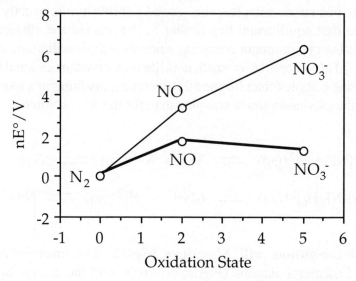

A portion of the Frost diagram for nitrogen. The points connected with the bold lines are for basic solution, and the other points are for acid solution.

S6.11 The possibility of finding Fe_2O_3 in a waterlogged soil? According to Figure 6.14, a typical waterlogged soil (organic rich and oxygen depleted) has a pH of about 4 and a potential of about –0.1 V. If you find the point on the Pourbaix diagram for naturally occurring iron species, Figure 6.13, you see that Fe_2O_3 is not stable and Fe^{2+} (aq) will be the predominant species. In fact, as long as the potential remained at -0.1 V, Fe^{2+} remains the predominant species below pH = 8. Above pH = 8, Fe^{2+} is oxidized to Fe_2O_3 at this potential. Note that for a potential of -0.1 V and a pH below 2, Fe^{2+} is still the predominant iron species in solution, but water is reduced to H_2.

S6.12 **Which couple has the higher $E°$ value, $[Ni(en)_3]^{2+}/Ni$ or $[Ni(NH_3)_6]^{2+}/Ni$?** Re-read the Section *The effect of complex formation on potentials*. The formation of a more stable complex when the metal has the higher oxidation state, in this case Ni^{2+}, favors oxidation and makes the reduction potential more negative. Due to the chelate effect (see Section 7.7(c)), ethylenediamine forms a more stable complex with Ni^{2+} than does ammonia. Therefore, the $[Ni(NH_3)_6]^{2+}/Ni$ couple has the higher $E°$, since the $[Ni(en)_3]^{2+}/Ni$ couple favors oxidation and has the lower (or more negative) $E°$. Thermodynamically (but not mechanistically), you can think of the process occurring by decomplexation followed by reduction of the nickel(II) aqua ion to nickel metal. For either complex, the second equilibrium is exactly the same. However, the first equilibrium lies farther to the left for the ethylenediamine complex than for the ammonia complex, since the ethylenediamine complex is more stable. Therefore, the *overall* equilibrium constant is smaller for the ethylenediamine complex than for the NH_3 complex, resulting in a more negative potential for the ethylenediamine complex than for the NH_3 complex.

$$[Ni(en)_3]^{2+}(aq) \rightleftharpoons 3\,en + Ni^{2+}(aq) \rightleftharpoons Ni(s)$$

$$[Ni(NH_3)_6]^{2+}(aq) \rightleftharpoons 6\,NH_3 + Ni^{2+}(aq) \rightleftharpoons Ni(s)$$

6.1 **Under what conditions will Al reduce MgO?** The lines for Al_2O_3 and MgO on the Ellingham diagram (Figure 6.3) represent the change in $\Delta G°$ with temperature for the following reactions:

$$1/3\,Al(l) + O_2(g) \rightarrow 2/3\,Al_2O_3(s) \quad \text{and} \quad 2\,Mg(l) + O_2(g) \rightarrow 2\,MgO(s)$$

At temperatures below about 1400°C, the free energy change for the MgO reaction is more negative than for the Al_2O_3 reaction. This means that under these conditions MgO is more stable with respect to its constituent elements than is Al_2O_3, and that Mg will react with Al_2O_3 to form MgO and Al. However, above about 1400°C the situation reverses, and Al will react with MgO to reduce it to Mg with the concomitant formation of Al_2O_3. This is a rather high temperature, achievable in an electric arc furnace (compare the extraction of silicon from its oxide, discussed in Section 6.1).

6.2   **Gaseous oxygen vs. gaseous chlorine?** Whenever a thermodynamically stable compound is not readily formed in a process that "should" form it, you can always suspect that there is a kinetic barrier to the reaction. In this case, the electrode process that produces gaseous chlorine, Cl_2, is the oxidation of Cl^- ions by one electron. On the other hand, the electrode process that produces gaseous oxygen, O_2, must be a more complicated process involving the removal of two electrons from H_2O or OH^-. One-electron processes are usually more rapid than multi-electron processes.

6.3   **Suggest chemical reagents for redox transformations?** **(a) Oxidation of HCl to Cl_2?** Assume that all of these transformations are occurring in acid solution (pH 0). Referring to Appendix 2, you will find that the oxidation potential (not the reduction potential) for the Cl^-/Cl_2 couple is -1.358 V. To oxidize Cl^- to Cl_2, you need a couple with a reduction potential more positive than 1.358 V, because then the net potential will be positive and ΔG will be negative. Examples include the $S_2O_8^{2-}/SO_4^{2-}$ couple ($E^\circ = 1.96$ V), the H_2O_2/H_2O couple ($E^\circ = 1.763$ V), and the α-PbO_2/Pb^{2+} couple ($E^\circ = 1.468$ V). Therefore, since the *oxidized* form of the couple will get *reduced* by *oxidizing* Cl^-, you would want to use either $S_2O_8^{2-}$, H_2O_2, or α-PbO_2 to oxidize Cl^- to Cl_2.

(b) Reducing Cr^{3+}(aq) to Cr^{2+}(aq)? In this case, the reduction potential is -0.424 V. Therefore, to have a net $E^\circ > 0$, you need a couple with an oxidation potential more positive than 0.424 V (or a couple with a reduction potential more negative than -0.424 V). Examples include Mn^{2+}/Mn ($E^\circ = -1.18$ V), Zn^{2+}/Zn ($E^\circ = -0.7626$ V), and NH_3OH^+/N_2 ($E^\circ = -1.87$ V). Remember that it is the reduced form of the couple that you want to use to reduce Cr^{3+} to Cr^{2+}, so in this case you would choose either metallic manganese, metallic zinc, or NH_3OH^+.

(c) Reducing Ag^+(aq) to Ag(s)? The reduction potential is 0.799 V. As above, the reduced form of any couple with a reduction potential less than (i.e., less positive than) 0.799 V will reduce Ag^+ to silver metal.

(d) Reducing I_2 to I^-? The reduction potential is 0.535 V. As above, the reduced form of any couple with a reduction potential less than 0.535 V will reduce iodine to iodide.

6.4 **Write balanced equations, if a reaction occurs, for the following species in aerated aqueous acid?** **(a) Cr^{2+}?** For all of these species, you must determine whether they can be oxidized by O_2. The standard potential for the reduction $O_2 + 4H^+ + 4e^- \rightarrow 2H_2O$ is 1.23 V. Therefore, only redox couples with a reduction potential less positive than 1.23 V will be driven to completion to the oxidized member of the couple by the reduction of O_2 to H_2O. Since the Cr^{3+}/Cr^{2+} couple has $E° = -0.424$ V, Cr^{2+} *will* be oxidized to Cr^{3+} by O_2. The balanced equation is:

$$4\,Cr^{2+}(aq) + O_2(g) + 4H^+(aq) \rightarrow 4\,Cr^{3+}(aq) + 2H_2O(l) \quad E° = 1.65\ V$$

(b) Fe^{2+}? Since the Fe^{3+}/Fe^{2+} couple has $E° = 0.771$ V, Fe^{2+} *will* be oxidized to Fe^{3+} by O_2. The balanced equation is:

$$4\,Fe^{2+}(aq) + O_2(g) + 4H^+(aq) \rightarrow 4\,Fe^{3+}(aq) + 2H_2O(l) \quad E° = 0.46\ V$$

(c) Cl^-? Both of the following couples have $E°$ values, shown in parentheses, more positive than 1.23 V, so there will be no reaction when O_2 is mixed with aqueous chloride ion in acid solution: ClO_4^-/Cl^- (1.287 V); Cl_2/Cl^- (1.358 V). The appropriate equation is $Cl^-(aq) + O_2(g) \rightarrow$ NR (NR = no reaction):

(d) HOCl? Since the $HClO_2/HClO$ couple has $E° = 1.701$ V, HClO *will not* be oxidized to $HClO_2$ by O_2. The standard potential for the oxidation of HOCl by O_2 is –0.47 V.

(e) $Zn(s)$? Since the Zn^{2+}/Zn couple has $E° = -0.763$ V, metallic zinc *will* be oxidized to Zn^{2+} by O_2. The balanced equation is:

$$2\,Zn(s) + O_2(g) + 4H^+(aq) \rightarrow 2\,Zn^{2+}(aq) + 2H_2O(l) \quad E° = 1.99\ V$$

A competing reaction is $Zn(s) + 2H^+(aq) \rightarrow Zn^{2+}(aq) + H_2(g)$ ($E° = 0.763$ V).

6.5 **Balanced equations for redox reactions?** **(a) Fe^{2+}?** When a chemical species is dissolved in aerated acidic aqueous solution, you need to think about four possible redox reactions. These are (i) the species might oxidize water to O_2, (ii) the species might reduce water (hydronium ions) to H_2, (iii) the species might be oxidized by O_2, and (iv) the species might undergo disproportionation. In the case of Fe^{2+}, the two couples of interest are Fe^{3+}/Fe^{2+} ($E° = 0.77$ V) and Fe^{2+}/Fe ($E° = -0.44$ V). Consider the four possible reactions (refer to Appendix

2 for the potentials you need): (i) The O_2/H_2O couple has $E° = 1.229$ V, so the oxidation of water would only occur if the Fe^{2+}/Fe potential were *greater* than positive 1.229 V. In other words, the net reaction below is not spontaneous because the net $E°$ is less than zero:

$$2Fe^{2+} + 4e^- \longrightarrow 2Fe \qquad\qquad E° = -0.44 \text{ V}$$

$$2H_2O \longrightarrow O_2 + 4H^+ + 4e^- \quad E° = -1.23 \text{ V}$$

$$2Fe^{2+} + 2H_2O \longrightarrow 2Fe + O_2 + 4H^+ \quad E° = -1.67 \text{ V}$$

Therefore, Fe^{2+} will not oxidize water. (ii) The H_2O/H_2 couple has $E° = 0$ V (by definition), so the reduction of water would only occur if the potential for the oxidation of Fe^{2+} to Fe^{3+} was positive, and it is -0.77 V (note that it is the *reduction* potential for the Fe^{3+}/Fe^{2+} couple that is *positive* 0.77 V). Therefore, Fe^{2+} will not reduce water. (iii) Since the O_2/H_2O has $E° = 1.229$ V, the reduction of O_2 would occur as long as the potential for the oxidation of Fe^{2+} to Fe^{3+} was less negative than -1.229 V. Since it is only -0.77 V (see above), Fe^{2+} will reduce O_2 and in doing so will be oxidized to Fe^{3+}. (iv) The disproportionation of a chemical species will occur if it can act as its own oxidizing agent and reducing agent, which will occur when the potential for reduction minus the potential for oxidation is *positive*. For the Latimer diagram

$$Fe^{3+} \xrightarrow{0.771} Fe^{2+} \xrightarrow{-0.44} Fe$$

the difference $(-0.44 \text{ V}) - (0.771 \text{ V}) = -1.21$ V, so disproportionation will not occur.

(b) Ru^{2+}? Consider the Ru^{3+}/Ru^{2+} ($E° = 0.249$ V) and Ru^{2+}/Ru ($E° = 0.81$ V) couples. Since the Ru^{2+}/Ru couple has a potential that is not more positive than 1.229 V, Ru^{2+} will not oxidize water. Also, since the Ru^{3+}/Ru^{2+} couple has a positive potential, Ru^{2+} will not reduce water. However, since the potential for the oxidation of Ru^{2+} to Ru^{3+}, -0.249 V, is less negative than -1.229 V, Ru^{2+} will reduce O_2 and in doing so will be oxidized to Ru^{3+}. Finally, since the difference $(0.81 \text{ V}) - (0.249 \text{ V})$ is positive, Ru^{2+} will disproportionate in aqueous acid to Ru^{3+} and metallic ruthenium. It is not possible to tell from the potentials whether the reduction of O_2 by Ru^{2+} or the disproportionation of Ru^{2+} will be the faster process.

(c) $HClO_2$? Consider the $ClO_3^-/HClO_2$ ($E° = 1.181$ V) and $HClO_2/HClO$ ($E° = 1.674$ V) couples. Since the $HClO_2/HClO$ couple has a potential that is more positive than 1.229 V, $HClO_2$ *will* oxidize water. However, since the $ClO_3^-/HClO_2$ couple has a positive potential, $HClO_2$ will not reduce water. Since the potential for the oxidation of $HClO_2$ to ClO_3^-, -1.181 V, is less negative than -1.229 V, $HClO_2$ will reduce O_2 and in doing so will be oxidized to ClO_3^-. Finally, since the difference (1.674 V) – (1.181 V) is positive, $HClO_2$ will disproportionate in aqueous acid to ClO_3^- and $HClO$. As above, it is not possible to tell from the potentials whether the reduction of O_2 by $HClO_2$ or the disproportionation of $HClO_2$ will be the faster process.

(d) Br_2? Consider the Br_2/Br^- ($E° = 1.065$ V) and $HBrO/Br_2$ ($E° = 1.604$ V) couples. Since the Br_2/Br^- couple has a potential that is not more positive than 1.229 V, Br_2 will not oxidize water. Also, since the $HBrO/Br_2$ couple has a positive potential, Br_2 will not reduce water. Since the potential for the oxidation of Br_2 to $HBrO$, -1.604 V, is more negative than -1.229 V, Br_2 will not reduce O_2. Finally, since the difference (1.065 V) – (1.604 V) is negative, Br_2 will not disproportionate in aqueous acid to Br^- and $HBrO$. In fact, the equilibrium constant for the following reaction is 7.2×10^{-9}.

$$Br_2(aq) + H_2O(l) \rightleftharpoons Br^-(aq) + HBrO(aq) + H^+(aq)$$

6.6   Write the Nernst equation for?   (a) The reduction of O_2? In general terms, the Nernst equation is given by the formula:

$$E = E° - (RT/NF)\ln Q \qquad \text{where } Q \text{ is the reaction quotient}$$

For the reduction of oxygen, $O_2(g) + 4\,H^+(aq) + 4\,e^- \rightarrow 2\,H_2O(l)$,

$$Q = 1/(p(O_2)[H^+]^4) \qquad \text{and} \qquad E = E° - (0.059/4)\left(\log(1/(p(O_2))[H_3O^+]^4)\right)$$

Therefore, the potential for O_2 reduction at pH = 7 and $p(O_2) = 0.20$ bar is

$$E = 1.229 \text{ V} - 0.42 = 0.81 \text{ V}$$

(b) The reduction of $Fe_2O_3(s)$? For the reduction of solid iron(III) oxide, $Fe_2O_3(s) + 6\,H^+(aq) + 6\,e^- \rightarrow 2\,Fe(s) + 3\,H_2O(l)$,

$$Q = 1/[H^+]^6 \qquad \text{and} \qquad E = E° - (RT/nF)(13.8\,\text{pH})$$

since pH $= -\log[H^+] = -2.3\ln[H^+]$ and $\log[H^+]^6 = 6\log[H^+]$.

6.7   Calculate $E, \Delta G°$, and K for the reduction of CrO_4^{2-} and $[Cu(NH_3)_4]^+$ by H_2 in basic solution? You must recognize that although the two values of $E°$ are so similar, the two reactions involve different numbers of electrons, N, and the expressions for $\Delta G°$ and K involve N whereas E does not:

$$\Delta G° = -NFE \qquad \text{and} \qquad \Delta G° = -RT\ln K$$

For the reduction of chromate ion:

$$\Delta G° = -nFE = -(3)(9.65 \times 10^4 \text{ C mol}^{-1})(-0.11 \text{ V}) = 31.8 \text{ kJ mol}^{-1}$$

(recall that 1 V = 1 J C^{-1}) and, since RT = 2.48 kJ mol^{-1} at 25 °C,

$$K = \exp((-31.8 \text{ kJ mol}^{-1})/(2.48 \text{ kJ mol}^{-1})) = 2.70 \times 10^{-6}$$

For the reduction of $[Cu(NH_3)_4]^+$:

$$\Delta G° = -(9.65 \times 10^4 \text{ C mol}^{-1})(-0.10) = 9.65 \text{ kJ/mol}^{-1}$$
$$\text{and } K = \exp((-9.65 \text{ kJ mol}^{-1})/(2.48 \text{ kJ mol}^{-1})) = 2.04 \times 10^{-2}$$

So, because the reductions differ by the number of electrons required (3 vs. 1), the equilibrium constants are different by a factor of about 8000.

6.8   Using Frost diagrams? (a) What happens when Cl_2 is dissolved in aqueous basic solution? The Frost diagram for chlorine in basic solution is shown in Figure 6.16 and is reproduced below. If the points for Cl$^-$ and ClO$_4^-$

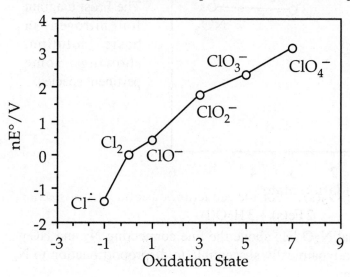

are connected by a straight line, Cl$_2$ lies above it. Therefore, Cl$_2$ is thermodynamically susceptible to disproportionation to Cl$^-$ and ClO$_4^-$ when it is dissolved in aqueous base. In practice, the oxidation of ClO$^-$ is slow, so a solution of Cl$^-$ and ClO$^-$ is formed when Cl$_2$ is dissolved in aqueous base.

(b) What happens when Cl_2 is dissolved in aqueous acid solution? The Frost diagram for chlorine in acidic solution is shown in Figure 6.16. If the points for Cl^- and any positive oxidation state of chlorine are connected by a straight line, the point for Cl_2 lies below it (if only slightly). Therefore, Cl_2 will not disproportionate. However, $E°$ for the Cl_2/Cl^- couple, 1.36 V, is more positive than $E°$ for the O_2/H_2O couple, 1.23 V. Therefore, Cl_2 is thermodynamically capable of oxidizing water as follows, although the reaction is very slow:

$$Cl_2(aq) + H_2O(l) \rightarrow 2\,Cl^-(aq) + 2\,H^+(aq) + 1/2\,O_2(l) \qquad E° = 0.13\ V$$

(c) Should $HClO_3$ disproportionate in aqueous acid solution? The point for ClO_3^- in acidic solution on the Frost diagram in Figure 6.16 lies above the single straight line connecting the points for Cl_2 and ClO_4^-. Therefore, since ClO_3^- is thermodynamically unstable with respect to disproportionation in acidic solution (i.e. it *should* disproportionate), the failure of it to exhibit any observable disproportionation must be due to a kinetic barrier.

6.9   **Write equations for the following reactions: (a) N_2O is bubbled into aqueous NaOH solution?** Part of the Frost diagram for nitrogen in basic solution is shown below (the entire diagram is shown in Figure 6.9).

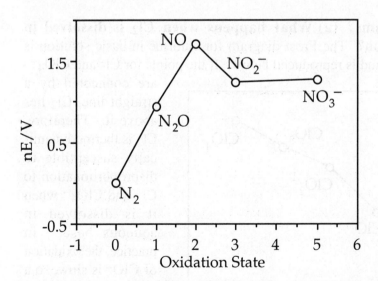

The Frost diagram for nitrogen in basic solution, showing some pertinent species.

Inspection of it shows that N_2O lies above the line connecting N_2 and NO_3^-. Therefore, N_2O is thermodynamically susceptible to disproportionation to N_2 and NO_3^- in basic solution:

$$5 N_2O(aq) + 2 OH^-(aq) \rightarrow 2 NO_3^-(aq) + 4 N_2(g) + H_2O(l)$$

However, the redox reactions of nitrogen oxides and oxyanions are generally slow. In practice, N_2O has been found to be inert.

(b) Zinc metal is added to aqueous acidic sodium triiodide? The overall reaction is:

$$Zn(s) + I_3^-(aq) \rightarrow Zn^{2+}(aq) + 3 I^-(aq)$$

Since $E°$ values for the Zn^{2+}/Zn and I_3^-/I^- couples are –0.76 V and 0.54 V, respectively (these potentials are given in Appendix 2), $E°$ for the net reaction above is 0.54 V + 0.76 V = 1.30 V. Since the net potential is positive, this *is* a favorable reaction, and it should be kinetically facile if the zinc metal is finely divided and thus well exposed to the solution.

(c) I_2 is added to excess aqueous acidic $HClO_3$? Since $E°$ values for the I_2/I^- and ClO_3^-/ClO_4^- couples are 0.54 V and –1.20 V, respectively (see Appendix 2), the net reaction involving the reduction of I_2 to I^- and the oxidation of ClO_3^- to ClO_4^- will have a net negative potential, $E° = 0.54$ V + (–1.20 V) = –0.66 V. Therefore, this net reaction will not occur. However, since $E°$ values for the IO_3^-/I_2 and ClO_3^-/Cl^- couples are 1.19 V and 1.47 V, respectively, the following net reaction will occur, with a net $E° = 0.28$ V:

half-reaction: $ClO_3^-(aq) + 6 H^+(aq) + 6 e^- \rightarrow Cl^-(aq) + 3 H_2O(l)$

half-reaction: $I_2(s) + 6 H_2O(l) \rightarrow 2 IO_3^-(aq) + 12 H^+(aq) + 10 e^-$

net rxn: $3 I_2(s) + 5 ClO_3^-(aq) + 3 H_2O(l) \rightarrow 6 IO_3^-(aq) + 5 Cl^-(aq) + 6 H^+(aq)$

6.10 **Will acid or base most favor the following half-reactions?** **(a) $Mn^{2+} \rightarrow MnO_4^-$?** You can answer questions such as this by applying Le Chatelier's principle to the complete, balanced half-reaction, which in this case is

$$Mn^{2+}(aq) + 4 H_2O(l) \rightarrow MnO_4^-(aq) + 8 H^+(aq) + 5 e^-$$

Since hydrogen ions are produced by this oxidation half-reaction, raising the pH will favor the reaction. (Alternatively, you could come to the same conclusion by using the Nernst equation.) Thus, this reaction is favored in basic solution. At a sufficiently high pH, Mn^{2+} will precipitate as $Mn(OH)_2$. The rate of oxidation for this solid will be slower than for dissolved species.

(b) $ClO_4^- \rightarrow ClO_3^-$? The balanced half-reaction is

$$ClO_4^-(aq) + 2H^+(aq) + 2e^- \rightarrow ClO_3^-(aq) + H_2O(l)$$

Since hydrogen ions are consumed by this reduction half-reaction, lowering the pH will favor the reaction. Thus, this reaction is favored in acidic solution. This is the reason that perchloric acid is a dangerous oxidizing agent, even though salts of the perchlorate ion are frequently stable in neutral or basic solution.

(c) $H_2O_2 \rightarrow O_2$? The balanced half-reaction is

$$H_2O_2(aq) \rightarrow O_2(g) + 2H^+(aq) + 2e^-$$

Since hydrogen ions are produced by this oxidation half-reaction, raising the pH will favor the reaction. Thus, the reaction is favored in basic solution. This means that hydrogen peroxide is a better reducing agent in base than in acid.

(d) $I_2 \rightarrow 2I^-$? Since protons are neither consumed nor produced in this reduction half-reaction, and since I^- is not protonated in aqueous solution (because HI is a very strong acid), this reaction has the same potential in acidic or basic solution, 0.535 V (this can be confirmed by consulting Appendix 2).

6.11 **Comment on the mechanisms of the following reactions:** **(a) HOI(aq) + I$^-$(aq) $\rightarrow$ I$_2$(s) + OH$^-$(aq)?** The transformation of reactants into products can be envisioned to take place by the simple transfer of an "I^+" ion from HOI to I^-, as shown below. Therefore, this is probably an atom transfer reaction.

$$HOI + I^- \rightarrow HOI\cdots I^- \rightarrow HO^-\cdots I^+\cdots I^- \rightarrow HO^-\cdots I_2 \rightarrow HO^- + I_2$$

(b) $[Co(phen)_3]^{3+}(aq) + [Cr(bipy)_3]^{2+}(aq) \rightarrow [Co(phen)_3]^{2+}(aq) + [Cr(bipy)_3]^{3+}(aq)$? This reaction takes place without any net change in the coordination spheres of the two metal ions, since Co^{3+} and Cr^{3+} are kinetically inert to substitution. The cobalt ion has three phenanthroline ligands whether it is +3 or +2 and the chromium ion has three bipyridine ligands whether it is +2 or +3. Therefore, this is probably a simple outer-sphere electron transfer reaction.

(c) $IO_3^-(aq) + 8I^-(aq) + 6H^+(aq) \rightarrow 3I_3^-(aq) + 3H_2O(l)$? Each of the I_3^- ions are formed by the reaction of I^- with I_2 (see the solution to Exercise 6.9(c)). It is unlikely that I_2 could be formed by a simple single-step reaction of I^- with IO_3^-, since IO_3^- must lose all three of its oxygen atoms. Therefore, this is probably a multistep mechanism.

6.12   Determine the standard potential for the reduction of ClO_4^- to Cl_2?
The Latimer diagram for chlorine in acidic solution is given in Appendix 2 and
the relevant portion of it is reproduced below:

$$ClO_4^- \xrightarrow{1.201 \text{ V}} ClO_3^- \xrightarrow{1.181 \text{ V}} HClO_2 \xrightarrow{1.674 \text{ V}} HClO \xrightarrow{1.630 \text{ V}} Cl_2$$

$$+7 \qquad\qquad +5 \qquad\qquad +3 \qquad\qquad +1 \qquad\qquad 0$$

To determine the potential for any couple, you must calculate the *weighted
average* of the potentials of intervening couples. In general terms it is

$$(n_1 E°_1 + n_2 E°_2 + \dots + n_n E°_n)/(n_1 + n_2 + \dots + n_n)$$

and in this specific case it is

$$((2)(1.201 \text{ V}) + (2)(1.181 \text{ V}) + (2)(1.674)$$
$$+ (1)(1.630 \text{ V}))/(2 + 2 + 2 + 1) = 1.392 \text{ V}$$

Thus, the standard potential for the ClO_4^-/Cl_2 couple is 1.392 V. The balanced
half-reaction for this reduction is

$$2\,ClO_4^-(aq) + 16\,H^+(aq) + 14\,e^- \rightarrow Cl_2(g) + 8\,H_2O(l)$$

Note that the point for ClO_4^- at pH 0 in the Frost diagram shown in Figure 6.16
has a y value of ~9.7, which is $NE°/V$ or 7(1.392).

**6.13   Calculate the equilibrium constant for $Au^+(aq) + 2\,CN^-(aq) \rightarrow$
$[Au(CN)_2]^-(aq)$?** To calculate an equilibrium constant using thermodynamic
data, you can make use of the expressions $\Delta G = -RT\ln K$ and $\Delta G = -NFE$. You
can use the given potential data to calculate ΔG for each of the two half-
reactions, then use Hess' Law to calculate ΔG for the overall process, then
calculate K from the net ΔG. The net reaction $Au^+(aq) + 2\,CN^-(aq) \rightarrow$
$[Au(CN)_2]^-(aq)$ is the following sum:

$$Au^+(aq) + e^- \longrightarrow Au(s)$$

$$Au(s) + 2CN^-(aq) \longrightarrow [Au(CN)_2]^-(aq)$$

$$Au^+(aq) + 2CN^-(aq) \longrightarrow [Au(CN)_2]^-(aq)$$

ΔG for the first reaction is $-nFE = -(1)(96.5 \text{ kJ mol}^{-1} \text{ V}^{-1})(1.69 \text{ V}) = -163 \text{ kJ mol}^{-1}$. ΔG for the second reaction is $-(1)(96.5 \text{ kJ mol}^{-1} \text{ V}^{-1})(0.6 \text{ V}) = -58 \text{ kJ mol}^{-1}$. The net ΔG is the sum of these two values, -221 kJ mol^{-1}. Therefore, assuming that $T = 298$ K:

$$K = \exp(-\Delta G/RT) = \exp[(221 \text{ kJ mol}^{-1})/(8.31 \text{ J K}^{-1} \text{ mol}^{-1})(298)]$$

$$K = \exp(89.2) = 5.7 \times 10^{38}$$

6.14 **Find the approximate potential of an aerated lake at pH = 6, and predict the predominant species? (a) Fe?** According to Figure 6.14, the potential range for surface water at pH 6 is $0.5 - 0.6$ V, so a value of 0.55 V can be used as the approximate potential of an aerated lake at this pH. Inspection of the Pourbaix diagram for iron (Figure 6.13) shows that at pH 6 and $E = 0.55$ V, the stable species of iron is $Fe(OH)_3$. Therefore, this compound of iron would predominate.

(b) Mn? Inspection of the Pourbaix diagram for manganese (Figure 6.15) shows that at pH 6 and $E = 0.55$ V, the stable species of manganese is Mn_2O_3. Therefore, this compound of manganese would predominate.

(c) S? At pH 0, the potential for the HSO_4^-/S couple is 0.387 V (this value was calculated using the weighted average of the potentials given in the Latimer diagram for sulfur in Appendix 2), so the lake will oxidize S_8 all the way to HSO_4^-. At pH 14, the potentials for intervening couples are all negative, so SO_4^{2-} would again predominate. Therefore, HSO_4^- is the predominant sulfur species at pH 6.

6.15 **What is the maximum E for an anaerobic environment rich in Fe^{2+} and H_2S?** Any species capable of oxidizing either Fe^{2+} or H_2S at pH 6 cannot survive in this environment. According to the Pourbaix diagram for iron (Figure 6.13), the potential for the $Fe(OH)_3/Fe^{2+}$ couple at pH 6 is approximately 0.3 V. Using the Latimer diagrams for sulfur in acid and base (see Appendix 2), the H_2S, S potential at pH 6 can be calculated as follows:

$$0.14 \text{ V} - (6/14)(0.14 \text{ V} - (-0.45 \text{ V})) = -0.11 \text{ V}$$

Any potential higher than this will oxidize hydrogen sulfide to elemental sulfur. Therefore, as long as H_2S is present, the maximum potential possible is approximately -0.1 V.

6.16 How will edta^{4-} complexation affect $M^{2+} \rightarrow M^0$ reductions? Since edta^{4-} forms very stable complexes with M^{2+}(aq) ions of period 4 d-block elements but *not* with the zerovalent metal atoms, the reduction of a $M(edta)^{2-}$ complex will be more difficult than the reduction of the analogous M^{2+} aqua ion. Since the reductions are more difficult, the reduction potentials become less positive (or more negative, as the case may be). This is represented in the equation below. The reduction of the $M(edta)^{2-}$ complex includes a decomplexation step, with a positive free energy change. The reduction of M^{2+}(aq) does not require this additional expenditure of free energy.

$$M(edta)^{2-} \quad \xleftarrow{\quad -edta^{4-} \quad} \quad M^{2+} \quad \xrightarrow{\quad +2\,e^- \quad} \quad M^0$$

A drawing of a metal complex of edta^{4-} is shown with the answer to Ex. 7.5.

6.17 Which of the boundaries depend on the choice of $[Fe^{2+}]$? Any boundary between a soluble species and an insoluble species will change as the concentration of the soluble species changes. For example, consider the line separating Fe^{2+}(aq) from $Fe(OH)_3$(s) in Figure 6.13. As shown in the text, $E = E° - (0.059\,V/2)\log([Fe^{2+}]^2/[H_3O^+]^6)$ (see Section 6.10(b), *Pourbaix diagrams*). Clearly, the potential at a given pH is dependent on the concentration of soluble Fe^{2+} (aq). The boundaries between the two soluble species, Fe^{2+}(aq) and Fe^{3+}(aq), and between the two insoluble species, $Fe(OH)_2$(s) and $Fe(OH)_3$(s), will not depend on the choice of $[Fe^{2+}]$.

6.18 Using cyclic voltammograms? If the metal ion is coordinated by only water molecules, then the potential measured from the cyclic voltammogram (i.e. the midpoint between the cathodic and anodic peak potentials) should be equal to the value derived from standard potentials, taking into account the pH of the solution. However, if other ligands are present, the measured potential will not correspond to the standard value. Instead, the reduction potential will be either higher or lower than the standard value depending on whether the ligands coordinate more strongly to the reduced metal ion or to the oxidized metal ion.

6.19 Why is aerated water containing CO_2 corrosive to iron? The oxidation of metallic iron by O_2 is a favorable process, even in neutral aqueous solution. However, the presence of dissolved CO_2 produces some acid (H_3O^+)

and some carbonate ion (CO_3^{2-}). The lower pH will make the oxidation more favorable. The presence of carbonate ion will also make the oxidation more favorable, since it forms complexes with iron cations in aqueous solution.

Guide to Solutions Quiz

1 Use reduction potentials for the NO_2^-/N_2O_4 (basic solution) and HNO_2/N_2O_4 (acidic solution) couples to determine K_a for HNO_2 at 25°C.

2 Consider the Latimer diagram for samarium, shown below. Calculate $[Sm^{3+}]$ when the equilibrium concentration of Sm^{2+} is 0.1 M at 25°C.

$$Sm^{3+} \xrightarrow{\;-1.55\;} Sm^{2+} \xrightarrow{\;-2.67\;} Sm$$

3 Calculate the potential for the following reaction in acid solution when the concentration of the products are both 0.1 M. Note: Al and I_2 are solids.

$$2\,Al + 3\,I_2 \rightarrow 2\,Al^{3+} + 6\,I^-$$

4 Sulfuric and nitric acids are oxidizing acids because HSO_4^- and NO_3^- have more positive reduction potentials than H^+ (i.e., H_3O^+). Which of the following are also oxidizing acids: HCl, H_3PO_4, HIO_3, H_2S?

5 (a) Construct one Frost diagram for the three p-block metals Ga, In, and Tl in aqueous acid. (b) Can excess gallium metal be used to reduce Tl^{3+} all the way to thallium metal? Explain carefully.

6 (a) Construct one Frost diagram for the three group 6 metals Cr, Mo, and W in aqueous acid. (b) Explain, using your diagram, why Cr(VI) is a very powerful oxidizing agent while W(VI) is not an oxidizing agent at all.

7 Comment on the effect of raising the pH on the reduction potentials for the following couples: S_8/S^{2-}; NO_3^-/NO_2^-; I_2/I^-. Be sure to consider any Brønsted acid–base reactions that might occur with any of the species present.

8 Write balanced equations for the following reactions in water: (a) An acidic solution of V^{3+} is added to excess acidic Fe^{3+} solution. (b) A basic solution of $FeSO_4$ is added to excess basic ClO^- solution. (c) Sulfur is added to a basic solution of Na_2SO_3. (d) Arsenic is added to an acidic solution of Ag^+.

9 Calculate the HPO_3^{2-}/P_4 E value when $[OH^-] = 2.50$ M and $[HPO_3^{2-}] = 1$ M.

10 Use the Latimer diagram for plutonium in acid solution (Appendix 2) to answer the following questions: (a) Will Pu metal react with H_2O? (b) Under what pH conditions would PuO_2^+ be least likely to disproportionate to Pu^{4+} and PuO_2^{2+}?

7   *d*-Metal complexes

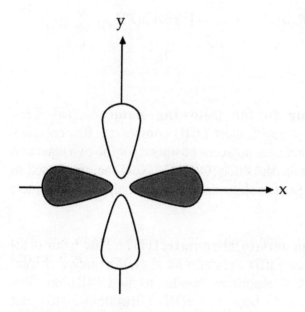

Many properties of *d*-metal complexes can be understood on the basis of the relative energies of the five *d* orbitals that are part of the valence shells of *d*-metal atoms or ions. These include their structure, their spectra, their magnetic behavior, and aspects of their thermodynamic and kinetic reactivity. Ligands in an octahedral complex are frequently placed on the *x*, *y*, and *z* axes. This results in the metal d_{x2-y2} orbital, shown at the left, becoming a metal–ligand σ antibonding MO.

S7.1   Identifying isomers?   The two square-planar isomers of [PtBrCl(PR$_3$)$_2$] are shown below. The NMR data indicate that isomer A is the *trans* isomer since the two trialkylphosphine ligands occupy opposite corners of the square plane. Isomer B is the *cis* isomer. Note that the two phosphine ligands in the *trans* isomer are related by symmetry elements that this C_{2v} molecule possesses, namely the C_2 axis (the Cl–Pt–Br axis) and the σ$_v$ mirror plane that is perpendicular to the molecular plane. Therefore, they exhibit the same chemical shift in the ^{31}P NMR spectrum of this compound. The two phosphine ligands in the *cis* isomer are not related by the σ mirror plane that this C_s molecule possesses. Since they are chemically nonequivalent, they give rise to separate groups of ^{31}P resonances.

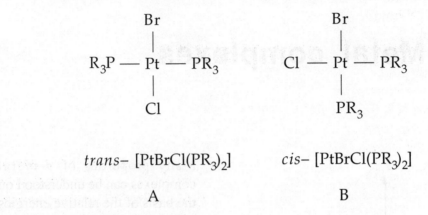

$$trans- [PtBrCl(PR_3)_2] \qquad cis- [PtBrCl(PR_3)_2]$$

$$A \qquad\qquad B$$

S7.2   **Give formulas corresponding to the following names?**   **(a) *Cis*-diaquadichloroplatinum(II)?** As with most Pt(II) complexes, this complex is square-planar. The *cis* prefix indicates adjacent positions for the two aqua and two chloro ligands. The formula is *cis*-[PtCl$_2$(OH$_2$)$_2$]. Note the order used in naming complexes of the *d*-block metals (ligands are listed in alphabetical order; then the metal ion is listed).

(b) Diammine*tetrakis*(isothiocyanato)chromate(III)? This is an octahedral complex of Cr(III) The *ate* suffix indicates an overall negative charge. Two NH$_3$ molecules and four NCS$^-$ anions are bonded to the Cr(III) ion. The NCS$^-$ ligands use their N atoms to bond to Cr(III) (-isothiocyanato-, not -thiocyanato-). The formula is [Cr(NCS)$_4$(NH$_3$)$_2$]$^-$. Note that a complex with this composition can exist as two structural isomers, *cis*-[Cr(NCS)$_4$(NH$_3$)$_2$]$^-$ or *trans*-[Cr(NCS)$_4$(NH$_3$)$_2$]$^-$.

(c) *Tris*(ethylenediamine)rhodium(III)? This is also an octahedral complex. The metal ion is Rh(III), there are three symmetric bidentate ligands, and the complex is not anionic (-rhodium(III), not -rhodate(III)). Since the accepted abbreviation for ethylenediamine (H$_2$NCH$_2$CH$_2$NH$_2$) is en, the formula may be written as [Rh(en)$_3$]$^{3+}$.

S7.3   **Sketches of the *mer* and *fac* isomers of [Co(gly)$_3$]?** The anion of glycine is an unsymmetrical bidentate ligand (it has a neutral amine nitrogen donor atom and a negatively charged carboxylate oxygen donor atom). In the *fac* isomer, the three possible N–Co–N bond angles are ~90°, while in the *mer* isomer, two N–Co–N bond angles are ~90° and one is 180°. If you imagine that the complex is a sphere, the three N atoms in the *mer* isomer lie on a *meridian* of the sphere (the largest circle that can be drawn on the surface of the sphere). In contrast, the three N atoms in the *fac* isomer form one of the eight triangular *faces* of the [Co(gly)₃] octahedron.

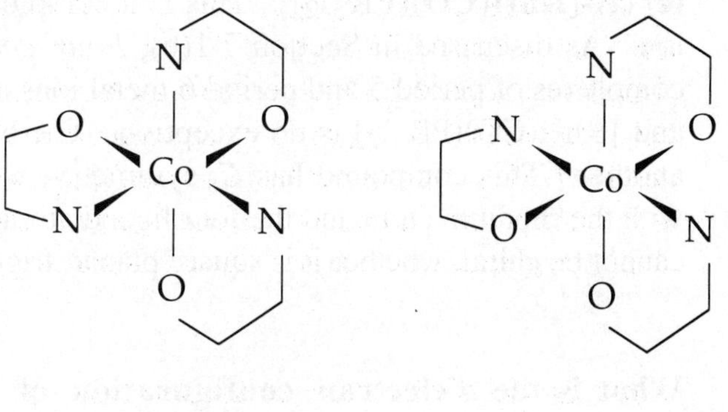

fac– [Co(gly)₃]   *mer*– [Co(gly)₃]

S7.4 **Which of the following are chiral?** (a) *cis*-[CrCl₂(ox)₂]³⁻? Drawings of two mirror images of this complex are shown below. They are not superimposable, and therefore represent two enantiomers. Therefore, this complex is chiral. Note that it does not possess an S_n axis, only a single C_2 axis. The point group of both enantiomers is C_2, so they are dissymmetric, not asymmetric.

(b) *trans*-[CrCl₂(ox)₂]³⁻? Drawings of two mirror images of this complex are also shown below. They *are* superimposable, and therefore do not represent two enantiomers but only a single isomer. Since this complex is achiral, it must possess at least one S_n axis. In fact, it possesses three different σ planes of symmetry, each of which is an S_1 axis.

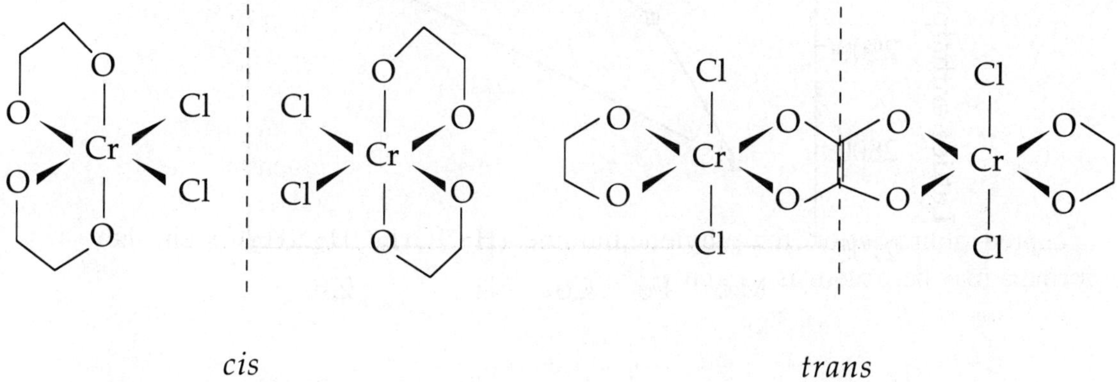

cis   *trans*

(c) *cis*-**[RhH(CO)(PR$_3$)$_2$]?** This is a complex of Rh(I), which is a d^8 metal ion. As discussed in Section 7.1(b), *Four coordination*, four-coordinate d^8 complexes of period 5 and period 6 metal ions are almost always square planar, and [RhH(CO)(PR$_3$)$_2$] is no exception. The bulky PR$_3$ ligands are *cis* to one another. This compound has C_s symmetry, with the mirror plane coincident with the rhodium atom and the four ligand atoms bound to it. A planar complex cannot be chiral, whether it is square planar, trigonal planar, etc.

S7.5   **What is the d electron configuration of [Mn(NCS)$_6$]$^{4-}$?** Since each isothiocyanate ligand has a single negative charge, the oxidation state of the manganese ion is II. Since Mn(II) is d^5, there are two possibilities for an octahedral complex, low-spin (t_{2g}^5), with one unpaired electron, or high-spin ($t_{2g}^3 e_g^2$), with five unpaired electrons. The observed magnetic moment of 6.06 μ_B is close to the spin-only value for five unpaired electrons, $(5 \times 7)^{1/2} = 5.92$ μ_B. Therefore, this complex is high-spin and has a $t_{2g}^3 e_g^2$ configuration.

S7.6   **Account for the variation in ΔH_L of octahedral difluorides?** If it were not for ligand field stabilization energy, LFSE, MF$_2$ lattice enthalpies would increase from Mn(II) to Zn(II). This is because the decreasing ionic radius, which is due to the increasing Z_{eff} as you cross through the d block from left to right, leads to decreasing M–F separations. Therefore, you expect that ΔH_L for MnF$_2$ (2780 kJ mol^{-1}) will be smaller than ΔH_L for ZnF$_2$ (2985 kJ mol^{-1}). In addition, as discussed for aqua ions and oxides, we expect additional

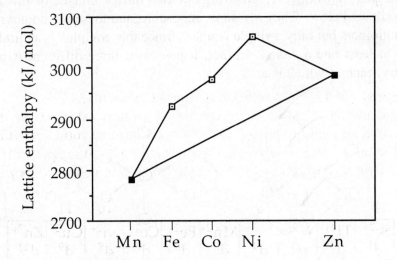

LFSE for these compounds. From Table 7.4, you see that LFSE = 0 for Mn(II), $0.4\Delta_o$ for Fe(II), $0.8\Delta_o$ for Co(II), $1.2\Delta_o$ for Ni(II), and 0 for Zn(II). The deviations of the observed values from the straight line connecting Mn(II) and Zn(II) are not quite in the ratio $0.4:0.8:1.2$, but note that the deviation for Fe(II) is smaller than that for Ni(II).

S7.7 **Photoelectron spectra of $[Fe(C_5H_5)_2]$ and $[Mg(C_5H_5)_2]$?** In the spectrum of $[Mo(CO)_6]$, the ionization energy around 8 eV was attributed to the t_{2g} electrons that are largely metal-based. The biggest difference between the photoelectron spectra of ferrocene and magnesocene is in the 6–8 eV region. Although Fe(II) has six *d* electrons (like Mo(0)), Mg(II) has no *d* electrons. Therefore, the differences in the 6–8 eV region can be attributed to the lack of *d* electrons for Mg(II).

S7.8 **Account for an increase of the second formation constant with respect to the first?** Normally, when a square-planar complex binds two ligands (one above and one below the square plane), $K_1 > K_2$. In this case, since the order is reversed, a structural and/or electronic rearrangement must be occurring during the second step. Since the metal is coordinated to the fairly rigid porphyrin macrocycle, a major structural change is unlikely. However, an electronic rearragement at the d^6 Fe(II) metal center does occur, as evidenced by the change from five-coordinate *high-spin* ($t_{2g}^4e_g^2$) to six-coordinate *low-spin* (t_{2g}^6). The change in ligand field stabilization energy (LFSE) on going from high-spin d^6 to low-spin d^6 is $0.4\Delta_o - 2.4\Delta_o = -2.0\Delta_o$. This negative enthalpy change makes ΔG for the second ligand binding equilibrium more negative than it would otherwise be, and $K_2 > K_1$.

7.1 **Tetrahedral $[MX_4]^{2-}$ complexes of the 3*d* elements?** The figure below shows the period 4 *d*-block metals as +2 cations, along with their d^n configuration and group number. The M^{2+} cations that form tetrahedral (T_d) $[MX_4]^{2-}$ tetrahalo complexes are the ones on the right of the series (re-read the Section 7.1(b), *Four-coordination*), and this is also indicated in the figure.

3	4	5	6	7	8	9	10	11	12
Sc^{2+}	Ti^{2+}	V^{2+}	Cr^{2+}	Mn^{2+}	Fe^{2+}	Co^{2+}	Ni^{2+}	Cu^{2+}	Zn^{2+}
d^1	d^2	d^3	d^4	d^5	d^6	d^7	d^8	d^9	d^{10}
					T_d	T_d	T_d	T_d	T_d

7.2   **Square-planar complexes? (a) Identify the elements?** The electron
configuration that is commonly found for four-coordinate square-planar
complexes is d^8. The metal ions that have this configuration are Rh^+, Ir^+, Ni^{2+},
Pd^{2+}, Pt^{2+}, and Au^{3+}. Note that Ni^{2+} also forms some four-coordinate
tetrahedral complexes (see the answer to Exercise 7.1, above).

3	4	5	6	7	8	9	10	11	12
							Ni^{2+}		
						Rh^+	Pd^{2+}		
						Ir^+	Pt^{2+}	Au^{3+}	

(b) Examples? Some examples of four-coordinate d^8 square-planar
complexes are $RhCl(PPh_3)_3$ (see Chapter 17), $IrCl(CO)(PPh_3)_2$, $Ni(CN)_4^{2-}$,
$PdCl_4^{2-}$, *cis*- and *trans*-$PtCl_2(NH_3)_2$, and $AuCl_4^-$. See also the answer to Self-
test S7.1. There are many other examples. Note that $NiCl_4^{2-}$ is tetrahedral, not
square-planar.

7.3   **Six-coordinate complexes?   (a) Sketch the two observed
structures?** Most six-coordinate complexes are either octahedral or trigonal-
prismatic. Drawings of these are shown below.

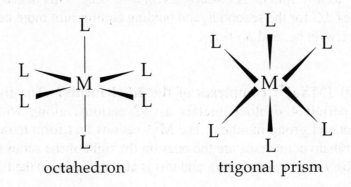

octahedron   trigonal prism

(b) Which one of these is rare? The trigonal prism is rare. Nearly all six-
coordinate complexes are octahedral.

(c) Examples? Some examples of six-coordinate octahedral complexes are
$[Cr(NH_3)_6)]^{3+}$, $[Cr(CO)_6]$, $[Cr(OH_2)_6]^{3+}$, $[Fe(CN)_6]^{3-}$, $[Fe(CN)_6]^{4-}$, and
$[RhCl_6]^{3-}$. There are many other examples.

7.4   **Name and draw structures of the following complexes?   (a)
[Ni(CO)4]?** The name of this complex is tetracarbonylnickel(0). Its structure,
which is tetrahedral, is shown below.

(b) [Ni(CN)4]$^{2-}$? The name of this complex is tetracyanonickelate(2–). Note
the use of the suffix -nickelate instead of -nickel, since the complex is negatively
charged. The structure, which is square-planar, is shown below.

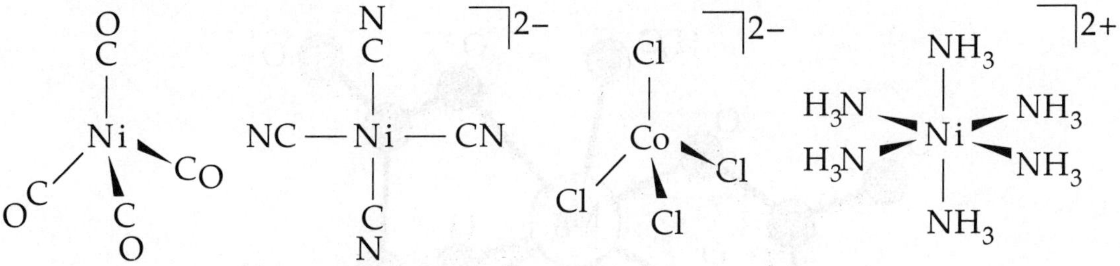

(c) [CoCl4]$^{2-}$? The name of this complex is tetrachlorocobaltate(2–). Its
structure, which is tetrahedral, is shown above.

(d) [Ni(NH3)6]$^{2+}$? The name of this six-coordinate octahedral complex is
hexaamminenickel(2+). Its structure is shown above.

7.5   **Draw structures of complexes that contain the ligands (a) en, (b)
ox, (c) phen, and (d) edta?** Generic drawings of octahedral complexes of
metal M for (a), (b), and (c) are shown below. In (a), the bidentate ligand
ethylenediamine (en = H$_2$NCH$_2$CH$_2$NH$_2$) takes up two coordination sites. For
clarity, the carbon atoms of the ethylene bridge are not explicitly shown. The
five-membered ring that is formed is not planar — one carbon atom is above and
one is below the plane formed by M and the two N atoms. In (b), the bidentate

ML$_4$(en)    ML$_4$(ox) 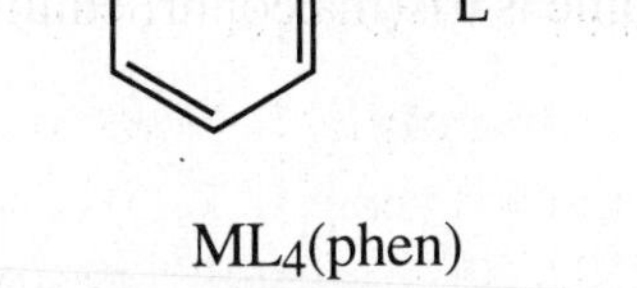   ML$_4$(phen)

ligand oxalate dianion (ox = $C_2O_4^{2-}$) also takes up two coordination sites. Once again, the carbon atoms of the ligand are shown simply as vertices. Due to the delocalized p system of the ligand, the five-membered ring in this case *is* planar. In (c), the bidentate ligand phenanthroline takes up two coordination sites. An example of a complex of ethylenediaminetetraacetate (edta^{4-} = $(O_2CCH_2)_2NCH_2CH_2N(CH_2CO_2)_2^{4-}$) is shown below. Note that the complex Mg(edta)(OH$_2$)$^{2-}$ is seven-coordinate.

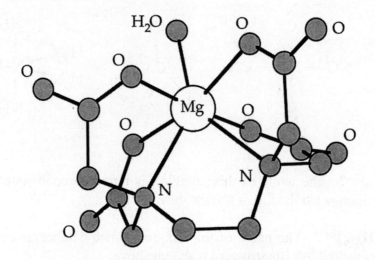

The structure of the [Mg(edta)(OH$_2$)]$^{2-}$ dianion. Hydrogen atoms have been omitted for clarity and the carbon atoms of the edta^{4-} ligand are left unlabelled. The hexadentate edta^{4-} ligand nearly surrounds the metal ion.

7.6   **Draw the structures of typical complexes?   (a) Square-planar four-coordinate?** Typical square-planar complexes are given in the answer to Exercise 7.2. The structure of one of them, [Ni(CN)$_4$]$^{2-}$, is shown in the answer to Exercise 7.4. Its name is tetracyanonickelate(2–).

(b) Trigonal-prismatic six-coordinate?   An example of a trigonal-prismatic complex is [Re(S$_2$C$_2$(CN)$_2$)$_3$], which has a structure very similar to that shown in Structure **12** in the text (see Section 7.1(d), *Six-coordination*). Its name is tris(maleonitriledithiolato)rhenium(0).

(c) Two-coordinate? Nearly all two-coordinate complexes are linear, including $[AgCl_2]^-$, $[Ag(NH_3)_2]^+$, $[Au(CN)_2]^-$, and $[Hg(CH_3)_2]$. The name of $[Au(CN)_2]^-$ is dicyanoaurate(1–). Its structure is shown below.

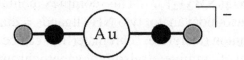

7.7   Draw structures?   (a) Pentaamminechlorocobalt(III) chloride?   (b) Hexaaquairon(III) nitrate? The structures of these two compounds are shown below.

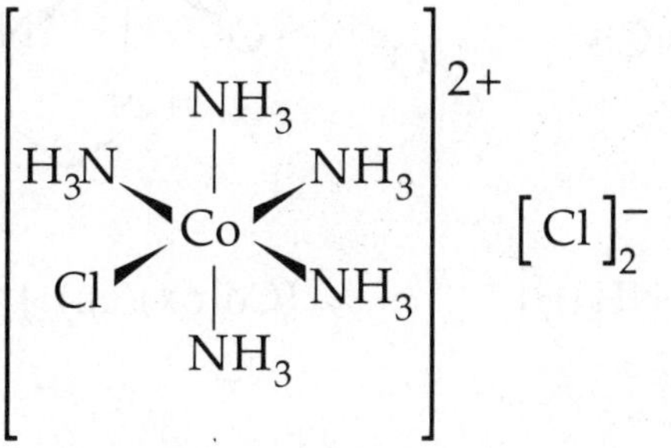

$$\left[\begin{array}{c} NH_3 \\ H_3N \quad | \quad NH_3 \\ Co \\ Cl \quad | \quad NH_3 \\ NH_3 \end{array}\right]^{2+} \quad \left[Cl\right]_2^-$$

$$\left[\begin{array}{c} OH_2 \\ H_2O \quad | \quad OH_2 \\ Fe \\ H_2O \quad | \quad OH_2 \\ OH_2 \end{array}\right]^{3+} \quad \left[NO_3\right]_3^-$$

$[CoCl(NH_3)_5][Cl]_2$ 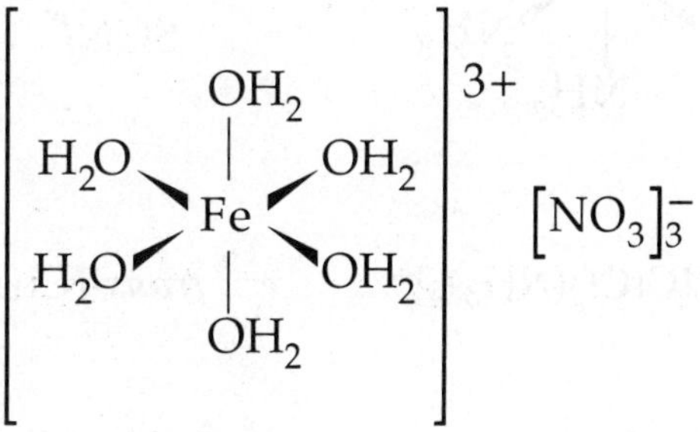 $[Fe(OH_2)_6][NO_3]_3$

(c) *cis*-dichloro*bis*(ethylenediamine)ruthenium(II)?   (d) µ-Hydroxo-*bis*(pentaamminechromium(III)) chloride? These structures are shown below.

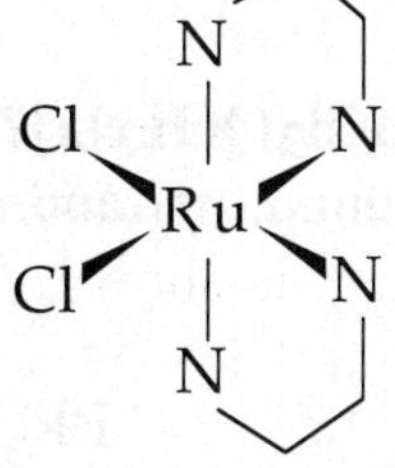

$$\begin{array}{c} N \\ Cl \quad | \quad N \\ Ru \\ Cl \quad | \quad N \\ N \end{array}$$

$$\left[\begin{array}{c} L \; L \quad H \quad L \; L \\ \backslash / \quad O \quad \backslash / \\ Cr \qquad Cr \\ L \diagup \; | \; \backslash \quad \diagup \; | \; \backslash L \\ L \; L \quad L \; L \end{array}\right]^{5+} \quad \left[Cl\right]_5^-$$

cis-[RuCl$_2$(en)$_2$] 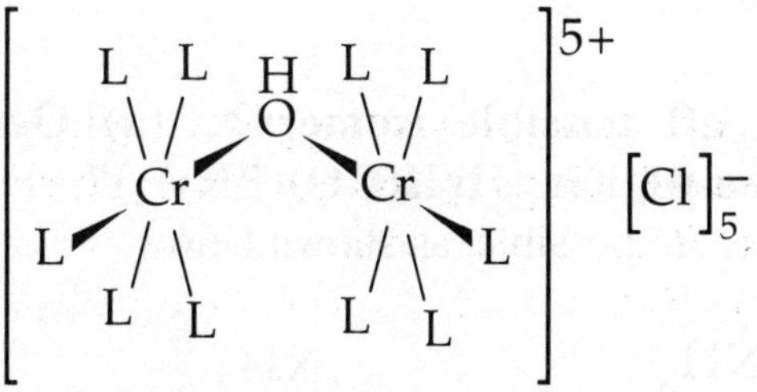 $[(H_3N)_5Cr(\mu\text{-}OH)Cr(NH_3)_5][Cl]_5$ (L = NH$_3$)

7.8   Name and draw structures?   (a) *cis*-[CrCl$_2$(NH$_3$)$_4$]$^+$? This complex contains (i) two Cl$^-$ ligands = dichloro, (ii) four NH$_3$ ligands = tetraammine (not tetraamine or tetramine or tetrammine), and (iii) Cr^{3+} in a complex that is not an anion = chromium(III). Putting (i), (ii), and (iii) together, the name of this complex is *cis*-tetraamminedichlorochromium(III). Note that the ligands are named in alphabetical order: tetr*a*ammine before di*c*hloro. The structure is shown below.

(b) *trans*-[Cr(NCS)$_4$(NH$_3$)$_2$]⁻? This complex contains four N-bonded NCS⁻ ligands = tetraisothiocyanato, two NH$_3$ ligands = diammine, and Cr^{3+} in a complex that is an anion (therefore, "chromate(III)" can be used). The name of this complex is *trans*-diamminetetraisothiocyanatochromate(III). Note the alphabetical ordering of the ligands. The structure is shown below.

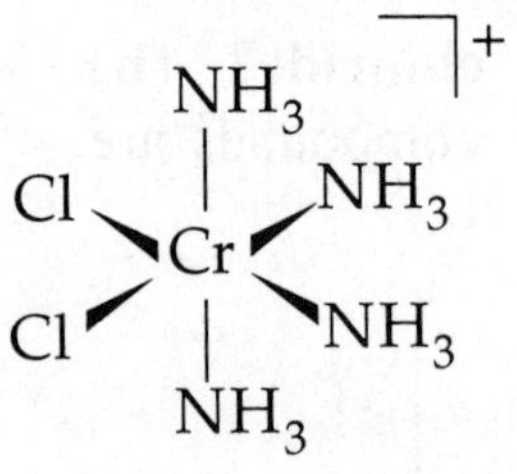

cis-[CrCl$_2$(NH$_3$)$_4$]⁺   *trans*-[Cr(NCS)$_4$(NH$_3$)$_2$]⁻   [Co(ox)(en)$_2$]⁺

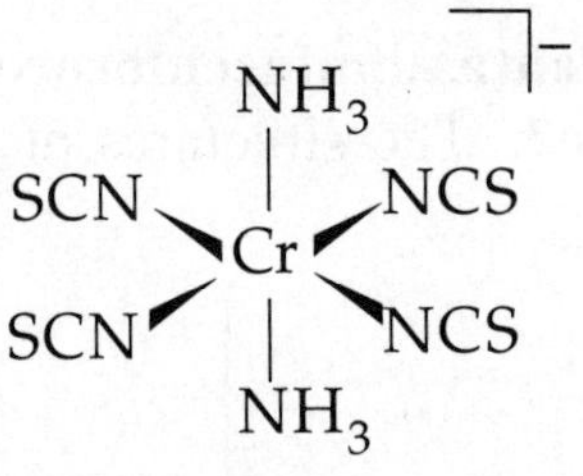

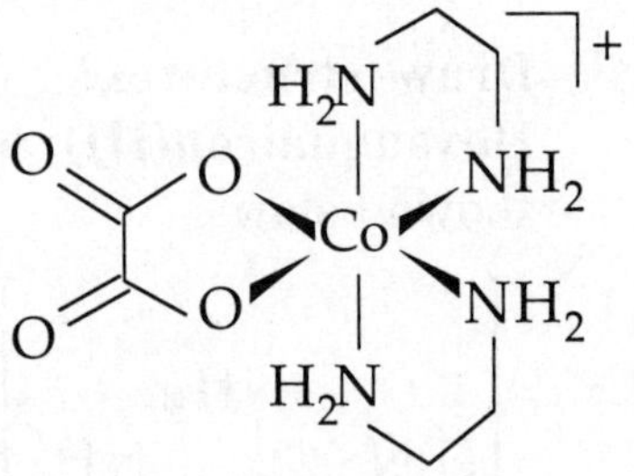

(c) [Co(ox)(en)$_2$]⁺? This complex contains (i) one oxalate dianion ligand (C$_2$O$_4$$^{2-}$) = oxalato, (ii) two ethylenediamine ligands = *bis*(ethylenediamine), and (iii) Co^{3+} in a complex that is not an anion = cobalt(III). Therefore, the name of this complex is *bis*(ethylenediamine)oxalatocobalt(III). The structure is shown above. Note that you do not need to use *cis*- or *trans*- for a complex of this composition, since neither of the two types of bidentate ligands in the complex can span two *trans* positions.

7.9   **Draw all possible isomers?   (a) Octahedral [RuCl$_2$(NH$_3$)$_4$]?   (b) Square-planar [IrH(CO)(PR$_3$)$_2$]?** For each compound, *cis* and *trans* isomers are possible, as shown below.

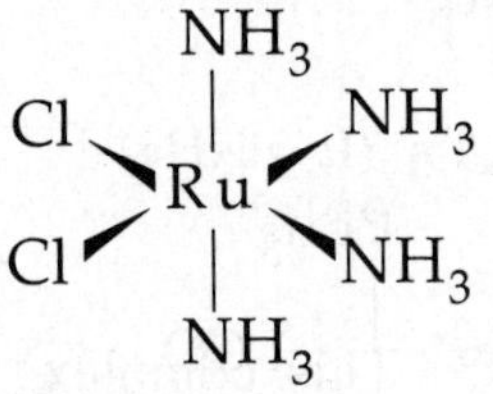

cis   *trans*   *trans*   *cis*

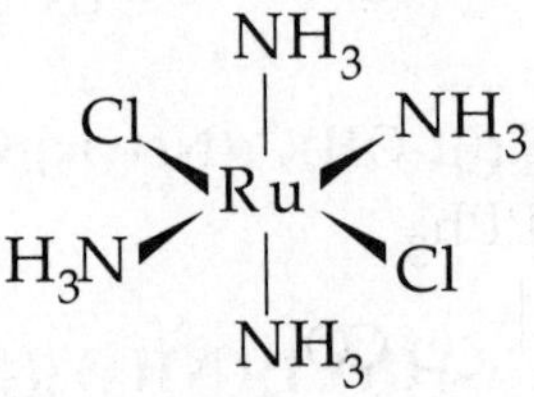

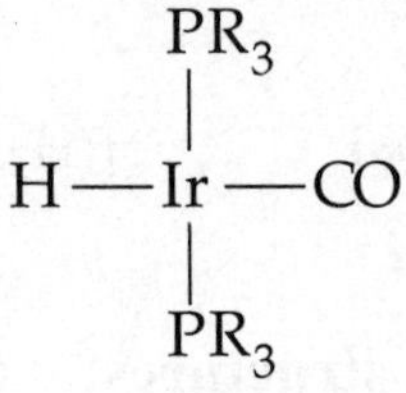

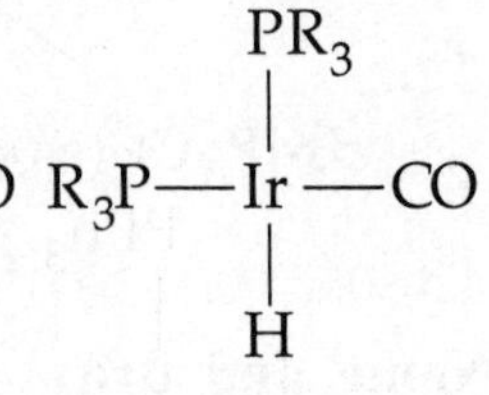

(c) Tetrahedral [CoCl₃(OH₂)]⁻?   (d) Octahedral [IrCl₃(PEt₃)₃]? There is only one isomer for (c), but two isomers, one *fac* (facial) and one *mer* (meridianal), for (d), as shown below.

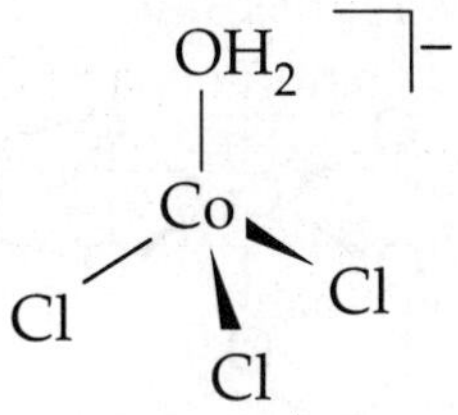

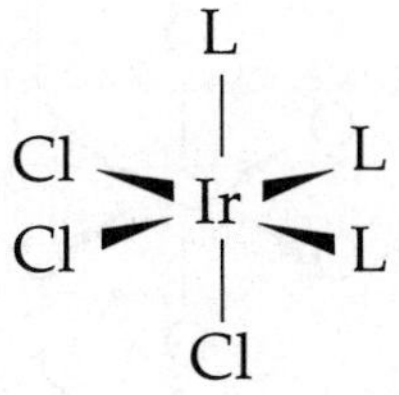

fac 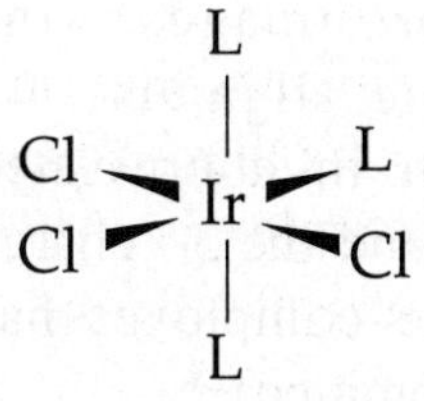   *mer*

L = PEt₃

(e) Octahedral [CoCl₂(NH₃)₂(en)]⁺? There are three isomers. In one, the two Cl⁻ ligands are *trans* to one another. In the other two, the two Cl⁻ ligands are *cis* to one another. The three isomers are shown below (the 1+ charge for each of these three isomers is not shown).

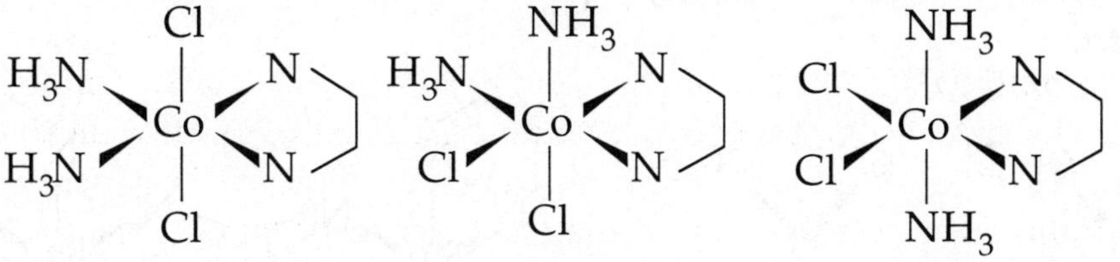

7.10   Draw and name iridium complexes? Structures of the three complexes are shown below. You should recognize the logic behind putting the two bulky triphenylphosphine ligands opposite one another in the two trigonal-bipyramidal complexes. The names of the three complexes, from left to right, are carbonyl(chloro)bis(triphenylphosphine)iridium(I), dicarbonyl(chloro)bis(triphenylphosphine)iridium(I), and dicarbonyl(hydrido)bis(triphenylphosphine)-iridium(I). Note the alphabetical ordering of the ligands, such as di*car*bonyl before *ch*loro.

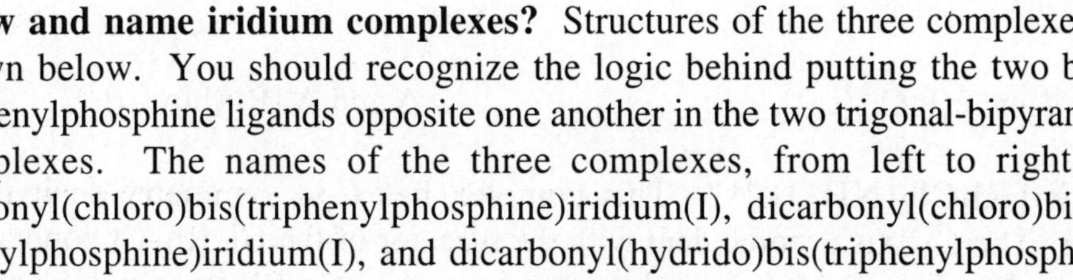

7.11   **Which of the following complexes are chiral?**   **(a) $[Cr(ox)_3]^{3-}$?** All octahedral *tris*(bidentate ligand) complexes are chiral, since they can exist as either a right-hand or a left-hand propeller, as shown in the drawings of the two nonsuperimposable mirror images (the oxalate ligands are shown in abbreviated form and the 3– charge on the complexes has been omitted).

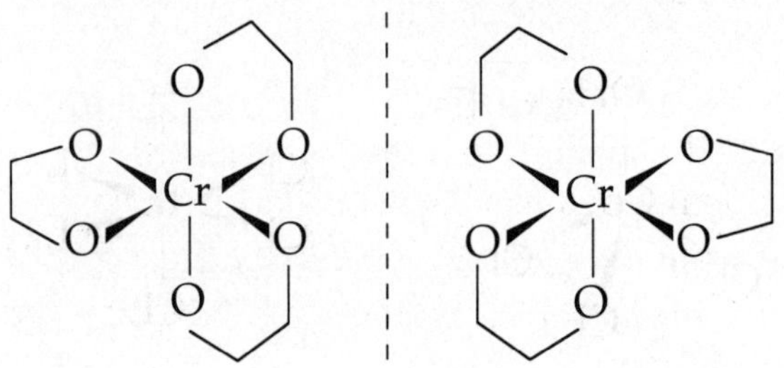

(b) *cis*-$[PtCl_2(en)]$?   This is a four-coordinate complex of a period 6 d^8 metal ion, so it is undoubtedly square-planar. You will recall from Chapter 3 that any planar complex contains at least one plane of symmetry and must be achiral.   In this case the five-membered chelate ring formed by the ethylenediamine ligand is not planar, so, strictly speaking, the complex is not planar. It *can* exist as two enantiomers, depending on the conformation of the chelate ring, as shown below. However, the conformational interconversion of the ethylene linkage is extremely rapid, so the two enantiomers cannot be separated.

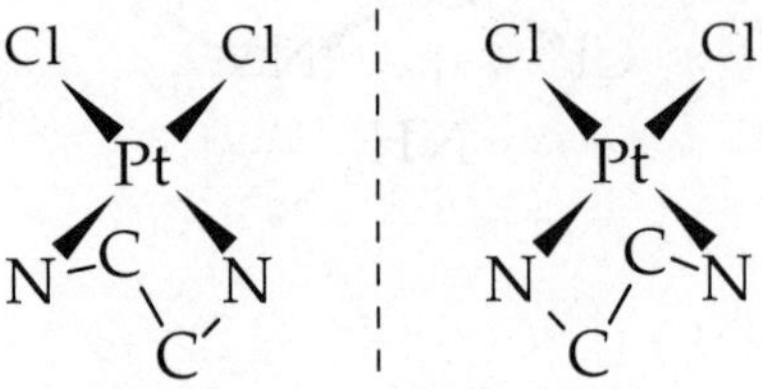

d and *l* $[PtCl_2(en)]$ Δ and Λ $[Ru(bipy)_3]^{3+}$

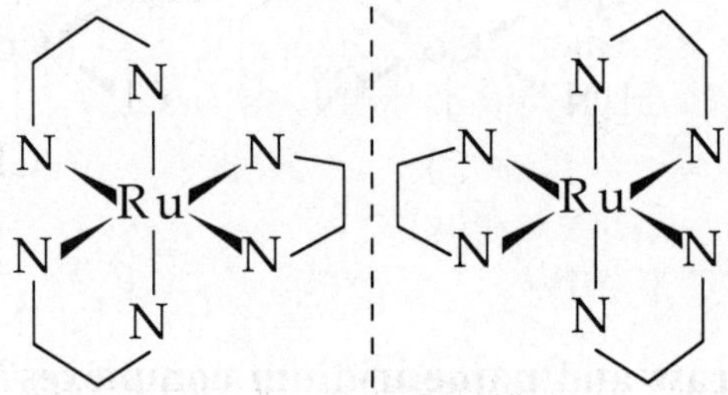

(c) *cis*-$[RhCl_2(NH_3)_4]^+$?   This complex has C_{2v} symmetry so it is not chiral. The C_2 axis is coincident with the bisector of the Cl–Rh–Cl bond angle, one σ_v plane is coincident with the plane formed by the Rh atom and the two Cl atoms, and the other σ_v plane is perpendicular to the first.

(d) $[Ru(bipy)_3]^{2+}$?   As stated in the answer to part (a), above, all octahedral *tris*(bidentate ligand) complexes are chiral, and this one is no exception. The two nonsuperimposable mirror images are shown above (the bipyridine ligands are shown in abbreviated form and the 3+ charge on the complex has been omitted).

(e) [Co(edta)]⁻?   A drawing of one of the enantiomers of this complex is shown in Structure **28** in the text (see Section 7.2(c), *Chelating ligands*). If you interchange the top and bottom oxygen atoms in that drawing, along with their respective acetate arms, you will form the other enantiomer.

(f) *fac*-[Co(NO₂)₃(dien)]?   If you ignore the conformations of the five-membered chelate rings, this complex, shown below, is not chiral. The plane perpendicular to the page that includes the central diethylenetriamine N atom and the Co atom is a symmetry plane for this complex (since the complex possesses C_s symmetry, this plane is the only symmetry element present).

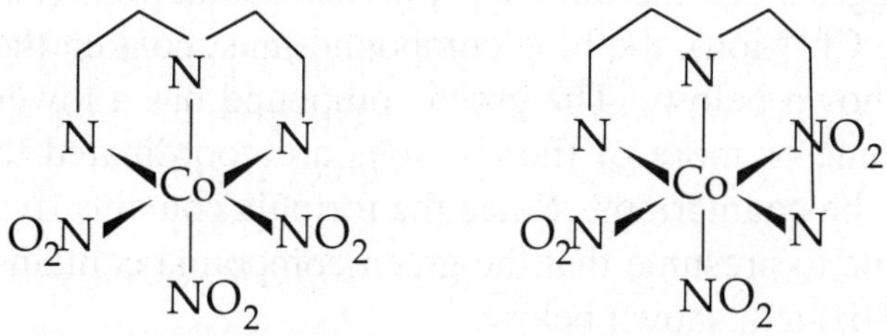

fac

mer

If you take into account the various conformations of the ethylene linkages, the stereo-isomer possibilities are much more complicated. As explained in the text for en, the ethylene linkages undergo *rapid* conformational inter-conversion.

(g) *mer*-[Co(NO₂)₃(dien)]?   As above, if you ignore the conformations of the five-membered chelate rings, this complex, which is shown above, is not chiral. The plane that contains the Co atom and all three dien N atoms is a symmetry plane of the complex, as is the plane that contains the Co atom and all three nitrite N atoms. Ignoring the chelate rings, this complex belongs to the C_{2v} point group.

7.12   **Deduce the structures of pink CoCl₃·5NH₃·H₂O and purple CoCl₃·5NH₃?**   Since the pink complex rapidly gives 3 mol AgCl on titration with AgNO₃ solution, the Cl⁻ ions cannot be coordinated to the Co^{3+} ion, which forms inert complexes. Instead, they must be counterions, and the five ammonia molecules and the water molecule must comprise the six ligands of this octahedral complex. The purple solid, on the other hand, does not contain any water, so one of the Cl⁻ ions must now be coordinated to Co^{3+}. The structures of the two complex ions are shown below. The name of the pink compound is pentaammineaquacobalt(III) chloride (the two neutral ligands are listed in alphabetical order, ammine before aqua). The name of the purple compound is pentaamminechlorocobalt(III) chloride. Dots are used in formulas for complexes where the structure is unknown or unspecified.

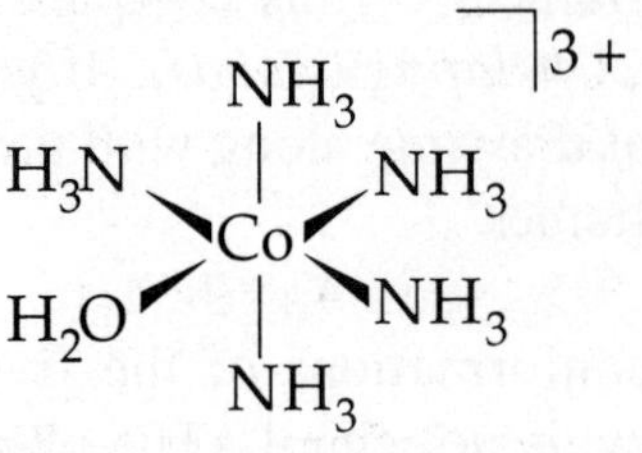

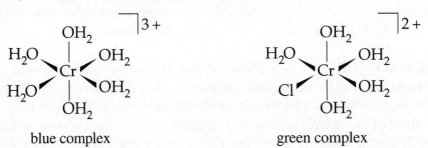

pink $[Co(NH_3)_5(OH_2)]^{3+}$ 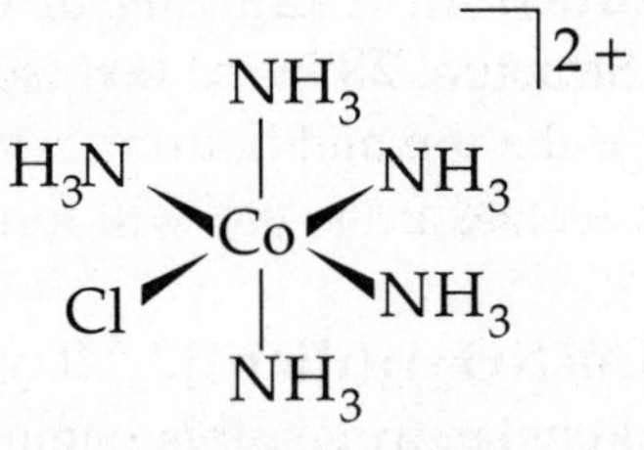   purple $[CoCl(NH_3)_5]^{2+}$

7.13   **Deduce the structures of blue $CrCl_3 \cdot 6H_2O$ and green $CrCl_3 \cdot 5H_2O$?**
Since the conductivity data suggests that the three Cl^- ions are counterions (i.e.
they are not coordinated to the Cr^{3+} ion), the blue compound must contain the
hexaaquachromium(III) ion, shown below. The green compound has a lower
conductivity, suggesting that one or more of the Cl^- ions are coordinated to
Cr^{3+}, leaving fewer Cl^- ions to be counterions. Since the formula contains five
water molecules, it is reasonable to presume that the green compound contains
the pentaaquachlorochromium(III) ion, shown below.

blue complex   green complex

7.14   **Name and draw the structure of the diaqua $[Pt(NH_3)_2(OH_2)_2]^{2+}$
complex?** The observation that this complex does not react with
ethylenediamine, which is unable to span two opposite positions in a square-
planar complex, suggests that this is the *trans* isomer. The name of the complex
ion is diamminediaquaplatinum(II) (alphabetical order requires *am*mine before
*aq*ua). Its structure is shown below, on the left.

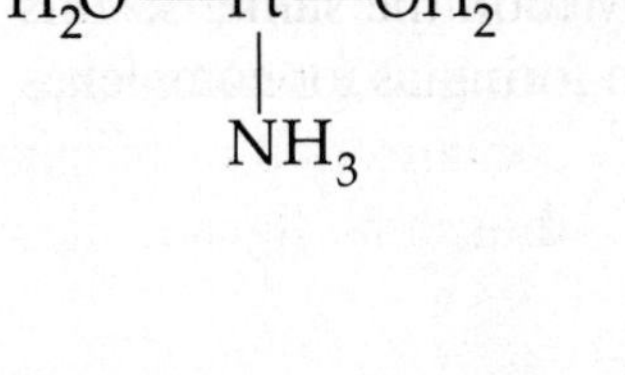

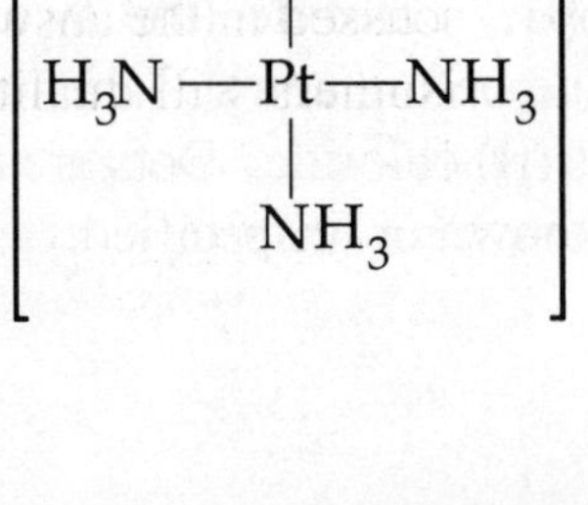

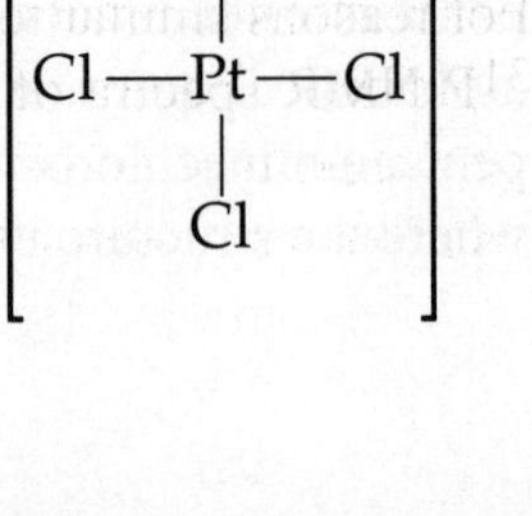

7.15   **Give the structure and name of the parent $PtCl_2 \cdot 2NH_3$ compound?** The metathesis reaction with $AgNO_3$ suggests the presence of $[Pt(NH_3)_4]^{2+}$ and $[PtCl_4]^{2-}$:

$$[Pt(NH_3)_4][PtCl_4] + 2AgNO_3 \rightarrow [Pt(NH_3)_4](NO_3)_2 \text{ and } Ag_2[PtCl_4]$$

This formulation would also explain the lack of formation of any AgCl. The name of the third isomer is tetraammineplatinum(II) tetrachloroplatinate(II), and the structures of the two square-planar complex ions that comprise this compound are shown above, in the middle and on the right. The names of the other two complexes are tetraammineplatinum(II) nitrate and silver(I) tetrachloroplatinate(II).

7.16   **Give the structures of Jensen's alkylphosphane and -arsane complexes?** The compounds do not have a dipole moment, so they are *trans* isomers. As discussed in Chapter 3, C_{2v} molecules like the *cis* isomers of these compounds will have dipole moments whereas D_{2h} molecules, like the *trans* isomers, will not. The structures of *trans*-$[PtCl_2(PR_3)_2]$ and *trans*-$[PtCl_2(AsR_3)_2]$ are shown at the right.

7.17   **^{31}P NMR spectra?** For both the *cis* and *trans* isomers of $[PtCl_2(PR_3)_2]$, symmetry elements make the two phosphine ligands equivalent. In the *cis* isomer, both the C_2 axis and one of the mirror planes interconverts the two ligands. In the *trans* isomer, the inversion center, two different C_2 axes, and a mirror plane interconvert them. Therefore, both isomers contain only one type of phosphine ligand and their ^{31}P NMR spectra will qualitatively look the same. There may be small differences in chemical shift, but this cannot be simply predicted.

7.18   **More on NMR spectroscopy?** (a) *cis* and *trans* $[W(CO)_4(PMe_3)_2]$? For reasons similar to those discussed in the answer to Exercise 7.17, above, the ^{31}P NMR spectra of the two isomers will qualitatively look the same, so this particular type of NMR experiment cannot be used to distinguish between the two isomers. However, the *cis* isomer (C_{2v} symmetry) contains two different pairs of carbonyl ligands, those that are 180° from the phosphine ligands and those that are 90° from the phosphine ligands. The *trans* isomer (D_{4h} symmetry) contains only one type of carbonyl ligand, because all four of them are interconverted by the C_4 axis. Therefore, ^{13}C NMR spectroscopy can be used to distinguish between the isomers, since the spectra will show two resonances for the *cis* isomer but only one resonance for the *trans* isomer.

(b) Trigonal-bipyramidal [M(CO)3(PR3)2]? In this case too, ^{31}P NMR spectra will just show one resonance in both cases. However, ^{13}C NMR spectra will show two resonances for the isomer with the phosphane ligands in the equatorial plane (this molecule has C_{2v} symmetry with two types of CO ligands) but only one resonance for the isomer with the phosphine ligands in the axial positions (this molecule has D_{3h} symmetry with all three CO ligands related by the C_3 axis). Therefore, as in part (a), ^{13}C NMR spectroscopy can be used to distinguish between these two isomers.

7.19   Draw   bonding   and   antibonding   σ   orbitals   for   *trans*-[W(CO)4(PR3)2]? This complex has D_{4h} symmetry.   The two linear combinations of P atom *s* orbitals are shown below. Spherical σ-type orbitals are used in the two linear combinations, which have A_{1g} and A_{2u} symmetries, but the P atom *s* orbitals could also be hybrid orbitals. The central W atom d_{z^2} orbital also has A_{1g} symmetry, and so can form bonding and antibonding combinations with the A_{1g} combination of phosphorus orbitals.

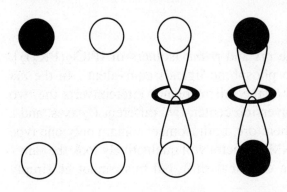

The combination on the far left is the A_{2u} linear combination of phosphorus σ orbitals. Moving to the right, you have the A_{1g} combination of phosphorus *s* orbitals, followed by the bonding and antibonding molecular orbitals formed by taking + and − combinations of the A_{1g} phosphorus combination and the metal d_{z^2} orbital.

7.20   Configuration, number of unpaired electrons, and LFSE? **(a) [Co(NH3)6]3+?** Since the NH_3 ligands are neutral, the cobalt ion in this octahedral complex is Co^{3+}, which is a d^6 metal ion. Ammonia is in the middle of the spectrochemical series but, since the cobalt ion has a 3+ charge, this is a strong field complex and hence is low-spin, with $S = 0$ and no unpaired electrons (the configuration is t_{2g}^6). The LFSE is $6(0.4\Delta_o) = 2.4\Delta_o$. Note that this is the largest possible value of LFSE for an octahedral complex.

(b) $[Fe(OH_2)_6]^{2+}$? The iron ion in this octahedral complex, which contains only neutral water molecules as ligands, is Fe^{2+}, which is a d^6 metal ion. Since water is lower in the spectrochemical series than NH_3 (i.e. it is a weaker field ligand than NH_3) *and* since the charge on the metal ion is only 2+, this is a weak field complex and hence is high-spin, with $S = 2$ and four unpaired electrons (the configuration is $t_{2g}^4 e_g^2$). The LFSE is $4(0.4\Delta_o) - 2(0.6\Delta_o) = 0.4\Delta_o$. Compare this small value to the large value for the low-spin d^6 complex in part (a) above.

(c) $[Fe(CN)_6]^{3-}$? The iron ion in this octahedral complex, which contains six negatively charged CN^- ion ligands, is Fe^{3+}, which is a d^5 metal ion. Cyanide ion is a very strong field ligand, so this is a strong field complex and hence is low-spin, with $S = 1/2$ and one unpaired electron. The configuration is t_{2g}^5 and the LFSE is $2.0\Delta_o$.

(d) $[Cr(NH_3)_6]^{3+}$? The complex contains six neutral NH_3 ligands, so chromium is Cr^{3+}, a d^3 metal ion. The configuration is t_{2g}^3, and so there are three unpaired electrons and $S = 3/2$. Note that, for octahedral complexes, only d^4–d^7 metal ions have the possibility of being either high-spin or low-spin. For $[Cr(NH_3)_6]^{3+}$, the LFSE $= 3(0.4\Delta_o) = 1.2\Delta_o$. (For d^1–d^3, d^8, and d^9 metal ions in octahedral complexes, only one spin state is possible.)

(e) $[W(CO)_6]$? Carbon monoxide (i.e. the carbonyl ligand) is neutral, so this is a complex of W(0). The W atom in this octahdral complex is d^6. Since CO is such a strong field ligand (it is even higher in the spectrochemical series than CN^-), $W(CO)_6$ is a strong field complex and hence is low-spin, with no unpaired electrons (the configuration is t_{2g}^6). The LFSE $= 6(0.4\Delta_o) = 2.4\Delta_o$.

(f) Tetrahedral $[FeCl_4]^{2-}$? The iron ion in this complex, which contains four negatively charged Cl^- ion ligands, is Fe^{2+}, which is a d^6 metal ion. All tetrahedral complexes are high-spin, since Δ_T is much smaller than Δ_o ($\Delta_T = (4/9)\Delta_o$, if the metal ion, the ligands, and the metal–ligand distances are kept constant) so for this complex $S = 2$ and there are four unpaired electrons. The configuration is $e^3 t^3$. The LFSE is $3(0.6\Delta_T) - 3(0.4\Delta_T) = 0.6\,\Delta_T$.

(g) Tetrahedral $[Ni(CO)_4]$? The neutral CO ligands require that the metal center in this complex is Ni^0, which is a d^{10} metal atom. Regardless of geometry, complexes of d^{10} metal atoms or ions will never have any unpaired electrons and will always have LFSE $= 0$, and this complex is no exception.

7.21   **What factors account for the ligand field strength of different ligands?** It is clear that π-acidity cannot be a requirement for a position high in the spectrochemical series, since H^- is a very strong field ligand but is not a π-acid (it has no *low energy* acceptor orbitals of local π symmetry). However, ligands that are very strong σ-bases will increase the energy of the e_g orbitals in an octahedral complex relative to the t_{2g} orbitals. Thus, there are two ways for a complex to develop a large value of Δ_o, by possessing ligands that are π-acids *or* by possessing ligands that are strong σ-bases (of course some ligands, like CN^-, exhibit both π-acidity and moderately strong σ-basicity). A class of ligands that are also very high in the spectrochemical series are alkyl anions, R^- (e.g. CH_3^-). These are not π-acids but, like H^-, are very strong bases.

7.22   **Estimate the spin-only contribution to the magnetic moment?** The formula for the spin-only moment is $\mu_{SO} = [(N)(N + 2)]^{1/2}$. Therefore, the spin-only contributions are:

complex	N	$\mu_{SO} = [(N)(N + 2)]^{1/2}$
$[Co(NH_3)_6]^{3+}$	0	0
$[Fe(OH_2)_6]^{2+}$	4	4.9
$[Fe(CN)_6]^{3-}$	1	1.7
$[Cr(NH_3)_6]^{3+}$	3	3.9
$[W(CO)_6]$	0	0
$[FeCl_4]^{2-}$	4	4.9
$[Ni(CO)_4]$	0	0

7.23   **Assign the colors pink, yellow, and blue to cobalt(II) complexes?** The colors of metal complexes are frequently due to ligand-field transitions involving electron promotion from one subset of d orbitals to another (e.g. from t_{2g} to e_g for octahedral complexes or from e to t_2 for tetrahedral complexes). Of the three complexes given, the lowest energy transition probably occurs for $[CoCl_4]^{2-}$, because it is tetrahedral ($\Delta_T = (4/9)(\Delta_O)$) and because Cl^- is a weak-field ligand. This complex is blue, since a solution of it will absorb low energy red light and reflect the complement of red, which is blue. Of the two complexes that are left, $[Co(NH_3)_6]^{2+}$ probably has a higher energy transition than $[Co(OH_2)_6]^{2+}$, since NH_3 is a stronger-field ligand than H_2O (see Table 7.3). The complex $[Co(NH_3)_6]^{2+}$ is yellow because only a small amount of visible light, at the blue end of the spectrum, is absorbed by a solution of this complex. By default, you should conclude that $[Co(OH_2)_6]^{2+}$ is pink.

7.24 Comment on the lattice enthalpies for CaO, TiO, VO, MnO, FeO, CoO, and NiO? As in the answer to Exercise S7.6, there are two factors that lead to the values given in this question and plotted below: decreasing ionic radius from left to right across the *d*-block, leading to a general increase in ΔH_L from CaO to NiO, and LFSE, which varies in a more complicated way for high-spin metal ions in an octahedral environment, increasing from d^0 to d^3, then decreasing from d^3 to d^5, then increasing again from d^5 to d^8, then decreasing again from d^8 to d^{10}. The straight line through the black squares is the trend expected for the first factor, the decrease in ionic radius (the last black square is not a data point, but simply the extrapolation of the line between ΔH_L values for CaO and MnO, both of which have LFSE = 0). The deviations of ΔH_L values for TiO, VO, FeO, CoO, and NiO from the straight line is a manifestation of the second factor, the nonzero values of LFSE for Ti^{2+}, V^{2+}, Fe^{2+}, Co^{2+}, and Ni^{2+}. You shall find in Chapter 18 that TiO and VO have considerable metal–metal bonding, and this factor also contributes to their stability.

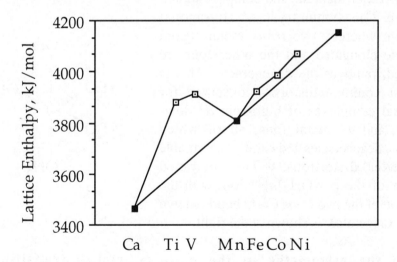

7.25 Diamagnetic and paramagnetic Ni(II) complexes? Perchlorate, ClO_4^-, is a very weakly basic anion (consider that $HClO_4$ is a very strong Brønsted acid). Therefore, in the compound containing Ni(II), the neutral macrocyclic ligand, and two ClO_4^- anions, there is probably a four-coordinate square-planar Ni(II) (d^8) complex and two noncoordinated perchlorate anions. Square-planar d^8 complexes are diamagnetic (see Figure 7.15), since they have a configuration $(xz,yz)^4(xy)^2(z^2)^2$. When SCN^- ligands are added, they coordinate to the nickel ion, producing a tetragonal (D_{4h}) complex which has two unpaired electrons (configuration $t_{2g}^6 e_g^2$).

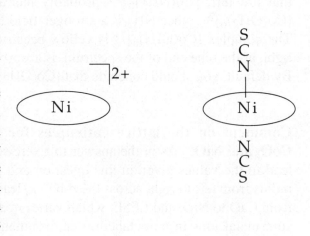

7.26 **Predict the structure of $[Cr(OH_2)_6]^{2+}$?** The main consequence of the Jahn–Teller theorem is that a nonlinear molecule or ion with an orbitally degenerate ground state is not as stable as a distorted version of the molecule or ion if the distortion removes the degeneracy. The high-spin d^4 complex $[Cr(OH_2)_6]^{2+}$ has the configuration $t_{2g}^3 e_g^1$, which is orbitally degenerate since the single e_g electron can be in either the d_{z^2} or the $d_{x^2-y^2}$ orbital. Therefore, by the Jahn–Teller theorem, the complex should not have O_h symmetry. A tetragonal distortion, whereby two *trans* metal–ligand bonds are elongated and the other four are shortened, removes the degeneracy. This is the most common distortion observed for octahedral complexes of high-spin d^4, low-spin d^7, and d^9 metal ions, all of which possess e_g degeneracies and exhibit measurable Jahn–Teller distortions. The predicted structure of the $[Cr(OH_2)_6]^{2+}$ ion, with the elongation of the two *trans* Cr–O bonds shown greatly exaggerated, is shown at the right.

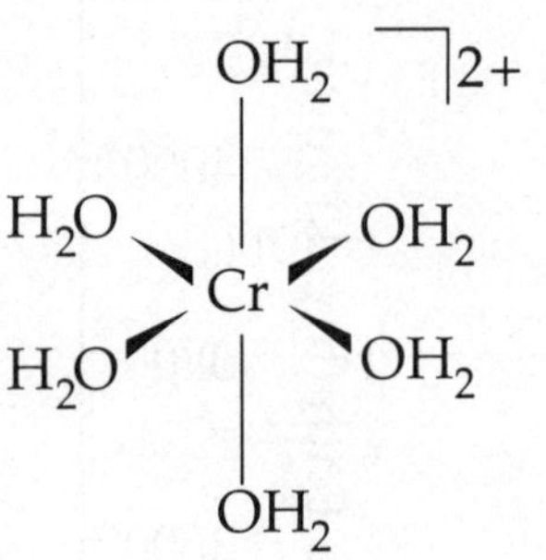

7.27 **Explain the asymmetry in the $e_g \leftarrow t_{2g}$ visible transition for Ti^{3+}(aq)?** It is suggested that you explain this observation using the Jahn–Teller theorem. The ground state of $Ti(OH_2)_6^{3+}$, which is a d^1 complex, is not one of the configurations that usually leads to an observable Jahn–Teller distortion (the three main cases are listed in the answer to Exercise 7.26, above). However, the electronic excited state of $Ti(OH_2)_6^{3+}$ has the configuration

$t_{2g}{}^0 e_g{}^1$, and so the excited state of this complex possesses an e_g degeneracy. Therefore, the "single" electronic transition is really the superposition of two transitions, one from an O_h ground-state ion to an O_h excited-state ion, and a lower energy transition from an O_h ground-state ion to a lower energy distorted excited-state ion (probably D_{4h}). Since these two transitions have slightly different energies, the unresolved superimposed bands result in an asymmetric absorption peak.

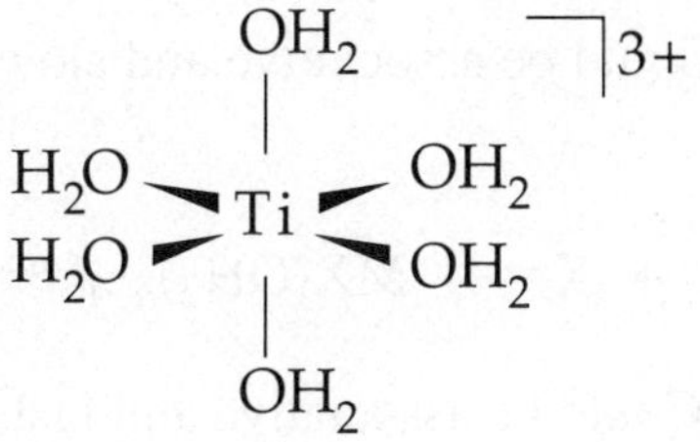

7.28   **What is the mechanism of formation of $[CoX(NH_3)_5]^{2+}$ from $[Co(NH_3)_5OH_2]^{3+}$?** The reaction in question is ($X^- = Cl^-$, Br^-, $N_3{}^-$, and SCN^-):

$$[Co(NH_3)_5OH_2]^{3+} + X^- \rightarrow [Co(NH_3)_5X]^{2+} + H_2O$$

Since the rate constant does not depend strongly on the nature of the incoming ligand, it is probably *not* associative. Therefore, it is likely to be dissociative.

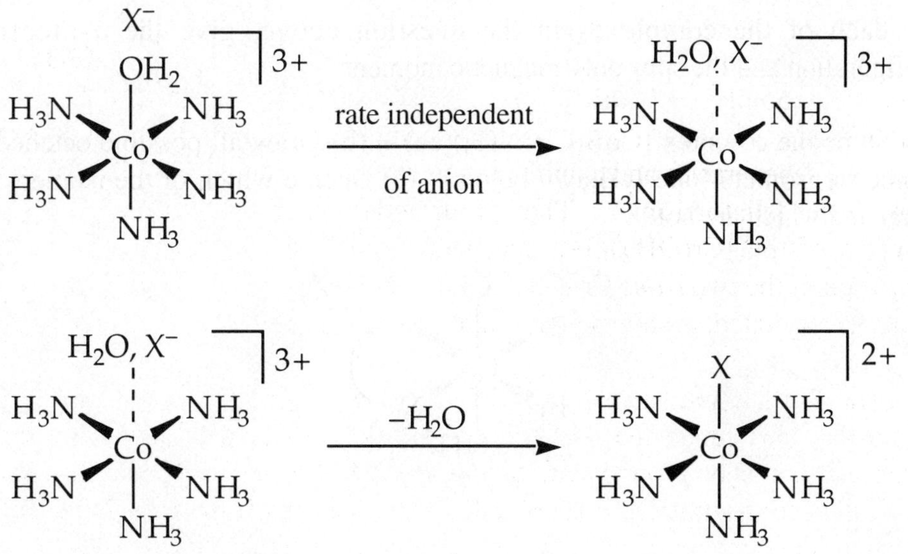

7.29   **If a substitution process is associative, why may it be difficult to characterize an aqua ion as labile or inert?** The classification of a complex as labile (reaction time < 1 minute) implies a judgement about its reaction with various partners. Since the rate of an associative reaction depends strongly on the entering ligand, the rates of substitution of an aqua ion may span many orders of magnitude.

$$M(OH_2)_n{}^{m+} + {}^*OH_2 \rightarrow M(OH_2)_{n-1}({}^*OH_2)^{m+} + OH_2$$

Could be associative and slow

$$M(OH_2)_n{}^{m+} + X^- \rightarrow MX(OH_2)_{n-1}{}^{(m-1)+} + OH_2$$

Could be associative and fast

Guide to Solutions Quiz

1   Give names and draw structures, including all possible isomers, of the following complexes: (a) $[ReO_4]^-$; (b) $[NbCl_6]^-$; (c) $[Co(CN)_2(trien)]^+$; (d) $[RhClH_2(PPh_3)_3]$; (e) $[MnCl(CO)_5]$; (f) $[TiCl_3(CH_3CN)_3]^+$; (g) $[PtCl_6]^{2-}$.

2   For each of the complexes in the question above, give the d-electron configuration and the spin-only magnetic moment.

3   (a) Name the complex $[CoBrCl_2(NH_3)(en)]$. (b) Draw all possible octahedral geometric isomers (one is drawn below). (c) Decide which of the isomers are chiral and which are not.

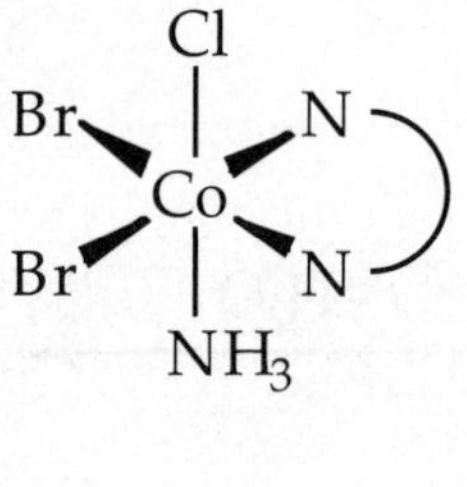

4 Many four-coordinate d^9 copper(II) complexes are neither square-planar nor tetrahedral but have a geometry somewhere in between. Which of the two complexes $[CuCl_4]^{2-}$ and $[CuBr_4]^{2-}$ is more likely to have a geometry closer to tetrahedral? Explain your answer.

5 Is it always possible to tell by magnetic measurements whether a new complex of Fe(II) is octahedral or tetrahedral? Explain your answer.

6 A particular six-coordinate octahedral complex of Cr(II) has six identical ligands and six equal Cr–L bond distances. What can you conclude about the spin-only magnetic moment of this complex? Explain your answer.

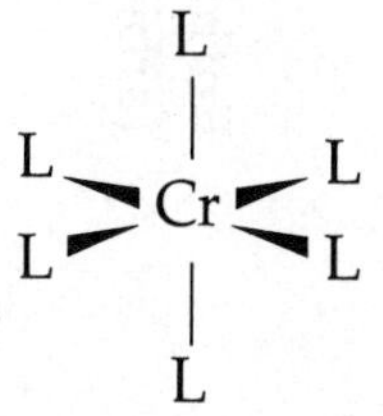

7 The absorption spectrum of $[Ti(OH_2)_6]^{3+}$ has a band centered at 20,300 cm^{-1} (see Figure 7.10). Do you expect v_{max} to be less than, greater than, or equal to 20,300 cm^{-1} in the spectrum of $[Ti(NCS)_6]^{3-}$? Explain your answer.

8 Consider the complex *trans*-$[W(CO)_4(PR_3)_2]$, which has approximate D_{4h} symmetry. Recall the convention that the major axis of a molecule is designated its z axis. Draw the metal–ligand π and π^* molecular orbitals involving the metal xy orbital and the carbonyl ligands. Be careful to indicate the relative sizes of the lobes of your orbitals.

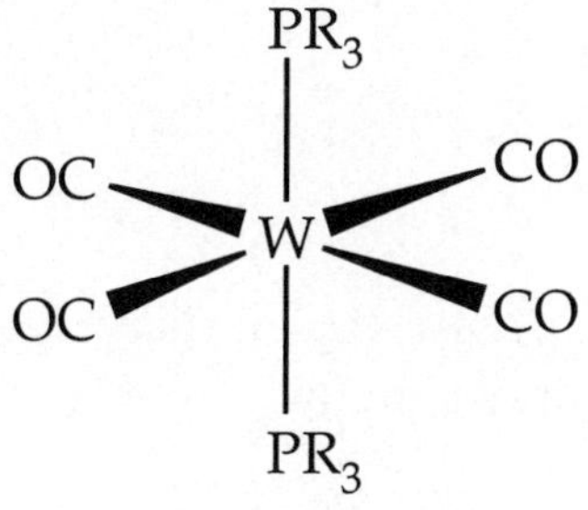

9 Graphs similar to the one below are obtained in many cases when ΔH values for *d*-block metals are plotted against the number of *d* electrons (e.g., see Figure 7.12). In these cases, the metal ions are invariably high-spin. Draw the graph that would result if the metal ions were all low-spin.

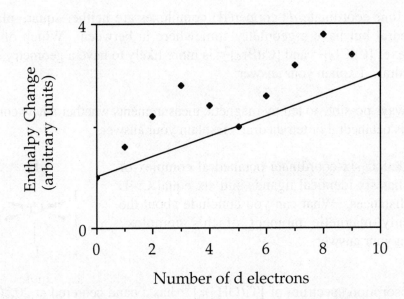

10   State which member of the following pairs of aqua complexes exchanges water
ligands faster, and why: (a) Al^{3+} or In^{3+}, (b) Ca^{2+} or Zn^{2+}, (c) Ni^{2+} or Cr^{3+},
(d) Cu^{2+} or Zn^{2+}.

8   Hydrogen

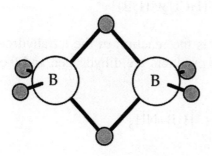

Diborane, B_2H_6, is the simplest member of a large class of compounds, the electron-deficient boron hydrides. Like all boron hydrides, it has a positive standard free energy of formation, and so cannot be prepared directly from boron and hydrogen. The bridge B–H bonds are longer and weaker than the terminal B–H bonds (1.32 vs. 1.19 Å).

S8.1   Reactions of hydrogen compounds?   (a) $Ca(s) + H_2(g) \rightarrow CaH_2(s)$. This is the reaction of an active *s*-metal with hydrogen, which is the way that saline metal hydrides are prepared.

(b) $NH_3(g) + BF_3(g) \rightarrow H_3N–BF_3(s)$. This is the reaction of a Lewis base and a Lewis acid. The product is a Lewis acid–base complex.

(c) $LiOH(s) + H_2(g) \rightarrow NR$. Although dihydrogen can behave as an oxidant (e.g. with Li to form LiH) or as a reductant (e.g. with O_2 to form H_2O), it does not behave as a Brønsted or Lewis acid or base. It does not react with strong bases, like LiOH, or with strong acids.

S8.2   **A procedure for making Et₃MeSn?** A possible procedure is as follows:

$$2Et_3SnH + 2Na \rightarrow Na^+Et_3Sn^- + H_2$$

$$Na^+Et_3Sn^- + CH_3Br \rightarrow Et_3MeSn + NaBr$$

S8.3   **Write equations for: The reaction of LiBH₄ with propene?** An analogous reaction is shown in Section 8.9(d) and reproduced below. The BH_4^- ion can also add a B–H bond across the C=C double bond by hydride transfer. The reaction product is an alkyltrihydroborate ion.

$$H_3B-OR_2 + H_2C=CH_2 \rightarrow CH_3CH_2BH_2 + OR_2$$

$$BH_4^- + CH_3CH=CH_2 \rightarrow CH_3CH_2CH_2BH_3^-$$

The reaction of LiBH₄ with NH₄Cl. This is the reaction of the tetrahydro-borate ion with a Brønsted acid, NH_4^+. The products are dihydrogen and the ammonia-borane Lewis acid–base complex.

$$BH_4^- + NH_4^+ \rightarrow H_2 + H_3B-NH_3$$.

8.1   **Assign oxidation numbers to elements?** **(a) H₂S?** When hydrogen is less electronegative than the other element in a binary compound (H is 2.20, S is 2.58; see Table 1.8), its oxidation number is chosen to be +1. Therefore, the oxidation number of sulfur in hydrogen sulfide (sulfane) is –2.

(b) KH? In this case, hydrogen (2.20) is more electronegative than potassium (0.82), so its oxidation number is chosen to be –1. Therefore, the oxidation number of potassium in potassium hydride is +1.

(c) [ReH₉]²⁻? The electronegativity of rhenium is not given in Table 1.8. However, since it is a metal, it is reasonable to conclude that it is less electronegative than hydrogen. Therefore, if hydrogen is counted as –1, then the rhenium atom in the $[ReH_9]^{2-}$ ion has an oxidation number of +7.

(d) H₂SO₄? The structure of sulfuric acid is shown below. Since the hydrogen atoms are bound to very electronegative oxygen atoms, their oxidation number is +1. Furthermore, since oxygen is always assigned an oxidation number of –2 (except for O_2 and peroxides), sulfur has an oxidation number of +6.

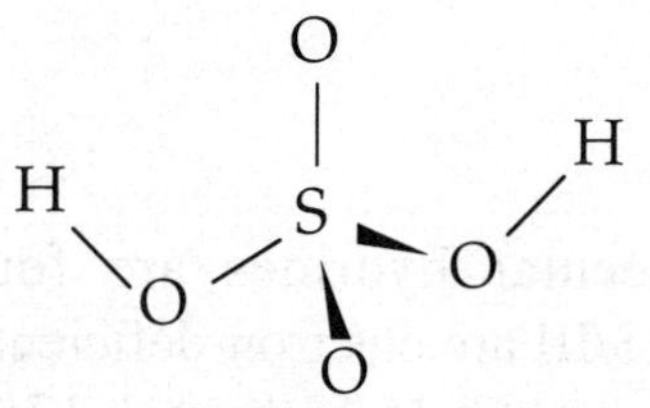

$$H_2SO_4$$

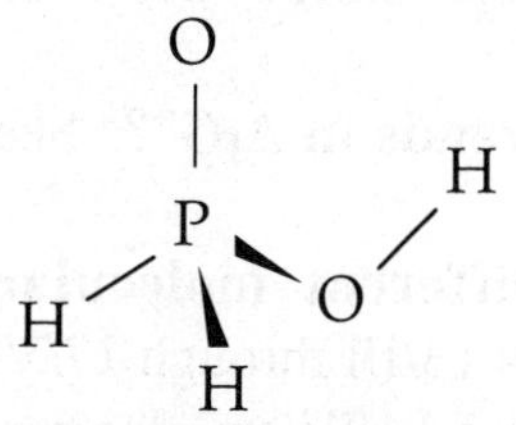

$$H_2PO(OH)$$

(e) H₂PO(OH)? The structure of hypophosphorous acid (also called phosphinic acid) is shown above. There are two types of hydrogen atoms. The one that is bonded to an oxygen atom has an oxidation number of +1. The two that are bonded to the phosphorus atom present a problem, since phosphorus (2.19) and hydrogen (2.20) have nearly equal electronegativities. If these two hydrogen atoms are assigned an oxidation number of +1, and oxygen, as above, is assigned an oxidation number of –2, then the phosphorus atom in H₂PO(OH) has an oxidation number of +1.

8.2 **Preparation of hydrogen gas?** As discussed in Section 8.3, *Properties and reactions of dihydrogen*, the three industrial methods of preparing H₂ are (i) steam reforming, (ii) the water-gas reaction, and (iii) the shift reaction (also called the water–gas shift reaction). The balanced equations are:

$$\text{(i)} \quad CH_4(g) + H_2O(g) \rightarrow CO(g) + 3\,H_2(g) \quad (1000\,°C)$$

$$\text{(ii)} \quad C(s) + H_2O(g) \rightarrow CO(g) + H_2(g) \quad (1000\,°C)$$

$$\text{(iii)} \quad CO(g) + H_2O(g) \rightarrow CO_2(g) + H_2(g)$$

These reactions are not very convenient for the preparation of small quantities of hydrogen in the laboratory. Instead, (iv) treatment of an acid with an active metal (such as zinc) or (v) treatment of a metal hydride with water would be suitable. The balanced equations are:

$$\text{(iv)} \quad Zn(s) + 2\,HCl(aq) \rightarrow Zn^{2+}(aq) + 2\,Cl^-(aq) + H_2(g)$$

$$\text{(v)} \quad NaH(s) + H_2O(l) \rightarrow Na^+(aq) + OH^-(aq) + H_2(g)$$

8.3   **Properties of hydrides of the elements?**   **(a) Position in the periodic table?**   See Figure 8.2.

(b) Trends in $\Delta_f G°$?   See Table 8.6.

(c) Different molecular hydrides?   Molecular hydrides are found in groups 13/III through 17/VII. Those in group 13/III are electron-deficient, those in group 14/IV are electron-precise, and those in group 15/V through 17/VII are electron-rich.

8.4   **Name and classify the following?**   **(a) BaH_2?**   This compound is named barium hydride. It is a saline hydride.

(b) SiH_4?   This compound is named silane. It is an electron-precise molecular hydride.

(c) NH_3?   This familiar compound is known by its common name, ammonia, rather than by the systematic names azane or nitrane. Ammonia is an electron-rich molecular hydride.

(d) AsH_3?   This compound is generally known by its common name, arsine, rather than by its systematic name, arsane. It is also an electron-rich molecular hydride.

(e) $PdH_{0.9}$?   This compound is named palladium hydride. It is a metallic hydride.

(f) HI?   This compound is known by its common name, hydrogen iodide, rather than by its systematic name, iodane. It is an electron-rich molecular hydride.

8.5   **Chemical characteristics of hydrides?** **(a) Hydridic character?** Barium hydride is a good example, since it reacts with proton sources such as H_2O to form H_2:

$$BaH_2(s) + 2\,H_2O(l) \rightarrow 2\,H_2(g) + Ba(OH)_2(s)$$

$$\text{net reaction:}\quad 2\,H^- + 2\,H^+ \rightarrow 2\,H_2$$

(b) Brønsted acidity? Hydrogen iodide is a good example, since it transfers its proton to a variety of bases, including pyridine (:py):

$$HI(g) + :py(g) \rightarrow [H:py]^+[I]^-(s)$$

(c) Variable composition? The compound $PdH_{0.9}$ is a good example.

(d) Lewis basicity? Ammonia is a good example, since it forms acid–base complexes with a variety of Lewis acids, including BF_3:

$$NH_3(g) + BF_3(g) \rightarrow F_3BNH_3(s)$$

8.6 **Phases of hydrides of the elements?** Of the compounds listed in Exercise 8.4, BaH_2 and $PdH_{0.9}$ are solids, none is a liquid, and SiH_4, NH_3, AsH_3, and HI are gases (see Figure 8.3). Only $PdH_{0.9}$ is likely to be a good electrical conductor.

8.7 **The structures of H_2Se, P_2H_4, and H_3O^+?** The Lewis structures of these three species are:

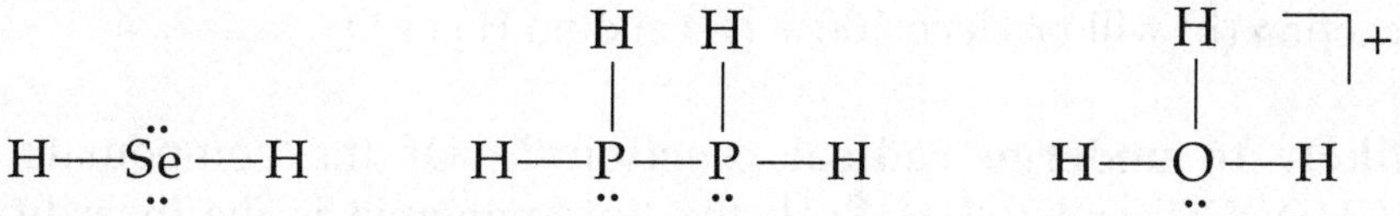

$$H\text{—}\overset{\cdot\cdot}{\underset{\cdot\cdot}{Se}}\text{—}H \qquad H\text{—}\overset{\overset{H}{|}}{\underset{\cdot\cdot}{P}}\text{—}\overset{\overset{H}{|}}{\underset{\cdot\cdot}{P}}\text{—}H \qquad \left[H\text{—}\overset{\overset{H}{|}}{\underset{\cdot\cdot}{O}}\text{—}H \right]^+$$

According to the VSEPR model (Section 3.3), H_2Se should be bent, and so it belongs to the C_{2v} point group; H_3O^+ should be trigonal pyramidal (like NH_3), and so it belongs to the C_{3v} point group; each phosphorus atom of P_2H_4 should have local pyramidal structure. If the molecule adopts the skew conformation (see the Newman diagram below), then it belongs to the C_2 point group (the C_2 axis bisects the P–P bond).

A drawing of the structure of P_2H_4. The P–P and P–H bond distances are 2.22 and 1.42 Å, respectively, and the P–P–H bond angles are all about 94° (cf. PH_3, in which the H–P–H bond angles are 93.8° (see Table 8.3)).

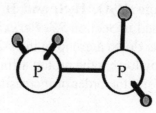

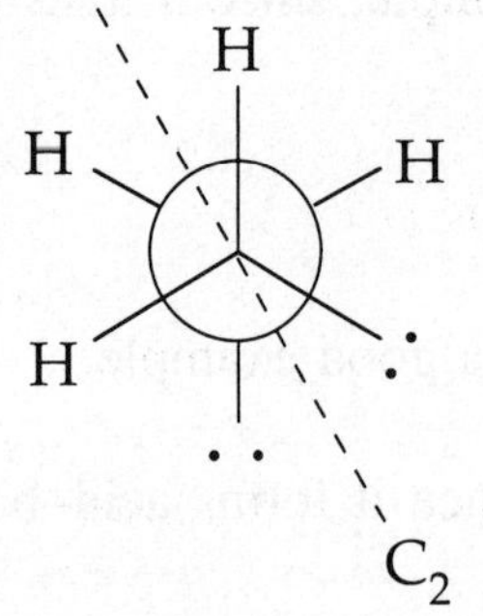

A Newman projection of the skew, or gauche, conformation of P_2H_4. The only element of symmetry that this structure possesses is a C_2 axis that bisects the P–P bond. Therefore, it has C_2 symmetry.

8.8   **The reaction that will give the highest proportion of HD?**   Reactions (a) and (c) both involve the production of both H and D atoms at the surface of a metal. The recombination of these atoms will give a statistical distribution of H_2 (25%), HD (50%), and D_2 (25%). However, reaction (b) involves a source of protons that is 100% $^2H^+$ (i.e. D^+) and a source of hydride ions that is 100% $^1H^-$:

$$D_2O(l) + NaH(s) \rightarrow HD(g) + NaOD(s)$$

$$\text{net reaction:} \quad D^+ + H^- \rightarrow HD$$

Thus, reaction (b) will produce 100% HD and no H_2 or D_2.

8.9   **Most likely to undergo radical reactions?**   Of the compounds H_2O, NH_3, $(CH_3)_3SiH$, and $(CH_3)_3SnH$, the tin compound is the most likely to undergo radical reactions with alkyl halides. This is because the Sn–H bond in $(CH_3)_3SnH$ is less polar *and* weaker than either O–H, N–H, or Si–H bonds. The formation of radicals involves the homolytic cleavage of the bond between the central element and hydrogen, and a weak nonpolar bond undergoes homolysis most readily.

8.10   **Arrange H_2O, H_2S, and H_2Se in order?**   **(a) Increasing acidity?**   As discussed in Section 5.3, *Periodic trends in aqua acid strength*, acidities of EH_n increase down a group in the *p*-block, mostly because the decrease in E–H bond enthalpy lowers the proton affinity of $[EH_{n-1}]^-$ (E is a generic *p*-block element). Therefore, the order of increasing acidity is $H_2O < H_2S < H_2Se$.

(b) Increasing basicity toward a hard acid?   In general, soft character increases down a group, so the hardest base of these three compounds is H_2O. The order of increasing basicity toward a hard acid is $H_2Se < H_2S < H_2O$.

8.11 **The synthesis of binary hydrogen compounds?** The three main methods of synthesis of binary hydrogen compounds are (i) direct combination of the elements, (ii) protonation of a Brønsted base, and (iii) metathesis using a compound such as LiH, $NaBH_4$, or $LiAlH_4$. The first method is limited to those binary hydrogen compounds that are exoergic. An example is:

$$\text{(i)} \quad 2\text{Li(s)} + \text{H}_2(\text{g}) \rightarrow 2\text{LiH(s)}$$

The second method can be used for the preparation of EH_n compounds when a source of the E^{n-} anion is available. An example is:

$$\text{(ii)} \quad \text{CaF}_2(\text{s}) + \text{H}_2\text{SO}_4(\text{l}) \rightarrow 2\text{HF(g)} + \text{CaSO}_4(\text{s})$$

Almost all of the hydrogen fluoride that is prepared industrially is made this way. The third method can be used to convert the chlorides of many elements E to the corresponding hydrides, as in the following example:

$$\text{PCl}_3(\text{l}) + 3\,\text{LiH(s)} \rightarrow \text{PH}_3(\text{g}) + 3\,\text{LiCl(s)}$$

8.12 **Give laboratory methods for synthesis?** **(a) H_2Se?** Since hydrogen selenide is endoergic (see Table 8.6), it cannot be prepared from elemental hydrogen and selenium. It can, however, be prepared by protonating a salt of the Se^{2-} ion, as in the following equations:

$$2\,\text{Na(s)} + \text{Se(s)} \rightarrow \text{Na}_2\text{Se(s)}$$

$$\text{Na}_2\text{Se(s)} + 2\,\text{H}_3\text{PO}_4(\text{l}) \rightarrow \text{H}_2\text{Se(g)} + 2\,\text{NaH}_2\text{PO}_4(\text{s})$$

(b) SiD_4? Both $SiCl_4$ and $LiAlH_4$ are exoergic compounds that can be prepared from their constituent elements. When reacted together, they form SiH_4. Therefore, the following reaction scheme can be used to prepare SiD_4:

$$2\,\text{Li(s)} + \text{D}_2(\text{g}) \rightarrow 2\,\text{LiD(s)}; \qquad 2\,\text{Al(s)} + 3\,\text{Cl}_2(\text{g}) \rightarrow 2\,\text{AlCl}_3(\text{s})$$

$$\text{AlCl}_3(\text{s}) + 4\,\text{LiD(s)} \rightarrow \text{LiAlD}_4(\text{s}) + 3\,\text{LiCl(s)}; \qquad \text{Si(s)} + 2\,\text{Cl}_2(\text{g}) \rightarrow \text{SiCl}_4(\text{l})$$

$$\text{SiCl}_4(\text{l}) + \text{LiAlD}_4(\text{s}) \rightarrow \text{LiAlCl}_4(\text{s}) + \text{SiD}_4(\text{g})$$

(c) Ge(CH$_3$)$_2$H$_2$? Once again, LiAlH$_4$ can be used to accomplish a H/Cl metathesis, as in the following reaction:

$$2\,Ge(CH_3)_2Cl_2(l)\ +\ LiAlH_4(s)\ \rightarrow\ 2\,Ge(CH_3)_2H_2(l)\ +\ LiAlCl_4(s)$$

(d) SiH$_4$ from Si and HCl? This reaction involves the oxidation of Si by HCl. To balance this redox process, something must be reduced, and the most likely candidate is H$^+$:

$$Si(s)\ +\ 3\,HCl(g)\ \rightarrow\ SiHCl_3(l)\ +\ H_2(g)$$

The trichlorosilane is then heated, whereupon it undergoes a redistribution reaction, as shown below:

$$4\,SiHCl_3(l)\ +\ heat\ \rightarrow\ SiH_4(g)\ +\ 3\,SiCl_4(l)$$

8.13   Is B$_2$H$_6$ stable in air? No, this compound reacts so vigorously with air that it spontaneously inflames when exposed to air (compounds that display this behavior are called pyrophoric). It can react with both oxygen and moisture in the air, according to the following reactions:

$$B_2H_6(g)\ +\ 3\,O_2(g)\ \rightarrow\ B_2O_3(s)\ +\ 3\,H_2O(l)\ \ (or\ 2\,B(OH)_3(s))$$

$$B_2H_6(g)\ +\ 3\,H_2O(l)\ \rightarrow\ 2\,B(OH)_3(s)\ +\ 3\,H_2(g)$$

In order to transfer diborane from a storage bulb, in which it exerts a pressure of 200 Torr, to a reaction vessel containing diethyl ether, you must first cool the reaction vessel with liquid nitrogen to –196°C (77 K). This will freeze the ether. Then you remove all of the noncondensable gases (e.g. N$_2$) from the reaction vessel with the vacuum system. Finally, with the reaction vessel still cooled to –196°C, you disconnect the vacuum system from the reaction vessel and connect the storage bulb to the reaction vessel. At this temperature, the sample of diborane will completely condense in the reaction vessel. Once you seal the vessel (by closing its valve), you can allow it to warm up to room temperature, at which point you will have a *solution* of diborane in diethyl ether.

8.14   Compare BH$_4^-$, AlH$_4^-$, and GaH$_4^-$? Since Al has the lowest electro-negativity of the three elements B (2.04), Al (1.61), and Ga (1.81, see Table 1.8), the Al–H bonds of AlH$_4^-$ are more hydridic than the B–H bonds of BH$_4^-$ or the Ga–H bonds of GaH$_4^-$. Therefore, since AlH$_4^-$ is more "hydride-like," it is the strongest reducing agent. The reaction of GaH_4^- with aqueous HCl is as follows:

$$GaH_4^-(aq) + 4\,HCl(aq) \rightarrow GaCl_4^-(aq) + 4\,H_2(g)$$

8.15 **Synthesis of boron compounds?** **(a) $B(C_2H_5)_3$?** A very convenient way to make boron alkyls is by hydroboration, the addition of olefins to B–H bonds. You would first prepare diborane, B_2H_6, by treating $NaBH_4$ with BF_3. This intermediate would be purified by trap-to-trap distillation on a vacuum line. It would then be treated with six equivalents of ethene (ethylene) to form triethylborane, which would also be purified using a vacuum line. The balanced equations for this two-step reaction are:

$$3\,NaBH_4(s) + 4\,BF_3(g) \rightarrow 2\,B_2H_6(g) + 3\,NaBF_4(s)$$

$$B_2H_6 + 6\,C_2H_4 \xrightarrow{\text{ether}} 2\,B(C_2H_5)_3$$

(b) Et_3NBH_3? In this case you want to cleave off one of the B–H bonds of tetrahydroborate(1–) and introduce an amine. You can do both of these things in one step if you added an alkylammonium salt to $NaBH_4$. The acidic hydrogen atom of the alkylammonium ion will cleave off the hydridic hydrogen atom of the BH_4^-, forming H_2, and the resulting Lewis acid (B_2H_6) and Lewis base (Et_3N) will form a complex, as follows:

$$NaBH_4 + Et_3NH^+Cl^- \rightarrow Et_3NBH_3 + NaCl + H_2$$

8.16 **The formation of pure Si from crude Si?** Crude silicon is treated with gaseous HCl (not aqueous HCl) to form $SiHCl_3$, which undergoes a redistribution reaction at moderately high temperature to form SiH_4 (silane) and $SiCl_4$. These two compounds are separated by fractional distillation (their normal boiling points are $-112°C$ and $58°C$, respectively), and the highly purified SiH_4 is decomposed at $500°C$ to pure Si and H_2. Balanced equations for this process are shown below:

$$Si(s) + 3\,HCl(g) \rightarrow SiHCl_3(l) + H_2(g)$$

$$4\,SiHCl_3(l) + \text{heat} \rightarrow SiH_4(g) + 3\,SiCl_4(l)$$

$$500°C: \quad SiH_4(g) \rightarrow Si(s) + 2\,H_2(g)$$

8.17   **Compare period 2 and period 3 hydrogen compounds?**   One important difference between period 2 and period 3 hydrogen compounds is their relative stabilities. The period 2 compounds, except for B_2H_6, are all exoergic (see Table 8.6). Their period 3 homologues are either much less exoergic or are endoergic (cf. HF and HCl, for which $\Delta_f G° = -273.2$ and -95.3 kJ mol^{-1}, and NH_3 and PH_3, for which $\Delta_f G° = -16.5$ and $+13.4$ kJ mol^{-1}). Another important chemical difference is that period 2 compounds tend to be weaker Brønsted acids and stronger Brønsted bases than their period 3 homologues. Diborane, B_2H_6 is a gas while AlH_3 is a solid. Methane, CH_4, is inert to oxygen and water while silane, SiH_4, reacts vigorously with both. The bond angles in period 2 hydrogen compounds reflect a greater degree of sp^3 hybridization than the homologous period 3 compounds (cf. the H–O–H and H–N–H bond angles of water and ammonia, which are 104.5° and 106.6°, respectively, to the H–S–H and H–P–H bond angles of hydrogen sulfide and phosphane, which are 92° and 93.8°, respectively). Several period 2 compounds exhibit strong hydrogen bonding, namely HF, H_2O, and NH_3, while their period 3 homologues do not (see Figure 8.3). As a consequence of hydrogen bonding, the boiling points of HF, H_2O, and NH_3 are all higher than their respective period 3 homologues.

8.18   **Describe the compound formed between water and Kr?**   This compound is called a clathrate hydrate (see Section 8.4(b)). It consists of cages of water molecules, all hydrogen bonded together, each surrounding a single krypton atom (cf. the structure of the clathrate hydrate of molecular chlorine, shown in Figure 8.8). Strong dipole–dipole forces hold the cages together, while weaker van der Waals forces hold the krypton atoms in the centers of their respective cages.

8.19   **Potential energy surfaces for hydrogen bonds?**   There are two important differences between the potential energy surfaces for the hydrogen bond between H_2O and Cl$^-$ ion and for the hydrogen bond in bifluoride ion, HF_2^-. The first difference is that the surface for the H_2O, Cl$^-$ system has a double minimum (as do most hydrogen bonds), since it is a relatively weak hydrogen bond, while the surface for the bifluoride ion has a single minimum (characteristic of only the strongest hydrogen bonds). The second difference is that the surface for the H_2O, Cl$^-$ system is not symmetric, since the proton is bonded to two different atoms (oxygen and chlorine), while the surface for bifluoride ion is symmetric. The two surfaces are shown below.

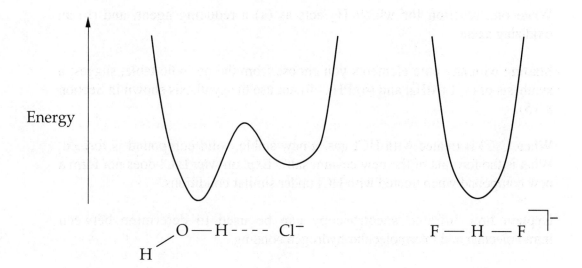

Guide to Solutions Quiz

1 What are the names and approximate natural abundances of the three isotopes of hydrogen? Which isotope is radioactive? By what process does it decay, and what is its half-life?

2 Write formulas for bismuthane, hydroxylamine, plumbane, hydrazine, and diborane.

3 Explain why only the most electropositive elements form saline hydrides. You may want to use a Born–Haber–type cycle in your answer.

4 Calculate the volume of palladium metal needed to store 1 mol of hydrogen gas. The density of palladium is 12 g cm^{-3}. Compare your answer with the volume of 1 mol of liquid hydrogen, the density of which is 0.07 g cm^{-3}.

5 Complete and balance the following equations: (a) $Ca + HNO_3 \rightarrow$; (b) $LiH + HOCl \rightarrow$; (c) $I_2 + H^- \rightarrow$; (d) $EtBH_2 + KH \rightarrow$.

6 Discuss the ways in which the following substances are more reactive than C_2H_6: (a) B_2H_6; (b) N_2H_4; (c) Si_2H_6.

7   Write one reaction for which H$_2$ acts as (a) a reducing agent, and (b) an oxidizing agent.

8   Starting with any pure elements you choose from the periodic table, suggest a synthesis of (a) LiAlH$_4$, and (b) PH$_3$ (do not use the synthesis shown in Section 8.15).

9   When CsCl is treated with HCl gas, a new stable, solid compound is formed. What is the formula of the new cesium salt? Explain why LiCl does not form a new compound when treated with HCl under similar conditions.

10   Explain how infrared spectroscopy can be used to determine between intramolecular and intermolecular hydrogen bonding.

9   The metals

1	2
Li	Be
Na	Mg
K	Ca
Rb	Sr
Cs	Ba
Fr	Ra

The s block of the periodic table, comprising the first two groups, is composed of the alkali metals (Group 1) and the alkaline earths (Group 2). All of these elements are metals, since they all have high electric and thermal conductivities, are malleable and ductile, and are electropositive. In addition, all of these elements form basic oxides and hydroxides. All of the elements in the d and f blocks of the periodic table are also metals, as are several of the elements in the p block.

S9.1   Choose the most appropriate ligand? (a) Li$^+$, Na$^+$, Rb$^+$, and Cs$^+$? Either the bicyclic cryptate ligand C2.2.2 or the acyclic polydentate ligand EDTA^{4-} would be preferred to OH$^-$, because both are polydentate and OH$^-$ is only monodentate (see Section 7.7, *Coordination equilibria*). The more appropriate ligand for alkali metals would be C2.2.2, since it forms very strong complexes with them (see Figure 9.7) and since EDTA^{4-} forms strong complexes with metal ions in 2+ or higher oxidation states. According to Figure 9.7, the order of stability would be Rb$^+$ > Na$^+$ ~ Cs$^+$ > Li$^+$.

(b) Cu^{2+}, Fe^{3+}? $EDTA^{4-}$ forms very strong complexes with 2+ and 3+ metal ions, so it would be the most appropriate choice here. Since it has a higher charge, Fe^{3+} would be expected to form a stronger complex with $EDTA^{4-}$ than would Cu^{2+}.

S9.2 **What happens when acidic aqueous V^{2+} is exposed to oxygen?** The Latimer diagram for vanadium in aqueous acid is shown below (see Appendix 2; reduction potentials are given in volts).

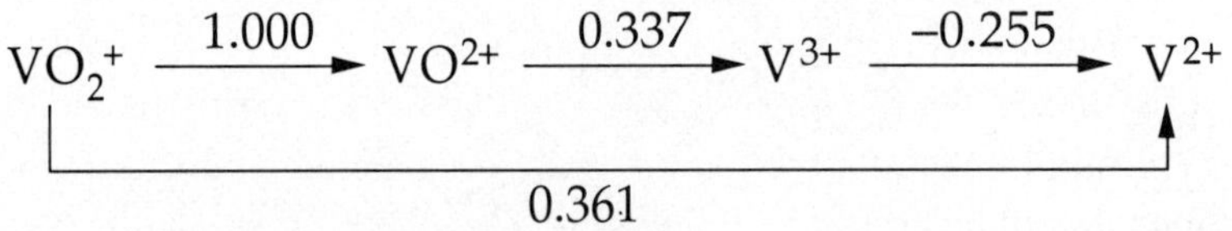

$$VO_2^+ \xrightarrow{1.000} VO^{2+} \xrightarrow{0.337} V^{3+} \xrightarrow{-0.255} V^{2+}$$
$$\underset{0.361}{\xleftarrow{\hspace{3cm}}}$$

Since the reduction potential for the O_2/H_2O couple is 1.229 V, V^{2+} will be oxidized all the way to VO_2^+ as long as sufficient O_2 is present. The net reaction is:

$$V^{2+} + O_2 + 2H^+ \rightarrow VO_2^+ + H_2O \qquad E° = 1.229\ V - 0.361\ V = 0.868\ V$$

S9.3 **The product of Re_3Cl_9 and PPh_3?** Figure 9.19 shows that Re_3Cl_9 molecules are linked in the solid state by weak intercluster chlorine bridges. When ligands are added, say Cl^- or PPh_3, these bridges are broken and discrete molecular species such as $Re_3Cl_{12}^{3-}$ or $Re_3Cl_9(PPh_3)_3$ are formed. Sterically, the most favorable place for each bulky triphenylphosphine ligand to go is in the terminal position in the Re_3 plane, as shown below.

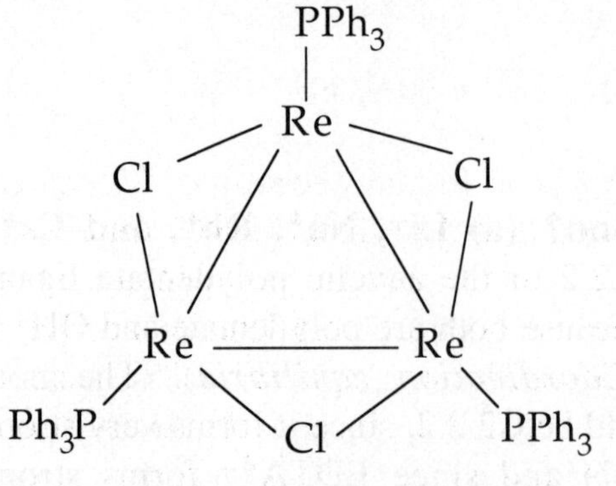

A portion of the structure of $Re_3Cl_9(PPh_3)_3$, showing the six ligands in the same plane as the Re_3 triangle. Six other Cl^- ligands are part of the molecule. Three terminal Re–Cl bonds are above the plane shown and three are below the plane shown.

S9.4 **Why is MoS_2 an effective lubricant?** The layered structure of molybdenum(IV) sulfide is the same as that shown in Figure 9.24 for CdI_2.

Although the Mo–S bonds are undoubtedly quite strong, the S···S interactions between adjacent S–Mo–S layers are weak and easily disrupted (these are best thought of as van der Waals interactions). The slipperiness of MoS_2 is due to the ease with which one layer can glide over another.

S9.5 **Propose chemical reactions?** **(a) $(CH_3)_2SAlCl_3$ and $GaBr_3$?** Toward a soft base such as dimethylsulfane (dimethylsulfide), the order of Lewis acidity is $GaBr_3 > AlCl_3$ because $GaBr_3$ is softer than $AlCl_3$ (see Section 5.12, *Hard and soft acids and bases*). This is for two reasons, the first of which is that softness increases down a group. The second reason is the presence of the soft bromine substituents on gallium relative to the harder chlorine substituents on aluminum (remember that soft ligands make for a soft metal center). Therefore, a good proposal is that the following exchange reaction will occur:

$$(CH_3)_2SAlCl_3 + GaBr_3 \rightarrow (CH_3)_2SGaBr_3 + AlCl_3$$

(b) $TlCl_3$ and formaldehyde in acidic water? Toward the bottom of groups 13/III and 14/IV, oxidation states two lower than the maximum oxidation state become increasingly more stable and hence more important. This being the case, it should come as no surprise that $TlCl_3$ is highly oxidizing (i.e. it is readily reduced to $TlCl$). Since formaldehyde is easily oxidized, the following redox reaction will occur:

$$2\,TlCl_3(aq) + CH_2O(aq) + H_2O(l) \rightarrow 2\,TlCl(s) + CO_2(aq) + 4\,HCl(aq)$$

This is the net reaction of the following two half-reactions:

$$2\,TlCl_3(aq) + 4\,e^- \rightarrow 2\,TlCl(s) + 4\,Cl^-(aq)$$

$$CH_2O(aq) + H_2O(l) \rightarrow CO_2(aq) + 4\,H^+(aq) + 4\,e^-$$

Note that $TlCl$ is insoluble in water, like $AgCl$. In many reactions, Tl^+ ion can be used to precipitate chloride, bromide, or iodide, just like Ag^+.

S9.6 **The most stable uranium ion in aqueous acid?** From the Frost diagram in Figure 9.32, it can be seen that the most stable oxidation state of uranium in aqueous acid is U^{4+} (that is, it has the most negative free energy of formation, the quantity plotted on the y axis). However, the reduction potentials for the UO_2^{2+}/U^{4+} and UO_2^{2+}/UO_2^+ couples are quite small, 0.380 and 0.170 V, respectively (the UO_2^+/U^{4+} potential, therefore, is 0.275 V). Therefore, since the O_2/H_2O reduction potential is 1.229 V, the most stable uranium ion is UO_2^{2+} if sufficient oxygen is present:

$$2\,U^{4+} + O_2 + 2\,H_2O \rightarrow 2\,UO_2^{2+} + 4\,H^+ \qquad E° = 1.229\ V - 0.275\ V = 0.954\ V$$

9.1   **Properties of the *s*-block metals?** A sketch of the *s* block is shown below, along with the trends in (a) melting point, (b) radii for common cations, and (c) the trend in the tendency of peroxides to decompose thermally to simple oxides:

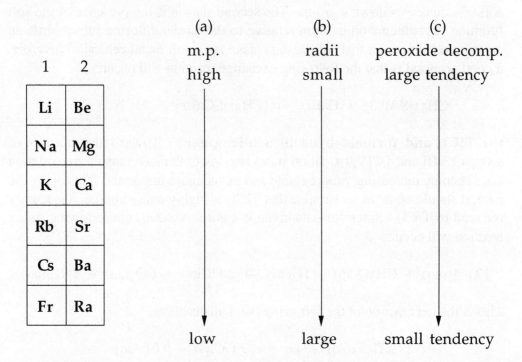

The trend in melting point can be gleaned from Figure 9.2, showing sublimation enthalpies for the metals. The only anomaly is that magnesium has a lower melting point (649°C) than calcium (839°C). (No weighable amount of francium has ever been isolated, but it is estimated that its melting point should be approximately 25°C.) One way to remember the trend is that bonds become weaker as you go down a group in the *s* and *p* blocks of the periodic table (the trend is just the opposite in the *d* block). The trend in ionic radii for Group 1 (charge = 1+) and Group 2 (charge = 2+) was presented in Section 1.8(a), *Atomic and ionic radii*. The trend in tendency for perioxides of these metals to decompose (to metal oxides and O_2) can be understood from the concepts presented in Section 2.12(a), *Thermal stabilities of ionic solids*.

9.2 **Which is more likely to lead to the desired result?** **(a) Cs^+ or Mg^{2+}, form an acetate complex?** The metal ion with the higher charge, Mg^{2+}, is more likely to form a complex with acetate ion than Cs^+. The reason is that the binding of ligands for these hard metal ions is governed by the electrostatic parameter, z^2/r. As stated in the text, the weakness with which s-block metal ions bind ligands explains why until recently very few s-metal complexes had been characterized.

(b) Be or Sr, dissolve in liquid ammonia? It is more likely that strontium will dissolve in liquid ammonia than beryllium. There are two ways to arrive at this conclusion. First, in Section 9.5, *Metal-rich oxides, electrides, and alkalides*, you can see that electropositive metals in low enthalpies of sublimation are prone to dissolve in liquid ammonia. Reference to Figure 9.2 will show that $\Delta_{sub}H(Sr) \sim (1/2)(\Delta_{sub}H(Be))$. The second reason is that the Be^{2+}/Be reduction potential, -1.97 V, is less negative than $E°$ for the Sr^{2+}/Sr potential, -2.89 V. Even though you were asked to focus on liquid ammonia solutions in this exercise and $E°$ values refer to redox reactions in *aqueous acid*, the principal steps are the same on going from $M^0 \to M^{2+}$, namely sublimation, ionization, and solvation. Therefore, you can expect a general agreement between $E°$ values and tendency to dissolve in liquid ammonia.

(c) Li^+ or K^+, form a complex with C2.2.2? Potassium ion is more likely to form a complex with the cryptate ligand C2.2.2 than Li^+. The difference has to do with the match between the interior size of the cryptate and the ionic radius of the alkali metal ion. According to Figure 9.7, K^+ forms a stronger complex, by about two orders of magnitude, with C2.2.2 than does Li^+.

9.3 **Explain the difference in coordination environment?** **(a) CaF_2 vs. MoS_2?** Calcium fluoride exhibits the fluorite structure, shown in Figure 2.13, while MoS_2 exhibits a variation of the layered CdI_2 structure, shown in Figure 9.24. In the fluorite structure, each Ca^{2+} ion is surrounded by eight F^- ions at the corners of a cube. The F^- ions fill all of the tetrahedral holes between close-packed planes of Ca^{2+} ions. The coordination numbers (Ca^{2+}:F^-) are 8:4. In layered MoS_2, the molybdenum ions sit in trigonal prismatic holes between two S^{2-} layers. The coordination numbers are 6:3. There are therefore two differences. The difference in coordination numbers can be attributed to degree of ionicity vs. degree of covalency. The high coordination numbers in CaF_2 are the hallmark of a large degree of ionic character, which makes sense for this compound containing hard cations and anions. The difference in distribution of ions in interstitial holes, i.e. the fact that MoS_2 is layered while CaF_2 is not, has a similar explanation. Layered structures like MoS_2 are only found when the compound has a large degree of covalent character, because only then will the interactions between atoms of the same type in adjacent layers (e.g. S···S interactions in MoS_2) be attractive instead of repulsive.

(b) CdI_2 vs. $MoCl_2$? The layered structure of CdI_2 is shown in Figure 9.24. The structure of $MoCl_2$ is described in Section 9.9, *Metal–metal bonded d-metal compounds*. It consists of octahedral Mo_6 clusters with strong metal–metal bonds linked by chlorine bridges. In general, only the *d*-block metals with partially filled *d* shells form strong metal–metal bonds, so it is not surprising that CdI_2, containing $4d^{10}$ Cd^{2+} ions, does not exhibit the $MoCl_2$ structure. Note that there is a weak Cd–Cd bond, involving *s–s* overlap, in the Cd_2^{2+} ion (see Section 9.13, *Redox reactions*).

(c) BeO vs. CaO? Beryllium oxide exhibits the wurtzite structure (i.e. the coordination number 4:4 hexagonal closest-packed ZnS structure shown in Figure 2.14), while CaO exhibits the 6:6 rock-salt structure (see Table 2.3). Size is the most important reason for the difference. Small, hard Be^{2+} is simply too small to have more than four nearest neighbors. On the other hand, a compound with a large degree of ionic character and reasonably sized ions, such as CaO, will be found to have as many nearest neighbors as possible, certainly more than four.

(d) $Mo_2(OAc)_4$ vs. $Be_4O(OAc)_6$? A drawing of $Mo_2(OAc)_4$ is shown in Structure 27, while a drawing of $Be_4O(OAc)_6$ is shown in Structure 5. The major difference in the two structures is that the former involves a strong metal–metal quadruple bond while the latter does not exhibit any metal–metal bonding. Divalent alkaline earth cations have a noble gas-like electron configuration and are incapable of metal–metal bonding.

9.4   The 3*d* metal ions with the group oxidation number? The metals from scandium through zinc are shown below, with the group numbers written above. Each metal has a notation indicating whether its group oxidation number is

3	4	5	6	7	8	9	10	11	12
Sc	Ti	V	Cr	Mn	Fe	Co	Ni	Cu	Zn
C	C	C	O	O	N	N	N	N	N

common and relatively unreactive (C), a strong oxidant (O), or not achieved (N). Examples of the first three are $ScCl_3$, TiO_2, and the VO_2^+ ion. Examples of the next three are CrO_3, MnO_4^-, and FeO_4^{2-}.

9.5 **The trend in stability of the group oxidation number?** In the d-block of the periodic table, the group oxidation number *increases* in stability as you descend a group. For example, CrO_3 is a strong oxidizing agent but WO_3 is not. In the p block, the group oxidation number *decreases* in stability as you descend a group. For example, Tl^{3+} is a strong oxidizing agent but Al^{3+} is not. Values for E° are shown below. Remember that the less positive (or the more negative) a reduction potential, the greater the stability of the oxidized form of the redox couple in question.

VO_2^+/V	−0.236 V	$Cr_2O_7^{2-}/Cr$	−0.320 V	Al^{3+}/Al	−1.676 V
Nb_2O_5/Nb	−0.65 V	H_2MoO_4/Mo	0.114 V	Ga^{3+}/Ga	−0.529 V
Ta_2O_5/Ta	−0.81 V	WO_3/W	−0.090 V	In^{3+}/In	−0.338 V
				Tl^{3+}/Tl	0.72 V

9.6 **Give balanced chemical equations and rationalize your answer?** **(a) $Cr^{2+} + Fe^{3+}$?** As you move from left to right across the d block, stable oxidation states in aqueous solution tend to get lower (for example, from TiO^{2+} and VO_2^+ on the left to Co^{2+}, Ni^{2+}, Cu^{2+}, and Zn^{2+} on the right). Based on this trend, you should expect Cr^{2+} to be oxidized to Cr^{3+} at the expense of Fe^{3+}, which will be reduced to Fe^{2+}. That is, iron is to the right of chromium, and the trend is left higher, right lower. In fact, the net potential for this redox reaction in aqueous acid is 1.195 V (see Appendix 2).

$$Cr^{2+}(aq) + Fe^{3+}(aq) \rightarrow Cr^{3+}(aq) + Fe^{2+}(aq) \quad E^\circ = 1.195 \text{ V}$$

(b) $CrO_4^{2-} + MoO_2$? Recall the trend that higher oxidation numbers become increasingly more stable as you descend a group in the d block. Since you are given two compounds containing Cr(VI) and Mo(IV), you should expect a redox reaction to occur, with the oxidation of MoO_2 and reduction of CrO_4^{2-}. The products should be Cr^{3+} and H_2MoO_4 (see Appendix 2). Note that Cr(V) and Cr(IV) are unstable with respect to disproportionation.

$$2\,CrO_4^{2-} + 3\,MoO_2(s) + 10\,H^+ \rightarrow 2\,Cr^{3+} + 3\,H_2MoO_4 + 2\,H_2O \quad E^\circ = 0.73 \text{ V}$$

(c) $MnO_4^- + Cr^{3+}$? This is another case where you should apply the trend left higher, right lower, predicting a reaction between these two species in which Cr^{3+} is oxidized and MnO_4^- is reduced. However, since these two elements are immediate neighbors, you should not expect the net $E°$ to be very large. In fact, the products are Mn^{2+} and $Cr_2O_7^{2-}$ and $E° = 0.12$ V.

$$6\,MnO_4^- + 10\,Cr^{3+} + 11\,H_2O \rightarrow 6\,Mn^{2+} + 5\,Cr_2O_7^{2-} + 22\,H^+ \qquad E° = 0.12\ V$$

9.7   **Is Ni^{2+} or Mn^{2+} more likely to form a sulfide?**   In general, hardness decreases and softness increases from left to right in the d-block, especially when comparing elements in the same oxidation state. Therefore, it is more likely for Ni^{2+} to form a sulfide than it is for Mn^{2+} to form a sulfide:

$$Ni^{2+}(aq) + H_2S(aq) \rightarrow NiS(s) + 2\,H^+(aq)$$

9.8   **Difluorides of the d-block metals?**   The d-block of the periodic table is shown below. Those metals that form difluorides having the rutile structure are indicated. The region in which metal–metal bonded halide compounds are found is indicated with a bold border. Examples of metal–metal bonded halides are Sc_5Cl_6, $ZrCl$, and Re_3Cl_9.

Sc	Ti	V	Cr	Mn	Fe	Co	Ni	Cu	Zn
	rutile	rutile	rutile	rutile	rutile	rutile	rutile		
Y	Zr	Nb	Mo	Tc	Ru	Rh	Pd	Ag	Cd
La/Lu	Hf	Ta	W	Re	Os	Ir	Pt	Au	Hg

9.9   **Reactions of cis-[RuLCl(OH$_2$)]$^+$?**   According to Figure 9.12, the Ru(II) complex cis-[RuLCl(OH$_2$)]$^+$ is stable at 0.2 V below about pH 8. What this means is that if there is a redox couple (Red and Ox) with $E° = 0.2$ V in solution with the ruthenium complex, no reaction will occur. Assume that $E°$ for Red/Ox does not change with pH. Now, if you raise the pH past 8, the ruthenium complex will be oxidized and Ox will be reduced:

$$cis\text{-}[Ru^{II}LCl(OH_2)]^+ + Ox + OH^- \rightarrow cis\text{-}[Ru^{III}LCl(OH)]^+ + Red + H_2O$$

Note that the charge on the two ruthenium complexes is the same, since the H_2O ligand in the Ru(II) complex has become a OH^- ligand in the Ru(III) complex.

If the pH is held at 6 and the potential is raised (by adding stronger and stronger oxidants to the solution), the original Ru(II) complex will be oxidized to a Ru(III) complex and then to a Ru(IV) complex, with a single deprotonation occurring at each oxidation:

$$cis\text{-}[Ru^{II}LCl(OH_2)]^+ + H_2O \rightarrow cis\text{-}[Ru^{III}LCl(OH)]^+ + H_3O^+ + e^-$$

$$cis\text{-}[Ru^{III}LCl(OH)]^+ + H_2O \rightarrow cis\text{-}[Ru^{IV}LCl(O)]^+ + H_3O^+ + e^-$$

As the oxidation state of the metal increases, its ability to accept electron density from an OH^- or O^{2-} ligand through π bonding also increases, and deprotonation of the original H_2O ligand occurs.

9.10 **Write plausible balanced chemical reactions?** **(a)** $MoO_4^{2-} + Fe^{2+}$? The only possible reaction that might occur here is a redox reaction. However, since the Fe^{3+}/Fe^{2+} reduction potential is 0.771 V and since there is no redox couple of molybdenum involving MoO_4^{2-} that is more positive than 0.771 V, no reaction will occur.

(b) The preparation of $[Mo_6O_{19}]^{2-}$ from K_2MoO_4? Acidification of solutions of MoO_4^{2-} will produce polyoxometallates such as $[Mo_6O_{19}]^{2-}$ (see Section 9.7(f), *Polyoxometallates*). The balanced equation is:

$$6\,MoO_4^{2-}(aq) + 10\,H^+(aq) \rightarrow [Mo_6O_{19}]^{2-}(aq) + 5\,H_2O(l)$$

(c) $ReCl_5 + KMnO_4$? In this case you have a compound of Re(V) and Mn(VII). Recall that the group oxidation number is more readily achieved by the heavier metals in a d-block group. Therefore, the rhenium compound will be oxidized and the manganese compound will be reduced. One plausible reaction is:

$$5\,ReCl_5(s) + 2\,MnO_4^-(aq) + 12\,H_2O(l) \rightarrow$$
$$5\,ReO_4^-(aq) + 2\,Mn^{2+}(aq) + 25\,Cl^-(aq) + 24\,H^+(aq)$$

(d) $MoCl_2$ + warm HBr? Aqueous Br^- is a reducing agent in that it can form Br_2 (although $E°$, at ~ 1 V, is high). Therefore, a plausible reaction is the following cluster formation:

$$6\,MoCl_2(s) + 2\,Br^-(aq) \rightarrow [Mo_6Cl_{12}]^{2-}(aq) + Br_2(aq)$$

9.11   **The structures of [Re(O)$_2$(py)$_4$]$^+$, [V(O)$_2$(ox)$_2$]$^{3-}$, [Mo(O)$_2$(CN)$_4$]$^{4-}$, and [VOCl$_4$]$^{2-}$?** The first three of these complexes are dioxo complexes and will have *cis*-dioxometal structural units, as reviewed in Section 9.7(d), *Mononuclear oxo complexes*. The reason that a *cis* geometry is observed rather than a *trans* geometry is that in the *cis* geometry the two oxo ligands only have to share a single *d* orbital for O→M π bonding. In the *trans* geometry, on the other hand, the two oxo ligands must share the same two *d* orbitals. The structures of the first three complexes are shown below.

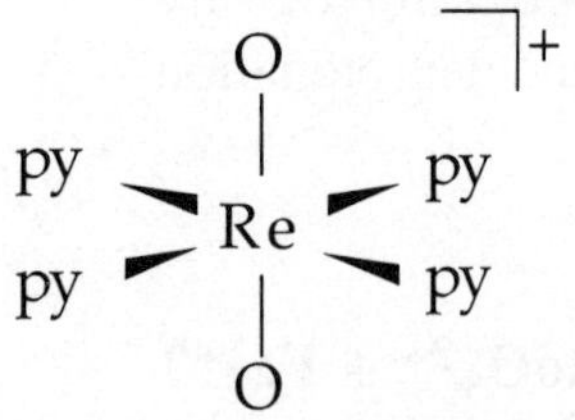

d^2 *trans* 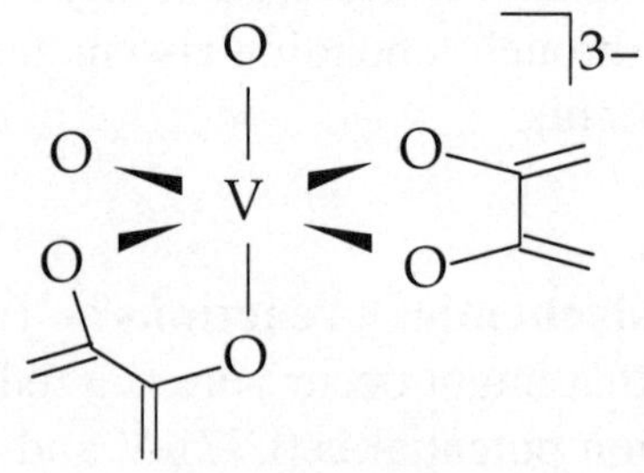 d^0 *cis* 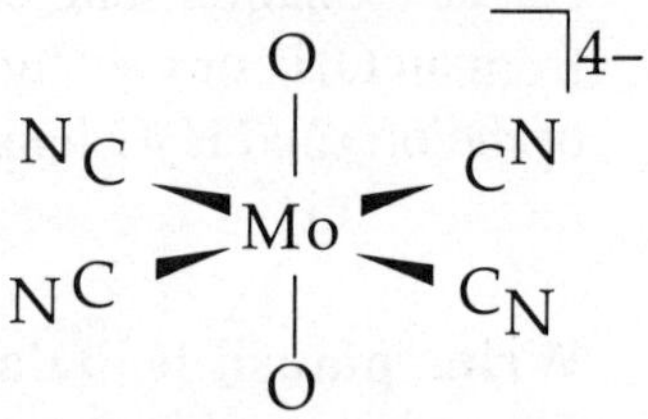 d^2 *trans*

The complex VOCl$_4{}^{2-}$, like VO(acac)$_2$, has a square-pyramidal structure with an apical oxo ligand. Its structure is shown at the right

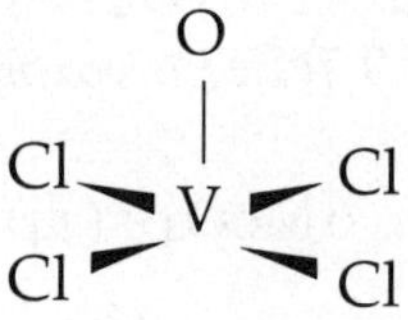

9.12   **Which are predominantly ionic, significantly covalent, and metal–metal bonded compounds? (a) NiI$_2$?** This is a typical example of a divalent metal diiodide. It should be an ionic compound with a significant degree of covalent character, since I$^-$ is such a soft and polarizable anion. You should expect NiI$_2$ to have a layered structure as opposed to a more ionic structure such as the rutile structure.

(b) NbCl$_4$? There are very few "ionic" compounds with metal ions in the 4+ oxidation state. You should expect NbCl$_4$ to be significantly covalent. You should not expect it to contain metal–metal bonds, since the oxidation number is too high. In addition, you should not expect it to be molecular, since a relatively large *d*-block metal would naturally be more than four-coordinate. Therefore, a structure with bridging halides is likely.

(c) FeF₂? A compound containing a divalent metal ion and hard fluoride ions is going to be ionic. Expect the rutile structure.

(d) PtS? With a soft metal ion like Pt^{2+} and a soft anion like S^{2-}, you should expect a significant amount of covalent character. Some metal monosulfides exhibit the nickel-arsenide structure (Figure 2.15), so that would be a good choice here. In fact, PtS exhibits a different type of covalent solid-state structure, with four-coordinate square-planar Pt^{2+} ions. Most of the d^8 ions, including Pd^{2+} and Pt^{2+}, tend to form structures containing four-coordinate square-planar metal ions, as discussed in Section 9.10, *Noble character*.

(e) WCl₂? This is a compound of a low valent period 4 or 5 early *d*-block metal. This is precisely the kind of situation in which metal–metal bonding occurs.

9.13 **Write plausible balanced chemical equations?** **(a) TiO + HCl(aq)?** Titanium in the 2+ oxidation state in aqueous acid is a strong reducing agent, strong enough even to reduce protons to H_2. The balanced equation is:

$$2\,TiO(s) \;+\; 6\,H^+(aq) \;\rightarrow\; 2\,Ti^{3+}(aq) \;+\; H_2(g) \;+\; 2\,H_2O(l) \qquad E° = 0.37\ V$$

(b) Ce⁴⁺ + Fe²⁺? Recall that all of the lanthanides are most stable in the 3+ oxidation state. Therefore, cerium in the 4+ oxidation state should be a strong oxidizing agent, and it will oxidize Fe^{2+} to Fe^{3+} as follows:

$$Ce^{4+}(aq) \;+\; Fe^{2+}(aq) \;\rightarrow\; Ce^{3+}(aq) \;+\; Fe^{3+}(aq) \quad E° = 0.99\ V$$

(c) Rb₉O₂ + H₂O? The compound Rb_9O_2 is an unusual suboxide of rubidium. The average oxidation state of rubidium in this compound is less than 1+, which is the common oxidation state of this alkali metal. Therefore, water will oxidize the compound:

$$2\,Rb_9O_2(s) \;+\; 14\,H_2O(l) \;\rightarrow\; 18\,Rb^+(aq) \;+\; 5\,H_2(g) \;+\; 18\,OH^-(aq)$$

(d) Na(am) + CH₃OH? A solution of sodium in liquid ammonia contains solvated Na^+ ions and solvated electrons. The electrons will react with any hydroxylic solvent (including H_2O and CH_3OH) to produce hydrogen gas and RO^- (R = H, CH_3, etc.):

$$2Na(am) \;+\; 2CH_3OH(am) \;\rightarrow\; 2Na^+(am) \;+\; 2OCH_3^-(am) \;+\; H_2(g)$$

**9.14   Probable   occupancies   of σ, π, and δ orbitals?   (a)
[Mo$_2$(O$_2$CCH$_3$)$_4$]?** This is a neutral molecule with two molybdenum ions
and four 1– acetate ions. Therefore, each of the two molybdenum ions has a 2+
charge and a d^4 configuration. Consequently, the Mo$_2$ fragment has eight
electrons and a $\sigma^2\pi^4\delta^2$ configuration. This molecule has a molybdenum–
molybdenum quadruple bond.

(b) [Cr$_2$(O$_2$CC$_2$H$_5$)$_4$]? This molecule is very similar to the one in part (a).
Instead of Mo^{2+} ions, in contains another divalent ion from Group 6, Cr^{2+}.
Instead of 1– acetate ions, it contains 1– propionate ions. Consequently, it has
the same electron configuration as far as the metal–metal bonding orbitals are
concerned, $\sigma^2\pi^4\delta^2$. This molecule has a chromium–chromium quadruple bond.

(c) [Cu$_2$(O$_2$CCH$_3$)$_4$]? With two copper ions and four 1– acetate ions, this
molecule contains two d^9 Cu^{2+} ions. With 18 electrons between the two Cu^{2+}
ions, all of the metal–metal bonding and antibonding orbitals are filled. Sixteen
of the electrons form the $\sigma^2\pi^4\delta^2\delta^{*2}\pi^{*4}\sigma^{*2}$ configuration. Each copper ion also
has an additional unpaired electron in its $d_{x^2-y^2}$ orbital (the one that points
directly at the acetate oxygen ligands). There is no metal–metal bond in this
molecule.

9.15   The reactions of Zn(NH$_2$)$_2$ in liquid ammonia? If you accept the
analogy between liquid ammonia and water, try to imagine what would happen
if Zn(OH)$_2$ were treated with H$_3$O$^+$ or OH$^-$ in aqueous solution. The
complexes Zn(OH$_2$)$_6{}^{2+}$ and Zn(OH)$_3{}^-$, respectively, would form. Therefore,
when Zn(NH$_2$)$_2$ is treated with NH$_4{}^+$ or NH$_2{}^-$ in liquid ammonia, the likely
products are Zn(NH$_3$)$_6{}^{2+}$ and Zn(NH$_2$)$_4{}^{2-}$. In actuality, the stable Zn(II)
ammine species is Zn(NH$_3$)$_4{}^{2+}$, not Zn(NH$_3$)$_6{}^{2+}$, but that is a detail that you
could not predict at this point.

9.16   Give balanced equations?   (a) Hg^{2+}(aq) + Cd(s)? Of the three metals
in group 12, mercury is by far the most noble. The consequence of this is that
either zinc or cadmium metal will reduce Hg^{2+} to metallic mercury:

$$Hg^{2+}(aq) + Cd(s) \rightarrow Hg(l) + Cd^{2+}(aq)$$

(b) Tl^{3+}(aq) + Ga(s)? The element thallium is one of those that exhibits the
inert pair effect. This is manifested in the fact that Tl$^+$ is more stable in aqueous
solution than Tl^{3+}, which is a strong oxidant. In contrast, the only oxidation

state of gallium that is stable in aqueous solution is Ga^{3+}. In addition, Ga is a strong enough reducing agent to reduce Tl^+ to $Tl(s)$. The balanced equation is:

$$Tl^{3+}(aq) + Ga(s) \rightarrow Tl(s) + Ga^{3+}(aq)$$

(c) $[AlF_6]^{3-}(aq) + Tl^{3+}(aq)$? As discussed above, Tl^{3+} is a strong oxidant. However, in this case, there is no species that can be readily reduced by it. The Al(III) complex $[AlF_6]^{3-}$ is very stable with respect to reduction to aluminum metal. Therefore, no reaction will occur in this case.

9.17 **The elements of Groups 13/III and 14/IV?** **(a) Summarize the trends?** As a general rule, only the group oxidation state is achieved in stable compounds of the lighter elements in the group. The maximum is 3+ and 4+ for the Group 13/III and 14/IV elements, respectively. (A few exceptions exist, such as B_2Cl_4 and CO.) Only as the heavier elements indium, thallium, tin, and lead are approached, does the oxidation state two less than the maximum become important. In aqueous solution, In^+ is unstable with respect to disproportionation and Sn^{2+} is oxidized by O_2. As far as the inert pair effect is concerned, only Tl^+ and Pb^{2+} are stable in an aerated, aqueous environment. In addition, the noble character of the metals increases down each group.

(b) Write balanced equations? **(i) $Sn^{2+}(aq) + PbO_2(s)$ (excess)?** Since Pb^{2+} is more stable relative to Pb^{4+} than Sn^{2+} is relative to Sn^{4+}, the following redox reaction will occur:

$$Sn^{2+}(aq) + PbO_2(s) \text{ (excess)} + 4H^+(aq) \rightarrow Sn^{4+}(aq) + Pb^{2+}(aq) + 2H_2O(l)$$

(ii) $Tl^{3+}(aq) + Al(s)$ (excess)? Since the noble character increases down a group, aluminum will be oxidized at the expense of thallium(III):

$$Tl^{3+}(aq) + Al(s) \text{ (excess)} \rightarrow Tl(s) + Al^{3+}(aq)$$

(iii) $In^+(aq)$? As stated in part (a), above, monovalent indium is unstable with respect to disproportionation in aqueous solution:

$$3In^+(aq) \rightarrow 2In(s) + In^{3+}(aq)$$

(iv) $Sn^{2+}(aq) + O_2(air)$? As stated in part (a), above, Sn^{2+} will be oxidized by oxygen in aqueous solution:

$$2Sn^{2+}(aq) + O_2(air) + 4H_3O^+(aq) \rightarrow 2Sn^{4+}(aq) + 6H_2O(l)$$

(v) $Tl^+(aq) + O_2(air)$? Thallium(I) is stable in aerated aqueous solution due to the inert pair effect. Therefore, no reaction will occur in this case.

9.18   **Potentials for reactions in Exercise 9.17(b)?   (i) $Sn^{2+}(aq)$ + $PbO_2(s)$ (excess) + $4H^+(aq) \rightarrow Sn^{4+}(aq) + Pb^{2+}(aq) + 2H_2O(l)$?** The potential for the Sn^{2+}/Sn^{4+} oxidation in acid solution is –0.15 V. The potential for the PbO_2/Pb^{2+} reduction in acid solution is 1.468 V. Therefore, $E°$ for the overall reaction is 1.32 V. Recall that $\Delta G° = -NFE°$. This means that there is a large driving force for this reaction.

(ii) $Tl^{3+}(aq)$ + Al(s) (excess) $\rightarrow$ Tl(s) + $Al^{3+}(aq)$? The potentials for the Tl^{3+}/Tl^0 and Al^0/Al^{3+} redox processes in acid solution are 0.72 V and 1.676 V, respectively. Therefore, $E°$ for the overall reaction is 2.40 V, a very large driving force.

(iii) $3In^+(aq) \rightarrow 2In(s) + In^{3+}(aq)$? The potentials for the In^+/In^0 and In^+/In^{3+} redox processes in acid solution are –0.126 and 0.444 V, respectively. Therefore, $E°$ for the overall reaction is 0.318 V. There is not as much driving force for this reaction as for the first two reactions, but 0.318 V is still enough to drive the disproportionation of monovalent indium almost to completion. At 25°C, the equilibrium constant for this reaction is approximately 2×10^5.

(iv) $2Sn^{2+}(aq) + O_2(g) + 4H_3O^+(aq) \rightarrow 2Sn^{4+}(aq) + 6H_2O(l)$? The potentials for the O_2/H_2O and Sn^{2+}/Sn^{4+} redox processes in acid solution are 1.23 and –0.15 V, respectively. Therefore, $E°$ for the overall reaction is 1.08 V. There is plenty of driving force for this redox reaction.

(v) $Tl^+(aq) + O_2(g) \rightarrow$ NR? If this reaction did occur, the balanced equation would be $2Tl^+(aq) + O_2(g) + 4H_3O^+(aq) \rightarrow 2Tl^{3+}(aq) + 6H_2O(aq)$. The potentials for the O_2/H_2O and Tl^+/Tl^{3+} redox processes in acid solution are 1.23 and –1.25 V, respectively. Therefore, $E°$ for the overall reaction is –0.02 V. There is no driving force for this redox reaction, since $E° < 0$.

9.19   **Lanthanide metals and water?   (a) Balanced equation?** All of the lanthanide metals are very electropositive and reduce water (or protons) to H_2 while being oxidized to the trivalent state. If Ln is used as the symbol for a generic lanthanide element, the balanced equation is:

$$2Ln(s) + 6H_3O^+(aq) \rightarrow 2Ln^{3+}(aq) + 3H_2(g) + 6H_2O(l)$$

(b) Redox potentials? The potentials for the Ln^0/Ln^{3+} oxidations in acid solution range from a low of 1.99 V for europium to 2.38 V for lanthanum, a remarkably self consistent set of values spanning 15 elements. In fact, europium is the only lanthanide with a potential lower than 2.22 V. Since the potential for the H_3O^+/H_2 reduction is 0 V, the $E°$ values for the equation shown in part (a) for all lanthanides range from 1.99 to 2.38 V, a very large driving force.

(c) Two unusual lanthanides? The usual oxidation state for the lanthanide elements in aqueous acid is 3+. There are two lanthanides which deviate slightly from this trend. The first is Ce^{4+}, which, while being a strong oxidizing agent, is kinetically stable in aqueous acid. Since Ce^{3+} is $4f^1$ and Ce^{4+} is $4f^0$, you can see that the special stability of tetravalent cerium is due to its stable $4f^0$ electron configuration. The second deviation is Eu^{2+}, which is a strong reducing agent but exists in a growing number of compounds. Since Eu^{2+} is $4f^7$ and Eu^{3+} is $4f^6$, you can see that the special stability of divalent europium is due to its stable $4f^7$ electron configuration.

9.20 **Why were cerium and europium the easiest lanthanides to isolate?** In Exercise 9.19 you explained why Ce^{4+} and Eu^{2+} are relatively stable with respect to all other tetravalent and divalent lanthanide ions. These unusual oxidation states were used in separation procedures, since their charge/radius ratios are very different than those of the typical Ln^{3+} ions. Tetravalent cerium can be precipitated as $Ce(IO_3)_4$, leaving all other Ln^{3+} ions in solution because their iodates are soluble. Divalent europium, which resembles Ca^{2+}, can be precipitated as $EuSO_4$, leaving all other Ln^{3+} ions in solution because their sulfates are soluble.

9.21 **Fission products of ^{235}U?** The thermal neutron fission of ^{235}U yields two fragment nuclei of unequal mass, i.e. the fission is unsymmetrical. Several elements in the neighborhood of mass 95 are formed in ~5% each, and similarly, several elements in the neighborhood of mass 135 are formed in ~5% each (see Figure 9.33). Therefore, in terms of abundance, the isotopes ^{39}Ar and ^{228}Th will not present a large radiation hazard because their masses are far from the two peaks around masses 95 and 135. On the other hand, the isotopes ^{90}Sr and ^{144}Ce, with masses close to 95 and 135, respectively, will be present in relatively large abundance and therefore will pose a significant radiation hazard.

9.22 **Electronic spectra of Eu^{3+} and Am^{3+} complexes?** There is a considerable difference in the interactions of the $4f$ orbitals of the lanthanide ions and the $5f$ orbitals of the actinide ions with ligand orbitals. In the case of the $4f$ orbitals, interactions with ligand orbitals are negligible. Therefore, splitting of the $4f$ subshell by the ligands is also negligible and does not vary as the ligands vary. Since the colors of lanthanide ions are due to $4f$–$4f$ electronic transitions, the colors of Eu^{3+} complexes are invariant as a function of ligand. In contrast, the $5f$ orbitals of the actinide ions interact strongly with ligand orbitals, and the splitting of the $5f$ subshell, as well as the color of the complex, varies as a function of ligand.

Guide to Solutions Quiz

1   Both $AuCl_3$ and $AlCl_3$ have solid-state structures with bridging chloride ions. Predict the coordination geometries of the two metal ions.

2   Based on its empirical formula, the compound TlSe appears to contain Tl^{2+} ions. However, these $5d^{10}6s^1$ ions would be paramagnetic, and TlSe is diamagnetic. Suggest a more likely explanation for the composition of TlSe.

3   The chemistries of Tl^+ and Ag^+ are similar in some respects and different in some. Both ions form slightly soluble halides. Discuss two important differences in their redox chemistry.

4   Make a plot of first ionization energy vs. atomic number for the alkaline earth elements. Using this information and other facts, predict which alkaline earth element is most likely to form covalent as opposed to ionic compounds. Explain your prediction.

5   Make a plot of standard reduction potential vs. atomic number for the alkali metals. Discuss the trend or lack of a trend that you observe.

6   Predict the structures of the following molecular complexes: OsO_4, $VOCl_3$, $Pd(en)_2{}^{2+}$, $Ag(NH_3)_2{}^+$, $Pt(PPh_3)_4$, Nb_2Cl_{10}, $Cr_2O_7{}^{2-}$, Ru_4F_{20}.

7   Would you expect the metal–metal bond distance to increase, decrease, or remain about the same on going from (a) $Mo_2(SO_4)_4{}^{4-}$ to $Mo_2(SO_4)_4{}^{3-}$ and (b) $Tc_2Cl_8{}^{3-}$ to $Tc_2Cl_8{}^{2-}$? Explain your answers.

8   In some very important ways, the chemistry of Y^{3+} resembles that of the heavier lanthanides. In contrast, the chemistry of La^{3+} resembles that of the lighter lanthanides. Explain the difference.

9   Suggest a specific scheme to separate $CrO_4{}^{2-}$ and $MoO_4{}^{2-}$ in basic aqueous solution using any reagent(s) of your choice

10   Use the Frost diagrams and data in Appendix 2 to determine the most stable neptunium ion in aqueous acid in the presence of air.

10 The boron and carbon groups

12	13	14	15
	B	C	N
	Al	Si	P
Zn	Ga	Ge	As
Cd	In	Sn	Sb
Hg	Tl	Pb	Bi

III   IV

The elements of the boron and carbon groups exhibit a wide range of chemical and physical properties. Examples of non-metals, semimetals, and metals can be found in the two groups. Many of the elements have two or more polymorphs, including B (several allotropes), C (graphite and diamond), and Sn (gray and white). The electronic properties of some of the elements and their compounds are of current technological importance.

S10.1 Balanced equations for the following reaction mixtures? (a) BCl₃ and ethanol? As mentioned in the example, boron trichloride is vigorously hydrolyzed by water. Therefore, a good assumption is that it will also react with protic solvents such as alcohols, forming HCl and B–O bonds:

$$BCl_3(g) + 3\,EtOH(l) \rightarrow B(OEt)_3(l) + 3\,HCl(g)$$

(b) BCl₃ and pyridine in hydrocarbon solution? Neither pyridine nor hydrocarbons can cause the protolysis of the B–Cl bonds of boron trichloride, so the only reaction that will occur is a complex formation reaction, such as the one shown below:

$$BCl_3(g) \ + \ py(l) \ \rightarrow \ Cl_3B–py(s)$$

Note that only a 1:1 complex is formed, even if excess pyridine (py) is used. Boron and the other period 2 atoms cannot become hypervalent, in contrast with the heavier atoms of periods 3 and beyond.

(c) BBr₃ and F₃BN(CH₃)₃? Since boron tribromide is a stronger Lewis acid than boron trifluoride, it will displace BF_3 from its complex with $N(CH_3)_3$:

$$BBr_3(l) \ + F_3BN(CH_3)_3(s) \ \rightarrow \ BF_3(g) \ + \ Br_3BN(CH_3)_3(s)$$

S10.2 **Synthesis of (CH₃)₃B₃N₃(CH₃)₃?** This compound is permethyl-borazine. As discussed in Section 10.2, the reaction of ammonium chloride with boron trichloride yields *B*-trichloroborazine, while the reaction of a primary ammonium chloride with boron trichloride yields *N*-alkyl substituted *B*-trichloroborazine, as shown below:

$$3\,RNH_3{}^+Cl^- \ + \ 3\,BCl_3 \ \rightarrow \ 9\,HCl \ + \ Cl_3B_3N_3R_3 \ (R = H, \text{alkyl})$$

Therefore, if you use methylammonium chloride you will produce $Cl_3B_3N_3(CH_3)_3$ (i.e. R = CH_3). This product can be converted to the desired one by treating it with an organometallic methyl compound of a metal that is more electropositive than boron. Either methyllithium or methylmagnesium bromide could be used, as shown below:

$$Cl_3B_3N_3(CH_3)_3 \ + \ 3\,CH_3MgBr \ \rightarrow \ (CH_3)_3B_3N_3(CH_3)_3 \ + \ 3\,Mg(Br, Cl)_2$$

The structure of *N*, *N'*, *N''*-trimethyl-
B, *B'*, *B''*-trimethylborazine.

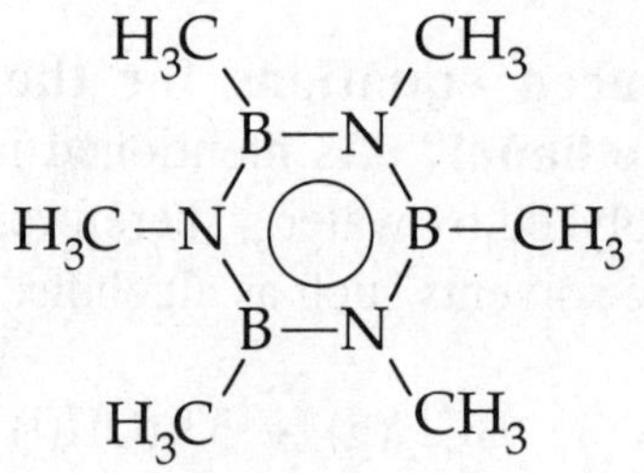

S10.3 How many skeletal electrons are present in B_5H_9? Five B–H units contribute $5 \times 2 = 10$ electrons and the four additional H atoms contribute four additional electrons, for a total of 14 electrons or 7 pairs of electrons.

S10.4 Use Wade's rules to determine the structure of B_5H_9? Five B–H units contribute $5 \times 2 = 10$ electrons and the four additional H atoms contribute four additional electrons, for a total of 14 electrons or 7 pairs of electrons. Boranes of formula B_nH_{n+4} have the *nido* structure, which is based on a *closo* structure with $n + 1$ vertices. In this case $n = 5$. The *closo* structure with 6 vertices is an octahedron, so the *nido* structure of B_5H_9 is based on an octahedron with one vertex missing, which is a square pyramid. In this compound, the four additional H atoms bridge the four B atoms that comprise the square plane of the square pyramid. Wade's rules do not predict the positions of the additional H atoms, just the framework structure of the compound. The C_{4v} structure of B_5H_9 is shown in Table 10.4.

S10.5 A plausible product for the interaction of $Li[B_{10}H_{13}]$ with $Al_2(CH_3)_6$? By analogy with the reaction of $[B_{11}H_{13}]^{2-}$ with $Al_2(CH_3)_6$, the plausible product would be $[B_{10}H_{11}(AlCH_3)]^-$, which would be formed as follows:

$$2\,[B_{10}H_{13}]^- + Al_2(CH_3)_6 \rightarrow 2\,[B_{10}H_{11}(AlCH_3)]^- + 4\,CH_4$$

S10.6 Propose a synthesis for $1,7\text{-}B_{10}C_2H_{10}(Si(CH_3)_2Cl)_2$? As in the example, you should consider attaching the $-Si(CH_3)_2Cl$ substituents to the carbon atoms of this carborane by using the dilithium derivative $1,7\text{-}B_{10}H_{10}C_2Li_2$. You can first prepare $1,2\text{-}B_{10}C_2H_{12}$ from decaborane as in the example. Then, this compound is thermally converted to a mixture of the 1,7- and 1,12-isomers, which can be separated by chromatography:

$$1,2\text{-}B_{10}C_2H_{12} \xrightarrow{\sim 500\,°C} 1,7\text{-}B_{10}C_2H_{12}\,(90\%) + 1,12\text{-}B_{10}C_2H_{12}\,(10\%)$$

The pure 1,7-isomer is lithiated with RLi and then treated with $Si(CH_3)_2Cl_2$:

$$1,7\text{-}B_{10}C_2H_{10}Li_2 + 2Si(CH_3)_2Cl_2 \rightarrow 1,7\text{-}B_{10}C_2H_{10}(Si(CH_3)_2Cl)_2 + 2LiCl$$

S10.7 **Describe how the electronic structure of graphite is altered?**
(a) With potassium? The extended π system for each of the planes of
graphite results in a band of π orbitals. The band is half filled in pure graphite,
that is, all of the bonding MOs are filled and all of the antibonding MOs are
empty. The HOMO–LUMO gap is ~0 eV, giving rise to the observed electrical
conductivity of graphite. Chemical reductants, like potassium, can donate their
electrons to the LUMOs (graphite π^* orbitals), resulting in a material with a
higher conductivity.

(b) With bromine? Chemical oxidants, like bromine, can remove electrons
from the π-symmetry HOMOs of graphite. This also results in a material with a
higher conductivity.

S10.8 **A synthesis of $D^{13}CO_2^-$?** As in the example, you would want to oxidize
^{13}CO to $^{13}CO_2$. However, unlike the example, you would then want to treat the
$^{13}CO_2$ with a source of deuteride ion, D^-. A good source would be LiD. The
entire synthesis would be:

$$^{13}CO(g) + 2MnO_2(s) \rightarrow {}^{13}CO_2(g) + Mn_2O_3(s)$$

$$2Li(s) + D_2 \rightarrow 2LiD(s)$$

$$^{13}CO_2(g) + LiD(et) \rightarrow Li^+D^{13}CO_2^-(et)$$

S10.9 **Determine the charge on $[Si_4O_{12}]^{n-}$?** Two views of the structure of this
cyclic silicate are shown below. It is an eight-membered ring of alternating Si
and O atoms with eight terminal Si–O bonds, two on each Si atom. Since each

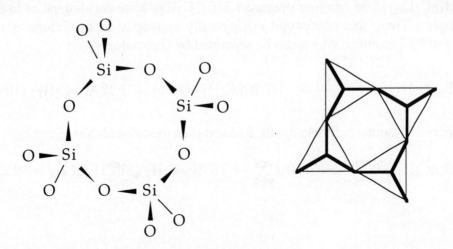

terminal O atom contributes −1 to the total charge on the cyclic ion, the overall charge is −8. The charge can also be determined from the oxidation numbers of the elements: $4(+4) + 12(−2) = −8$.

S10.10 How many Si and Al atoms are in one sodalite cage? A sodalite cage is based on a truncated octahedron. You can imagine an octahedron with four of its six vertices in the plane of this page. You would then look down upon one of the remaining vertices. This is shown in the drawing below and on the left. The truncation you should make is parallel to the plane of the page.

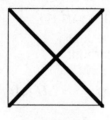

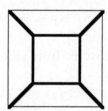

Octahedron   Octahedron truncated
                                   along one of its C_4 axes

The top vertex is removed, and in its place is a square plane parallel to the plane of the page. This is shown in the drawing on the right. Four new vertices have been created for the one that was removed. If you imagine performing this procedure six times, once for each of the vertices of the original octahedron, you will remove six vertices and create 24 new ones (6×4). Thus, a sodalite cage has 24 Si and Al atoms.

10.1   General properties of the elements?

element	type of element	diamond structure?	primarily occurs as oxide(s)?
B	nonmetal	no	yes
Al	metal	no	yes
Ga	metal	no	yes
In	metal	no	no
Tl	metal	no	no
C	nonmetal	yes	yes
Si	nonmetal	yes	yes
Ge	metalloid	yes	yes
Sn	metal	yes	yes
Pb	metal	no	no

10.2   **Draw the B$_{12}$ unit and find a C_2 axis?**   See the structure of *closo*-[B$_{12}$H$_{12}$]$^{2-}$ in Table 10.4. One of the many two-fold rotation axes is a horizontal line in the plane of the paper, bisecting opposite pairs of B–B linkages.

10.3   **Give balanced equations and conditions for the recovery of boron, silicon, and germanium from their ores?**   Boron is recovered from the mineral borax, Na$_2$B$_4$O$_5$(OH)$_4$·8H$_2$O, by formation of B$_2$O$_3$ followed by treatment with magnesium:

$$B_2O_3 + 3Mg \rightarrow 2B + 3MgO \quad \Delta H < 0$$

Silicon is recovered from silica by reduction with carbon:

$$SiO_2(s) + C(s) \rightarrow Si(s) + CO_2(g) \quad \Delta H < 0$$

Germanium is recovered from its oxide by reduction with hydrogen:

$$GeO_2(s) + 2H_2(g) \rightarrow Ge(s) + 2H_2O(g) \quad \Delta H < 0$$

The recovery of germanium is the most energy efficient, since it is less oxophilic than either silicon or boron. Oxophilicity decreases down a group.

10.4   **Lewis acidity of Group 13 and Group 14 halides?**   **(a) Arrange in order of increasing Lewis acidity toward hard Lewis bases: BF$_3$, BCl$_3$, SiF$_4$, AlCl$_3$?**   For a given halogen, the order of acidity for Group 13 halides toward hard Lewis bases like dimethylether or trimethylamine is BX$_3$ > AlX$_3$ > GaX$_3$, while the order toward soft Lewis bases such as dimethylsulfide or trimethylphosphine is BX$_3$ < AlX$_3$ < GaX$_3$. This fact establishes the order BCl$_3$ > AlCl$_3$ for the four Lewis acids in question. For boron halides, the order of acidity is BF$_3$ < BCl$_3$ < BBr$_3$, exactly opposite to the order expected from electronegativity trends. This is discussed in Section 10.2. While you can now predict with some confidence that both BF$_3$ and AlCl$_3$ are both weaker Lewis acids than BCl$_3$ towards hard Lewis bases, it is not possible to predict from the information given in the text whether BF$_3$ is stronger or weaker than AlCl$_3$. Finally, silicon tetrahalides are only mild Lewis acids, so SiF$_4$ is the weakest of the four acids given. So, the order of increasing Lewis acidity towards hard Lewis bases is SiF$_4$ < BF$_3$ ~ AlCl$_3$ < BCl$_3$.

(b) Predict the course of the following reactions?

(i) $F_4Si-N(CH_3)_3 + BF_3 \rightarrow F_3B-N(CH_3)_3 + SiF_4$ ($BF_3 > SiF_4$)

(ii) $F_3B-N(CH_3)_3 + BCl_3 \rightarrow Cl_3B-N(CH_3)_3 + BF_3$ ($BCl_3 > BF_3$)

(iii) $BH_3CO + BBr_3 \rightarrow NR$ (BH_3 softer than BBr_3; CO is a soft base)

10.5   Synthesis of $F_2BCH_2CH_2BF_2$? By analogy with the addition of B_2Cl_4 to ethene (ethylene) described in Section 10.2, you can prepare this compound by adding B_2F_4 to ethene. Starting with BCl_3, you prepare B_2Cl_4 and then convert it to B_2F_4 with a double replacement reagent such as AgF or HgF_2, as follows:

$$2\,BCl_3 + Hg \xrightarrow{\text{electron impact}} B_2Cl_4 + HgCl_2$$

$$B_2Cl_4 + 4\,AgF \rightarrow B_2F_4 + 4\,AgCl$$

$$B_2F_4 + C_2H_4 \rightarrow F_2BCH_2CH_2BF_2$$

10.6   The coordination numbers of boron, carbon, and silicon oxoanions? The formulas for these oxoanions are BO_3^{3-}, CO_3^{2-}, and SiO_4^{4-}. Therefore, the coordination numbers are three, three, and four, respectively. The Lewis structures of borate(3–) and carbonate(2–) involve multiple element–oxygen bonding, whereas in silicate(4–) there are only silicon–oxygen single bonds. Multiple bonding is relatively strong for period 2 atoms like boron and carbon, and relatively weak for period 3 atoms like silicon. That is why silicon forms four σ bonds in SiO_4^{4-} instead of three σ bonds and one π bond, as in BO_3^{3-} and CO_3^{2-}.

10.7   Which is more stable, B_6H_{10} or B_6H_{12}? Inspection of Table 10.5 shows that B_6H_{10} is an example of a *nido* borane, since it has the formula B_6H_{6+4}. The compound B_6H_{12}, which has the formula B_6H_{6+6}, is an *arachno* borane. In general, *arachno* boranes are much less stable than either *closo* or *nido* boranes. Therefore, B_6H_{10} would be expected to be more stable than B_6H_{12}.

10.8   The combustion of pentaborane(9)?   (a) Balanced equation? The balanced equation for the reaction of B_5H_9 with O_2 is:

$$2B_5H_9(l) + 12O_2(g) \rightarrow 5B_2O_3(s) + 9H_2O(g)$$

Note that the product water is listed as a gas rather than a liquid, since this combustion reaction will produce temperatures in excess of 100°C.

(b) Probable disadvantages? A serious drawback to using pentaborane(9), or any other borane for that matter, is that the boron-containing product of combustion is a solid, B_2O_3. If an internal combustion engine is used, the solid will eventually coat the internal surfaces, increasing friction, and will clog the exhaust valves. A similar problem would arise if silanes, Si_xH_y, were used as a fuel, since the silicon-containing product, SiO_2, is also a solid. Note that the carbon-containing product of hydrocarbon combustion, CO_2, is a gas.

10.9   Classify $B_{10}H_{14}$ and discuss its structure and bonding with respect to Wade's rules? This compound is an example of a B_nH_{n+4} compound with $n = 10$, so it is a *nido* borane. According to Wade's rules, ten B–H units contribute $10 \times 2 = 20$ electrons, and the four additional H atoms contribute four additional electrons (24 electrons = 12 pairs of skeletal electrons). The structure of $B_{10}H_{14}$ is shown in Table 10.4. The total number of valence electrons for $B_{10}H_{14}$ is $(10 \times 3) + (14 \times 1) = 44$. Since there are 10 (2c,2e) B–H bonds, which account for 20 of the valence electrons, the number of cluster valence electrons is the remainder, $44 - 20 = 24$.

10.10   The synthesis of Fe(*nido*-$B_9C_2H_{11}$)$_2$? The starting material, $B_{10}H_{14}$, is converted to the *closo* carborane 1,2-$B_{10}C_2H_{12}$ by treatment with acetylene in the presence of a Lewis base, usually diethylsulfide:

$$B_{10}H_{14} + C_2H_2 \xrightarrow{\text{SEt}_2} B_{10}C_2H_{12} + 2H_2$$

This compound is fragmented by the removal of a B atom as $B(OEt)_3$:

$$B_{10}C_2H_{12} + Na^+OEt^- + 2\,EtOH \rightarrow Na^+[B_9C_2H_{12}]^- + B(OEt)_3 + H_2$$

The salt is deprotonated and treated with $FeCl_2$ to form the product:

$$Na^+[B_9C_2H_{12}]^- + NaH \rightarrow (Na^+)_2[B_9C_2H_{11}]^{2-} + H_2$$

$$2(Na^+)_2[B_9C_2H_{11}]^{2-} + FeCl_2 \rightarrow (Na^+)_2[Fe(B_9C_2H_{11})_2]^{2-} + 2NaCl$$

10.11 Compare BN and graphite? (a) Their structures? Both of these substances have layered structures. The planar sheets in boron nitride and in graphite consist of edge-shared hexagons such that each B or N atom in BN has three nearest neighbors that are the other type of atom and each C atom in graphite has three nearest neighbor C atoms. The B–N and C–C distances within the sheets, 1.45 Å and 1.42 Å, respectively, are much shorter than the perpendicular interplanar spacing, 3.33 Å and 3.35 Å, respectively. In BN, the B_3N_3 hexagonal rings are stacked directly over one another so that B and N atoms from alternating planes are 3.33 Å apart, while in graphite the C_6 hexagons are staggered (see Figure 10.17) so that C atoms from alternating planes are either 3.35 Å or 3.64 Å apart (you should determine this yourself using trigonometry).

(b) Their reactivity with Na and Br_2? Graphite reacts with alkali metals and with halogens (see the answer to Self-test S10.7). In contrast, boron nitride is quite unreactive.

(c) Explain the differences? The large HOMO–LUMO gap in BN, which causes it to be an insulator, suggests an explanation for the lack of reactivity: since the HOMO of BN is a relatively low energy orbital, it is more difficult to remove an electron from it than from the HOMO of graphite, and since the LUMO of BN is a relatively high energy orbital, it is more difficult to add an electron to it than to the LUMO of graphite.

10.12 Devise a synthesis for the following borazines? (a) $Ph_3N_3B_3Cl_3$? The reaction of a primary ammonium salt with boron trichloride yields *N*-substituted *B*-trichloroborazines:

$$3\,PhNH_3{}^+Cl^- + 3\,BCl_3 \rightarrow Ph_3N_3B_3Cl_3 + 9\,HCl$$

(b) $Me_3N_3B_3H_3$? You first prepare $Me_3N_3B_3Cl_3$ using $MeNH_3{}^+Cl^-$ and the method described above, and then perform a Cl^-/H^- metathesis reaction using LiH as the hydride source:

$$3\,MeNH_3{}^+Cl^- + 3\,BCl_3 \rightarrow Me_3N_3B_3Cl_3 + 9\,HCl$$

$$Me_3N_3B_3Cl_3 + 3\,LiH \rightarrow Me_3N_3B_3H_3 + 3\,LiCl$$

The structures of $Ph_3N_3B_3Cl_3$ and $Me_3N_3B_3H_3$ are shown below:

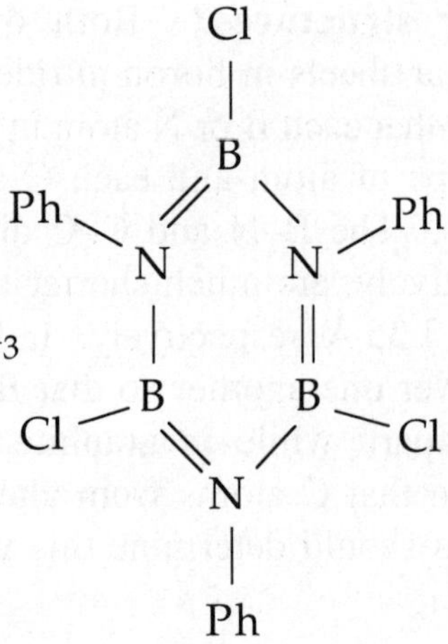

$Ph_3N_3B_3Cl_3$

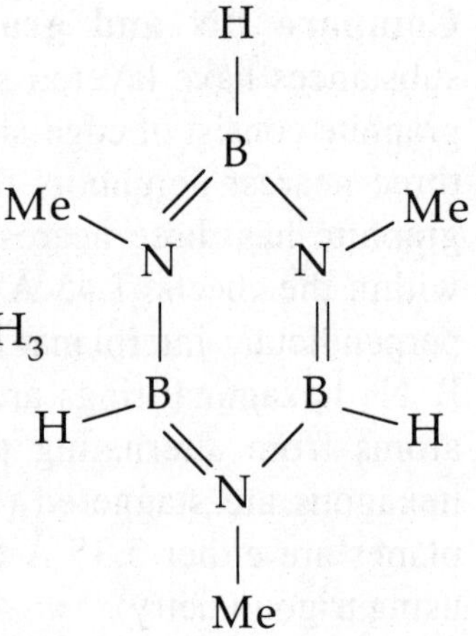

$Me_3N_3B_3H_3$

10.13 **Give the structural type and describe the structures of B_4H_{10}, B_5H_9, and $1,2\text{-}B_{10}C_2H_{12}$?** Simple polyhedral boranes come in three basic types, $B_nH_n^{2-}$ *closo* structures, B_nH_{n+4} *nido* structures, and B_nH_{n+6} *arachno* structures. The first compound given in this exercise, B_4H_{10}, is an example of a B_nH_{n+6} compound with n = 4, so it is an *arachno* borane. Its name is tetraborane(10) and its structure is shown in Table 10.4: two B–H units are joined by a (2c,2e) B–B bond; this B_2H_2 unit is flanked by four hydride bridges to two BH_2 units (the B–H–B bridge bonds are (3c,2e) bonds). The second compound, B_5H_9, is an example of a B_nH_{n+4} compound with $n = 5$, so it is a *nido* borane. Its name is pentaborane(9) and its structure is also shown in Table 10.4: four B–H units are joined by four hydride bridges; the resulting B_4H_8 unit, in which the four boron atoms are coplanar, is capped by an apical B–H unit that is bonded to all four of the coplanar boron atoms. The third compound is an example of a carborane in which two C–H units substitute for two B–H⁻ units in $B_{12}H_{12}^{2-}$, resulting in $1,2\text{-}B_{10}C_2H_{12}$ that retains the *closo* structure of the parent $B_{12}H_{12}^{2-}$ ion. Its name is $1,2\text{-}closo\text{-}$dodecaborane(12) and its structure is shown in Structure **24**: a B_5H_5 pentagonal plane is joined to a B_4CH_5 pentagonal plane that is offset from the first plane by 36°; The first plane is capped by a B–H unit while the second is capped by a C–H unit.

10.14 **Brønsted acidity of boranes?** In a series of boranes, the acidity increases as the size of the borane increases. This is because the negative charge, formed upon deprotonation, can be better delocalized over a large anion with many boron atoms than over a small one. Therefore, the Brønsted acidity increases in the order $B_2H_6 < B_5H_9 < B_{10}H_{14}$.

10.15 **Electrical conductivities?**   **(a) Describe the trend in E_g for C (diamond) through Sn (gray), and for cubic BN, AlP, and GaAs?** Reference to Table 10.6 will show that the bandgap decreases considerably from diamond (5.47 eV) to gray tin (~0 eV). In harmony with this trend of decreasing bandgap down a group, the bandgaps of the III–V compounds BN, AlP, and GaAs decrease in that order (both B and N are from period 2, both Al and P are from period 3, and both Ga and As are from period 4).

(b) Temperature dependence of σ(Si)? Silicon is a semiconductor. This type of material always experiences an increase in conductivity as the temperature is raised (see Section 3.15). Therefore, the conductivity of Si will be greater at 40°C than at 20°C. You should recall that the conductivity of a metal decreases as the temperature is raised.

(c) Temperature dependence of AlP and GaAs? The functional dependence of conductivity on temperature is:

$$\sigma = \sigma_0 e^{-Eg/2kT}$$

Accordingly, the greater the bandgap E_g, the more σ will change for a given change in temperature. Since AlP has a larger bandgap than GaAs, the conductivity of the former compound will be more sensitive to temperature.

10.16 **The elements that form saline, metallic, and metalloid carbides?** The distribution of carbides in the periodic table is shown in Figure 10.24. If you were not able to completely construct this diagram, try to remember that the saline carbides are formed by the same elements that form saline hydrides and ionic organometallic compounds, namely the alkali metals and the alkaline earths. In addition, aluminum carbide is saline. Metallic carbides are formed by the early and middle *d*-block metals. The metalloid carbides are formed by those two nonmetallic elements that also form organometallic compounds, namely boron and silicon.

10.17 **Describe the preparation, structure, and classification?** **(a) KC$_8$?** This compound is formed by heating graphite with potassium vapor or by treating graphite with a solution of potassium in liquid ammonia. The potassium atoms are oxidized to K$^+$ ions; their electrons are added to the LUMO π^* orbitals of graphite. The K$^+$ ions intercalate between the planes of the reduced graphite, so that there is a layered structure of alternating sp^2 carbon atoms and potassium ions. The structure of KC_8, which is an example of a saline carbide, is shown in Figure 10.25.

(b) CaC_2? There are two ways of preparing calcium carbide, and both require very high temperatures ($\geq 2000\ °C$). The first is the direct reaction of the elements, while the second is the reaction of calcium oxide with carbon:

$$Ca(l)\ +\ 2\,C(s)\ \rightarrow\ CaC_2(s)$$

$$CaO(s)\ +\ 3\,C(s)\ \rightarrow\ CaC_2(s)\ +\ CO(g)$$

The structure is quite different from that of KC_8. Instead of every carbon atom bonded to three other carbon atoms, as in graphite and KC_8, calcium carbide contains discrete C_2^{2-} ions with carbon–carbon triple bonds.

(c) K_3C_{60}? A solution of C_{60}, perhaps in toluene, can be treated with elemental potassium to form this compound. It is ionic, that is, it contains discrete K^+ ions and C_{60}^{3-} ions (see Section 10.8(a), *Carbon clusters*).

10.18 List four examples of amorphous or partially crystalline solids? Three forms of carbon have a low degree of crystallinity. These are carbon black, activated carbon, and carbon fibers, and are discussed in Section 10.8(d). Carbon black is used as a pigment and as an additive to automobile tires to improve their strength and wear resistance. Activated carbon is used as an adsorbent. Carbon fibers have great tensile strength, and so they increase the strength of materials to which they are added. A material referred to as amorphous silicon is really a solid silicon hydride SiH_x ($x \leq 0.5$), and was mentioned in Section 10.11 and discussed in Section 8.12(a). It is used in photovoltaic devices. A fifth amorphous material mentioned in this chapter is fused quartz (amorphous SiO_2, see Section 10.13), which is used to manufacture laboratory glassware that is transparent to UV radiation.

10.19 Differences between B and Al and between C and Si? (a) Structures and electrical properties? The various allotropes of elemental boron consist of icosahedral B_{12} cages connected together in various ways. In contrast, aluminum is a cubic close-packed metal at all temperatures. In harmony with the more localized bonding in boron, it is a semiconductor whereas aluminum is a metal. The only important form of silicon has the diamond structure, named after one of the naturally occurring forms of carbon. Despite the similar structures, silicon is a semiconductor and the diamond allotrope of carbon is an insulator (however, recall that a semiconductor and an insulator are merely extremes of the same type of conductivity behavior: see Section 3.15). The slightly more stable form of carbon, graphite, has no silicon homologue. Graphite is a semiconductor in the direction perpendicular to the planes of sp^2 carbon atoms and a metal in the directions parallel to the planes.

(b) Physical properties and structures of CO_2 and SiO_2? Carbon dioxide is a gas at ambient conditions while silicon dioxide is a hard solid with a high melting point (1710 °C). Both compounds contain four covalent bonds to oxygen atoms, but the structures are quite different, as expected from the different physical properties. Whereas CO_2 possesses two C=O double bonds, resulting in a triatomic molecule, SiO_2 possesses four Si–O single bonds, resulting in an extended three-dimensional covalent lattice. The reasons behind this difference were discussed in Section 3.10(b): Two C–O single bonds are weaker than one C=O double bond, while two Si–O single bonds are stronger than one Si=O double bond.

The different structures of CO_2 and SiO_2

(c) The Lewis acid/base properties of CX_4 and SiX_4? Carbon tetrahalides are not Lewis acids, despite the fact that the central carbon atom has four electronegative ligands. With its valence s and p orbitals taken up by bonding with halogen atoms, carbon has no low-lying empty orbitals with which it can accept an electron pair from a Lewis base. In contrast, silicon has valence $3d$ orbitals and can form complexes with one or two Lewis bases, yielding structures in which silicon is 5- or 6-coordinate, respectively.

(d) The structures of BX_3 and AlX_3? The boron halides are trigonal planar molecules that are gases or volatile liquids at ambient conditions. In contrast, aluminum halides are solids with extended lattices for X = F and Cl and with dimeric Al_2X_6 molecules for X = Br and I. The differences between the halides of boron and aluminum can be traced to their relative sizes. Since boron is so small it (i) resists formation of the halide bridges present in the AlX_3 structures and (ii) can form bonds with halogens that have partial double bond character, in marked contrast with Al–X bonds, which have little or no double bond character (as in the discussion of CO_2 vs. SiO_2, above, double bonds to period 3 atoms are very weak).

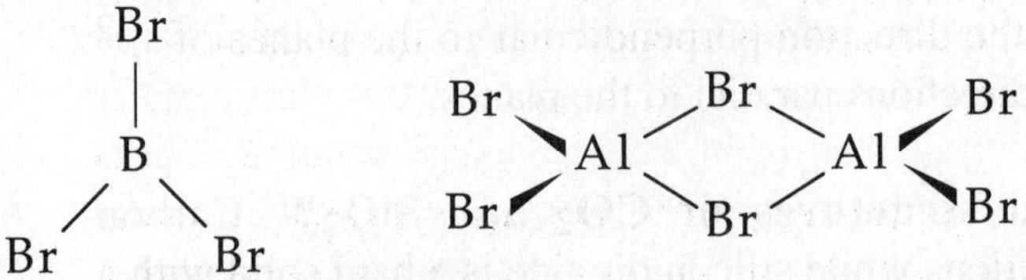

The different structures of BBr_3 and $AlBr_3$

10.20 **The reactions of K_2CO_3 and Na_4SiO_4 with acid?** Both of these compounds react with acid to produce the oxide, which for carbon is CO_2 and for silicon is SiO_2. The balanced equations are:

$$K_2CO_3(aq) + 2\,HCl(aq) \rightarrow 2\,KCl(aq) + CO_2(g) + H_2O(l)$$

$$Na_4SiO_4(aq) + 4\,HCl(aq) \rightarrow 4\,NaCl(aq) + SiO_2(s) + 2\,H_2O(l)$$

The second equation represents one of the ways that silica gel is produced.

10.21 **The nature of $[SiO_3{}^{2-}]_n$ ions and layered aluminosilicates?** In contrast with the extended three-dimensional structure of SiO_2, the structures of jadeite and kaolinite consist of extended one- and two-dimensional structures, respectively. The $[SiO_3{}^{2-}]_n$ ions in jadeite are a linear polymer of SiO_4 tetrahedra, each one sharing a bridging oxygen atom with the tetrahedron before it and the tetrahedron after it in the chain (this is referred to as a chain metasilicate; see Structure **43**). Each silicon atom has two bridging oxygen atoms and two terminal oxygen atoms. The two-dimensional aluminosilicate layers in kaolinite represent another way of connecting SiO_4 tetrahedra (see Figure 10.27). Each silicon atom has three oxygen atoms that bridge to other silicon atoms in the plane and one oxygen atom that bridges to an aluminum atom.

10.22 **Molecular sieves?** **(a) How many bridging O atoms are in a single sodalite cage?** A sodalite cage is based on a truncated octahedron. Refer to the answer to Self-test S10.10. Each of the heavy lines in the drawing of an octahedron truncated along one of its C_4 axes represents an M–O–M linkage

(M = Si or Al). There are eight such lines in the drawing, and since the truncation procedure is carried out six times to produce a sodalite cage, there are $8 \times 6 = 48$ bridging oxygen atoms.

(b) Describe the polyhedron at the center of the zeolite A structure? As shown in Figure 10.30, eight sodalite cages are linked together to form the large a cage of zeolite A. The polyhedron at the center has six octagonal faces (the one in the front of the diagram is the most obvious) and eight smaller square faces. If you look at Figure 10.29, you will see that a truncated octahedron has 8 hexagonal faces and 4 smaller square faces. Thus, the fusing together of 8 truncated octahedra produces a central cavity that is different than a truncated octahedron.

10.23 **Compare pyrophyllite and muscovite mica?** The generic structure of the 2:1 aluminosilicates that include pyrophyllite and muscovite mica is shown in Figure 10.28. The discrete O–Si–O–Al–O–Si–O layers are composed of a layer of AlO_6 octahedra sandwiched between layers of SiO_4 tetrahedra. In pyrophyllite, which has the formula $Al_2(OH)_2Si_4O_{10}$ (note: 2 Si per 1 Al, hence the name 2:1 aluminosilicate), the O–Si–O–Al–O–Si–O layers are electrically neutral, leading to weak van der Waals bonds between layers despite the strong Al–O and Si–O bonding within each layer. In muscovite mica, which has the formula $KAl_2(OH)_2Si_3AlO_{10}$, one Al(III) ion replaces a Si(IV) ion in the layers of SiO_4 tetrahedra (i.e. there are some AlO_4 tetrahedra within these layers). The result of this is that the layers are now negatively charged, and the potassium ions compensate the negative charge. They are accommodated between the layers, so the layers in muscovite mica are held together by relatively strong K–O ionic bonds. This accounts for the fact that mica is much harder than talc.

Guide to Solutions Quiz

1 Boron trichloride can be prepared from B_2O_3 and CCl_4 at high temperatures. Write a balanced equation for the reaction. What do you think provides most of the driving force for this reaction?

2 Predict the structures of $B_2H_7^-$ and $B_3H_8^-$.

3 Draw a likely structure for the $B_5O_6(OH)_4^-$ anion, which contains two six-membered rings.

4   Explain why borazine is more reactive than benzene toward addition of HCl.

5   Predict the oxidation numbers of the elements in TlI_3 and $GaCl_2$.

6   There is only one *stable* form of carbon at 25°C and 1 atm, and that form is graphite. Explain the existence of diamond and C_{60} under these conditions.

7   Dithiocarbamate anions, $R_2NCS_2{}^-$, are important ligands. Write a plausible synthesis for the preparation of the sodium salt of dimethyl-dithiocarbamate starting with carbon disulfide.

8   The Si–H bond is only slightly stronger than the Ge–H bond. However, the Si–F bond is much stronger than the Ge–F bond. Explain the difference.

9   Draw structures of several repeat units of $[SiO_3{}^{2-}]_n$, $[Si_4O_{11}{}^{6-}]_n$, $[Si_4O_{10}{}^{4-}]_n$, and SiO_2. What pattern do you see emerging?

10   Explain how molecular sieves work.

11 The nitrogen and oxygen groups

14	15	16	17
C	N	O	F
Si	P	S	Cl
Ge	As	Se	Br
Sn	Sb	Te	I
Pb	Bi	Po	At
	V	VI	

The elements of the nitrogen and oxygen groups contain no true metals, unlike the elements of Groups 13/III and 14/IV. They form compounds in a wide variety of oxidation states (e.g. N_2O, NO, and NO_2; $S_2O_4^{2-}$, SO_3^{2-}, and SO_4^{2-}), and therefore have a rich redox chemistry. The heavier members form ring and cluster compounds. Some of the industrially most important inorganic chemicals come from these two groups, including sulfuric acid and ammonia.

S11.1 The structure of bismuth? Considering the three nearest neighbors only, the structure around each Bi atom is trigonal pyramidal, like NH_3. A side view of the structure of bismuth is shown to the right. VSEPR theory predicts this geometry for an atom that has a Lewis structure with three bonding pairs and a lone pair.

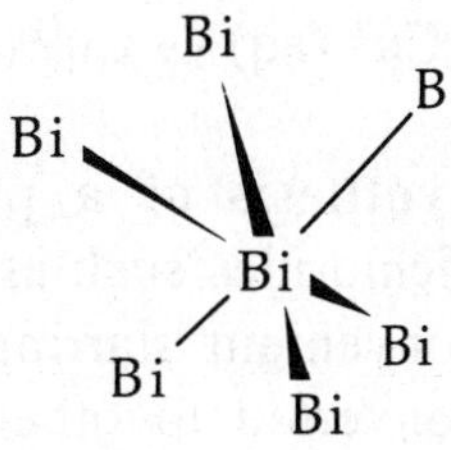

S11.2 Is phosphorus or sulfur a stronger oxidizing agent? The key to estimating the oxidizing strength of a substance is to determine how readily it is reduced. From the table of standard reduction potentials in Appendix 2 you can obtain the following information:

$$P(s) + 3\,e^- + 3\,H^+(aq) \rightarrow PH_3(aq) \qquad E° = -0.063 \text{ V}$$

$$S(s) + 2\,e^- + 2\,H^+(aq) \rightarrow H_2S(aq) \qquad E° = 0.144$$

Since $\Delta G = -NFE°$, sulfur is more readily reduced and is therefore the better oxidizing agent. This is in harmony with the higher electronegativity of sulfur ($\chi = 2.58$) than phosphorus ($\chi = 2.19$). As you will see in Chapter 12, the most electronegative element, fluorine, is also the most potent elemental oxidizing agent.

S11.3 The syntheses of hydrazine and hydroxylamine? The reactions that are employed to synthesize hydrazine are as follows:

$$NH_3 + ClO^- + H^+ \rightarrow [\,H_3N: \longrightarrow Cl–O^-\,] \rightarrow H_2NCl + H_2O$$

$$H_2NCl + NH_3 \rightarrow H_2NNH_2 + HCl$$

Both of these can be thought of as redox reactions since the formal oxidation state of the N atoms changes (from –3 to –1 in the first reaction and from –1 and –3 to –2 and –2 in the second reaction). Mechanistically, both reactions appear to involve nucleophilic attack by NH_3 (a Lewis base) on either ClO^- or NH_2Cl (acting as Lewis acids). The reaction employed to synthesize hydroxylamine, which produces the intermediate $N(OH)(SO_3)_2$, probably involves the attack of HSO_3^- (acting as a Lewis base) on NO_2^- (acting as a Lewis acid), although, since the formal oxidation state of the N atom changes from +3 in NO_2^- to +1 in $N(OH)(SO_3)_2$, this can also be seen as a redox reaction. Whether one considers reactions such as these to be redox reactions or nucleophilic substitutions may depend on the context in which the reactions are being discussed. It is easy to see that these reactions do not involve simple electron transfer, such as in $2\,Cu^+(aq) \rightarrow Cu^0(s) + Cu^{2+}(aq)$.

S11.4 Synthesis of a polyphosphazene high polymer? Cyclic phosphazene dichlorides such as $(Cl_2PN)_3$ (Structure **17**) and $(Cl_2PN)_4$ (Structure **18**) are convenient starting materials for polydichlorophosphazene, which can be converted to other polyphosphazenes by nucleophilic substitution of the

P–Cl bonds. In this case, you use the $N(CH_3)_2^-$ anion as the nucleophile (its lithium salt can be prepared from dimethylamine and lithium metal):

$$3 PCl_5 + 3 NH_4Cl \rightarrow (Cl_2PN)_3 + 12 HCl$$

$$(Cl_2PN)_3 \xrightarrow{290°} (Cl_2PN)_n$$

$$(Cl_2PN)_n + 2n\,LiN(CH_3)_2 \rightarrow (((CH_3)_2N)_2PN)_n + 2n\,LiCl$$

S11.5 Can Cl^- or Br^- catalyze the decomposition (disproportionation) of hydrogen peroxide? You must determine if the standard potentials for the reduction of Cl_2 to Cl^- or Br_2 to Br^- are within the range bounded by the reduction of O_2 to H_2O_2 (0.695 V) and the reduction of H_2O_2 to H_2O (1.76 V). From Appendix 2 you can get the following information:

$$Cl_2(aq) + 2e^- \rightarrow 2 Cl^-(aq) \qquad E° = 1.358\ V$$

$$Br_2(aq) + 2e^- \rightarrow 2 Br^-(aq) \qquad E° = 1.087\ V$$

Since both of the reduction potentials for these reactions are within the prescribed range, both chloride and bromide can catalyze the decomposition of H_2O_2. Great care must be exercised when storing H_2O_2 for long periods of time, since trace amounts of these common anions will slowly destroy the compound.

S11.6 Structures of SO_3 and SO_3F^-? The Lewis structures are shown below (only one of the three possible resonance structures for SO_3 are shown). The central S atom in sulfur trioxide is trigonal planar and, since all of the S–O bonds are equivalent, the point group is D_{3h}. The central S atom in fluorosulfate ion is tetrahedral, but the ion is C_{3v}, not T_d.

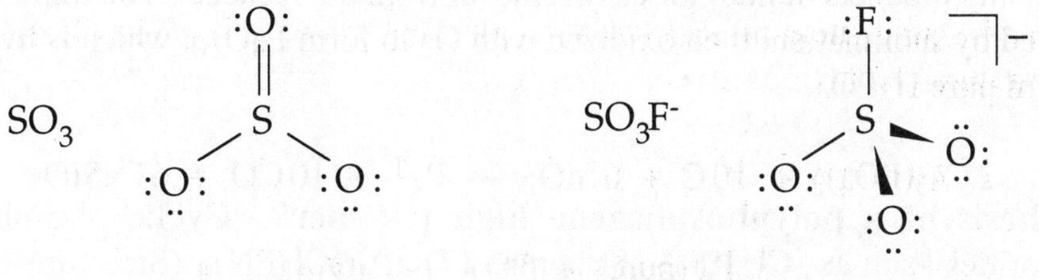

S11.7 The structure of Ge_9^{2-}? Each of the germanium atoms in this cluster has four valence electrons, so the cluster has $(9 \times 4) + 2 = 38$ electrons or 19

electron pairs. If you assume that each atom has a lone pair that is directed away from the other atoms in the cluster (and hence is unavailable for Ge–Ge framework bonding), there are 10 pairs of skeletal electrons, or one more than the number of cluster atoms. This suggests a *closo* structure, and the structure of $Ge_9{}^{2-}$ shown in Structure **38** has the same framework structure as the structure of *closo*-$B_9H_9{}^{2-}$ shown in Figure 10.12.

11.1   General properties of the elements?   In the list below, the properties of polonium have been omitted since the chemistry of this element has been little studied owing to its radioactivity.

	type of element	diatomic gas?	achieves maximum oxidation state?	displays inert pair effect?
N	nonmetal	yes	yes	no
P	nonmetal	no	yes	no
As	nonmetal	no	yes	no
Sb	metalloid	no	yes	no
Bi	metalloid	no	yes	yes
O	nonmetal	yes	no	no
S	nonmetal	no	yes	no
Se	nonmetal	no	yes	no
Te	nonmetal	no	yes	no

11.2   The synthesis of phosphoric acid?   (a) High purity H_3PO_4? The starting point is hydroxyapatite, $Ca_5(PO_4)_3OH$, which is converted to crude $Ca_3(PO_4)_2$. This compound is treated with (i) carbon to reduce phosphorus from P(V) in $PO_4{}^{3-}$ to P(0) in P_4 and with (ii) silica, SiO_2, to keep the calcium-containing products molten for easy removal from the furnace. The impure P_4 is purified by sublimation, then oxidized with O_2 to form P_4O_{10}, which is hydrated to form pure H_3PO_4.

$$2\,Ca_3(PO_4)_2 + 10\,C + 6\,SiO_2 \rightarrow P_4\uparrow + 10\,CO + 6\,CaSiO_3$$

$$P_4 \text{ (pure)} + 5\,O_2 \rightarrow P_4O_{10}$$

$$P_4O_{10} + 6\,H_2O \rightarrow 4\,H_3PO_4 \text{ (pure)}$$

(b) Fertilizer grade H_3PO_4? In this case, hydroxyapatite is treated with sulfuric acid, producing phosphoric acid that contains small amounts of many impurities.

$$Ca_5(PO_4)_3OH + 5 H_2SO_4 \rightarrow 3 H_3PO_4 \text{ (impure)} + 5 CaSO_4 + H_2O$$

(c) Account for the difference in cost? The synthesis of fertilizer grade (i.e. impure) phosphoric acid involves a single step and gives a product that requires little or no purification. In contrast, the synthesis of pure phosphoric acid involves several synthetic steps and a time-consuming and expensive purification step, the sublimation of white phosphorus, P_4.

11.3 Synthesis of ammonia? (a) Hydrolysis of Li_3N? Lithium nitride is one of two binary metal nitrides that can be prepared directly from the elements (the other is Mg_3N_2). Since it contains N^{3-} ions, which are extremely basic, it can be hydrolyzed with water to produce ammonia.

$$6 Li + N_2 \rightarrow 2 Li_3N$$

$$2 Li_3N + 3 H_2O \rightarrow 2 NH_3 + 3 Li_2O$$

(b) Reduction of N_2 by H_2? This reaction requires high temperatures and pressures to proceed at an appreciable rate.

$$N_2 + 3H_2 \rightarrow 2NH_3$$

(c) Account for the difference in cost? The second process is considerably cheaper than the first, even though the second process must be carried out at high temperature and pressure. This is because lithium, like many very electropositive metals (including aluminum), is very expensive.

11.4 Contrast the formulas and stabilities of nitrogen and phosphorus chlorides? The only isolable nitrogen chloride is NCl_3, and it is thermodynamically unstable with respect to its constituent elements (i.e. it is endoergic). The compound NCl_5 is unknown. In contrast, both PCl_3 and PCl_5 are stable and can be prepared directly from phosphorus and chlorine.

11.5 **Lewis structures and probable structures?** **(a) PCl_4^+?** The Lewis structure is shown below. With four bonding pairs of electrons around the central phosphorus atom, the structure is a tetrahedron.

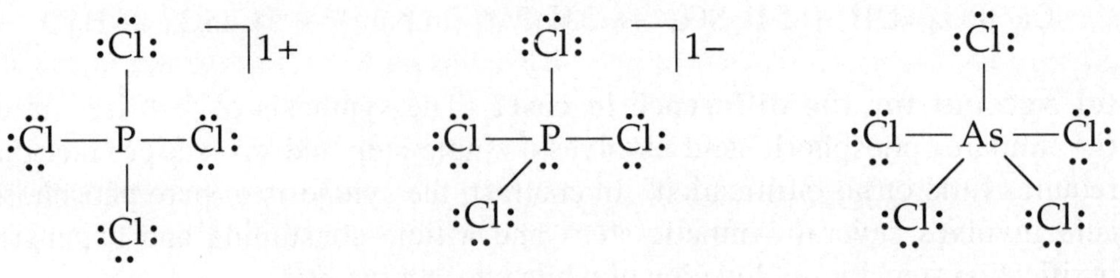

(b) PCl_4^-? The Lewis structure is shown above. With four bonding pairs of electrons and one lone pair of electrons around the central phosphorus atom, the structure is a see-saw (trigonal bipyramidal array of electron pairs with the lone pair in the equatorial plane).

(c) $AsCl_5$? The Lewis structure is shown above. With five bonding pairs of electrons around the central arsenic atom, the structure is a trigonal bipyramid.

11.6 **Give balanced equations?** **(a) Oxidation of P_4 with excess O_2?** The balanced equation is:

$$P_4 + 5\,O_2 \rightarrow P_4O_{10}$$

(b) Reaction of the product from part (a) with excess H_2O? The balanced equation is:

$$P_4O_{10} + 6\,H_2O \rightarrow 4\,H_3PO_4$$

(c) Reaction of the product from part (b) with $CaCl_2$? The products of this reaction would be calcium phosphate and a solution of hydrochloric acid. The balanced equation is:

$$2\,H_3PO_4(l) + 3\,CaCl_2(aq) \rightarrow Ca_3(PO_4)_2(s) + 6\,HCl(aq)$$

11.7 **Describe the relative oxidizing strengths and Lewis acid basicities of isoelectronic species?** Of the species listed, O_2^{2-} and $N_2H_5^+$ are isoelectronic. O_2^+ and NO are electronic, and CN^-, N_2, and NO^+ are isoelectronic. Lewis acid basicity increases and oxidizing strength decreases with increasing negative charge. Considering the first pair, the anionic species O_2^{2-}, is the stronger Lewis base; the cationic species will be easier to reduce and is therefore the stronger oxidant. Of the second pair, the neutral species NO can be expected to be the stronger Lewis base, but the cationic species will be the stronger reductant and therefore the weaker oxidant. In the third group, the species with the positive charge ($N_2H_5^+$) is most easily reduced and is therefore the strongest oxidant.

11.8 **Give the formula and name of a carbon-containing species that is isoelectronic and isostructural with the following?** **(a) NO_3^-?** The isoelectronic species is the carbonate dianion, CO_3^{2-}.

(b) NO_2^-? The isoelectronic and isostructural species is the CO_2^{2-} dianion.

(c) N_2O_4? The isoelectronic and isostructural species is oxalate dianion, $C_2O_4^{2-}$.

(d) N_2? Three examples are the cyanide ion, CN^-, carbon monoxide, CO, and the acetylide dianion, C_2^{2-} (formerly known as the carbide dianion).

(e) NH_3? The isoelectronic/isostructural species is CH_3^-.

11.9 **Give equations and conditions for the synthesis of the following?** **(a) HNO_3?** The synthesis of nitric acid, an example of the most oxidized form of nitrogen, starts with ammonia, the most reduced form:

$$4\,NH_3(aq) + 7\,O_2(g) \rightarrow 6\,H_2O(g) + 4\,NO_2(g)$$

High temperatures are obviously necessary for this reaction to proceed at a reasonable rate: ammonia is a flammable gas but at room temperature it does not react rapidly with air. The second step in the formation of nitric acid is the high-temperature disproportionation of NO_2 in water:

$$3\,NO_2(aq) + H_2O(l) \rightarrow 2\,HNO_3(aq) + NO(g)$$

(b) NO_2^-? Whereas the disproportionation of NO_2 in acidic solution yields NO_3^- and NO (see part (a)), in basic solution nitrite ion is formed:

$$2\,NO_2(aq) + 2\,OH^-(aq) \rightarrow NO_2^-(aq) + NO_3^-(aq) + H_2O(l)$$

(c) NH$_2$OH? The protonated form of hydroxylamine is formed in a very unusual reaction between nitrite ion and bisulfite ion in cold aqueous acidic solution (the NH$_3$OH$^+$ ion can be deprotonated with base):

$$NO_2^-(aq) + 2\,HSO_3^-(aq) + H_2O(l) \rightarrow NH_3OH^+(aq) + 2\,SO_4^{2-}(aq)$$

(d) N$_3^-$? The azide ion can be prepared from anhydrous molten sodium amide (m.p. ~200°C) and either nitrate ion or nitrous oxide at elevated temperatures:

$$3\,NaNH_2(l) + NaNO_3 \rightarrow NaN_3 + 3\,NaOH + NH_3(g)$$
$$2\,NaNH_2(l) + N_2O \rightarrow NaN_3 + NaOH + NH_3$$

11.10 The balanced chemical equation for the formation of P$_4$O$_{10}$(s)? This compound is formed by the complete combustion of white phosphorus, as shown below:

$$P_4(s) + 5\,O_2(g) \rightarrow P_4O_{10}(s)$$

The structure of P$_4$ is a tetrahedron of phosphorus atoms with six P–P σ bonds. As discussed in Section 11.1, white phosphorus (P$_4$) is adopted as the reference phase for thermodynamic calculations even though it is not the most stable phase of elemental phosphorus.

11.11 Frost diagrams for phosphorus and bismuth? The main points you will want to remember are (i) Bi(III) is much more stable than Bi(V) and (ii) P(III) and P(V) are both about equally stable (i.e. Bi(V) is a strong oxidant but P(V)

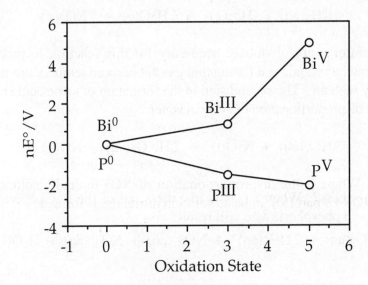

is not). This suggests that the point for Bi(III) on the Frost diagram lies *below* the line connecting Bi(0) and Bi(V), while the point for P(III) lies very close to the line connecting P(0) and P(V). The essential parts of the Frost diagrams for these two elements are shown above (see Figure 11.5 for the complete Frost diagrams).

11.12 **The pH dependence of the rate of reduction of NO_2^-?** The rates of reactions in which nitrite ion is reduced (i.e. in which it acts as an oxidizing agent) are increased as the pH is lowered. That is, acid enhances the rate of oxidations by NO_2^-. The reason is that NO_2^- is converted to the nitrosonium ion, NO^+, in strong acid:

$$HNO_2(aq) + H^+(aq) \rightarrow NO^+(aq) + H_2O(l)$$

This cationic Lewis acid can form complexes with the Lewis bases undergoing oxidation (species that are oxidizable are frequently electron rich and hence are basic). Therefore, at low pH the oxidant (NO^+) is a different chemical species than at higher pH (NO_2^-).

11.13 **Explain why NO at low concentrations in automobile exhaust reacts slowly with oxygen (air)?** The given observations suggest that more than one NO molecule is involved in the activated complex (i.e. in the rate-determining step), since the rate of reaction is dependent on the concentration of NO. Therefore, the rate law must be more than first order in NO concentration. It turns out to be second order: an equilibrium between NO and its dimer N_2O_2 precedes the rate-determining reaction with oxygen. At high concentrations of NO, the concentration of the dimer is higher.

11.14 **Balanced chemical equations for reactions of PCl_5?** **(a) H_2O?** In this case, water will form a strong P=O double bond, releasing two equivalents of HCl and forming "tetrahedral" $POCl_3$ (which actually has C_{3v} symmetry).

$$PCl_5 + H_2O \rightarrow POCl_3 + 2\,HCl$$

(b) H_2O in excess? When water is in excess, all of the P–Cl bonds will be hydrolyzed, and phosphoric acid will result:

$$2\,PCl_5(g) + 8\,H_2O(l) \rightarrow 2\,H_3PO_4(aq) + 10\,HCl(aq)$$

(c) AlCl$_3$? By analogy to the reaction with another group 13/III Lewis acid, BCl$_3$, PCl$_5$ will donate a chloride ion and a salt will result. Both the cation and the anion are tetrahedral.

$$PCl_5 + AlCl_3 \rightarrow [PCl_4]^+[AlCl_4]^-$$

(d) NH$_4$Cl? A phosphazene, containing cyclic molecules or linear chain polymers of (=P(Cl)$_2$–N=P(Cl)$_2$–N=P(Cl)$_2$–N=) bonds will be formed:

$$n\,PCl_5 + n\,NH_4Cl \rightarrow -[(N=P(Cl)_2)_n]- + 4n\,HCl$$

11.15 **The reaction of H$_3$PO$_2$ with Cu^{2+}?** From the standard potentials for the following two half-reactions, the standard potential for the net reaction can be calculated:

$$H_3PO_2(aq) + H_2O(l) \rightarrow H_3PO_3(aq) + 2\,e^- + 2\,H^+(aq) \qquad E° = 0.499 \text{ V}$$

$$Cu^{2+}(aq) + 2\,e^- \rightarrow Cu^0(s) \qquad E° = 0.340 \text{ V}$$

Therefore the net reaction is:

$$H_3PO_2(aq) + H_2O(l) + Cu^{2+}(aq) \rightarrow H_3PO_3(aq) + Cu^0(s) + 2\,H^+(aq)$$

and the net standard potential is $E° = 0.839$ V. To determine whether HPO$_3{}^{2-}$ and H$_2$PO$_2{}^-$ are useful as oxidizing or reducing agents, you must compare E values for their oxidations and reductions at pH 14 (these potentials are listed in Appendix 2):

$$HPO_3{}^{2-}(aq) + 3\,OH^-(aq) \rightarrow PO_4{}^{3-}(aq) + 2\,e^- + 2\,H_2O(l) \quad E(\text{pH }14) = 1.12 \text{ V}$$

$$HPO_3{}^{2-}(aq) + 2\,e^- + 2\,H_2O(l) \rightarrow H_2PO_2{}^-(aq) + 3\,OH^-(aq)\; E(\text{pH }14) = -1.57 \text{ V}$$

$$H_2PO_2{}^-(aq) + e^- \rightarrow P(s) + 2\,OH^-(aq) \qquad E(\text{pH }14) = -2.05 \text{ V}$$

Since a positive potential will give a negative value of ΔG, the oxidations of HPO$_3{}^{2-}$ and H$_2$PO$_2{}^-$ are much more favorable than their reductions, so these ions will be much better reducing agents than oxidizing agents.

11.16 **Disproportionations?** **(a) The disproportionation of H$_2$O$_2$ and HO$_2$?** To calculate the standard potential for a disproportionation reaction, you must sum the potentials for the oxidation and reduction of the species in question. For hydrogen peroxide in acid solution, the oxidation and reduction are:

$$H_2O_2(aq) \rightarrow O_2(g) + 2e^- + 2H^+(aq) \qquad E^\circ = -0.695 \text{ V}$$

$$H_2O_2(aq) + 2e^- + 2H^+(aq) \rightarrow 2H_2O(l) \qquad E^\circ = 1.763 \text{ V}$$

Therefore, the standard potential for the net reaction $2H_2O_2(aq) \rightarrow O_2(g) + 2H_2O(l)$ is $(-0.695 \text{ V}) + (1.763 \text{ V}) = 1.068 \text{ V}$.

(b) Catalysis by Cr^{2+}? As discussed in Self-test S11.5, Cr^{2+} can act as a catalyst for the decomposition of hydrogen peroxide if the Cr^{3+}/Cr^{2+} reduction potential falls between the values for the reduction of O_2 to H_2O_2 (0.695 V) and the reduction of H_2O_2 to H_2O (1.76 V). Reference to Appendix 2 reveals that the Cr^{3+}/Cr^{2+} reduction potential is -0.424 V, so Cr^{2+} is *not* capable of decomposing H_2O_2.

(c) The disproportionation of HO_2? The oxidation and reduction of superoxide ion (O_2^-) in acid solution are:

$$HO_2(aq) \rightarrow O_2(g) + e^- + H^+(aq) \qquad E^\circ = 0.125 \text{ V}$$

$$HO_2(aq) + e^- + H^+(aq) \rightarrow H_2O_2(aq) \qquad E^\circ = 1.51 \text{ V}$$

Therefore, the standard potential for the net reaction $2HO_2(aq) \rightarrow O_2(g) + H_2O_2(aq)$ is $(0.125 \text{ V}) + (1.51 \text{ V}) = 1.63 \text{ V}$. Since 1 V is equivalent to 96.5 kJ, $\Delta_r G^\circ = 157$ kJ. For the disproportionation of H_2O_2 (part (a)), $\Delta_r G^\circ = 103$ kJ.

11.17 **Is ethylenediamine or sulfur dioxide a better solvent for Na_2S_4, K_2Te_3, and $Cd_2(Al_2Cl_7)_2$?** As explained in the answer to Exercise 5.27, anionic species such as S_4^{2-} and Te_3^{2-} are intrinsically basic species that cannot be studied in solvents that are Lewis acids because complex formation will destroy the independent identity of the anion. Therefore, since the basic solvent ethylenediamine will not react with Na_2S_4 or with K_2Te_3, it is a better solvent for them than sulfur dioxide. On the other hand, the $Al_2Cl_7^-$ anion is not basic, despite its negative charge. This is because it can readily dissociate into two species, one of which is strongly acidic ($AlCl_3$) and the other of which is only weakly basic ($AlCl_4^-$). In order to inhibit the formation of an acid–base complex with the strong Lewis acid $AlCl_3$, you should choose a solvent that is itself a Lewis acid, in this case SO_2.

11.18 The strongest reducing agent and oxidizing agent? The three values of $E°$ for the $S_2O_8^{2-}/SO_4^{2-}$, the SO_4^{2-}/SO_3^{2-}, and the $SO_3^{2-}/S_2O_3^{2-}$ couples are 1.96 V, 0.158 V, and 0.400 V, respectively. Peroxydisulfate, $S_2O_8^{2-}$, is very easily reduced, so it is the strongest oxidizing agent. Sulfate dianion, SO_4^{2-}, is neither strongly oxidizing nor strongly reducing. Sulfite dianion, SO_3^{2-}, on the other hand, is relatively readily oxidized to sulfate, and hence is a moderate reducing agent. Two plausible redox reactions are:

$$S_2O_8^{2-}(aq) + NO_2^-(aq) + H_2O(l) \rightarrow 2\,HSO_4^-(aq) + NO_3^-(aq) \qquad E° = 1.02\ V$$

$$SO_3^{2-}(aq) + Cu^{2+}(aq) + H_2O(l \rightarrow SO_4^{2-}(aq) + Cu(s) + 2\,H^+(aq)\ E° = 0.182\ V$$

11.19 Compare Te(VI) and S(VI) in acidic solution? (a) Formulas? The formulas for these ions in acidic solution are $H_5TeO_6^-$ and HSO_4^- (the parent acids are H_6TeO_6, or $Te(OH)_6$, and H_2SO_4).

(b) An explanation? Tellurium is a much larger element than sulfur, and readily can increase its coordination number. This trend is found in other groups in the p block. An example involving perchlorate and periodate will be discussed in Chapter 12.

(c) A precedent in Group 15/V? There is such a precedent. The formulas for the oxyanions of the group 15/V elements in their highest oxidation state in basic solution are PO_4^{3-}, AsO_4^{3-}, and $Sb(OH)_6^-$.

11.20 Use isoelectronic analogies to infer the probable structures of the following? (a) $[Sb_4]^{2-}$? Antimony is in Group 15/V, as is bismuth. Therefore, $[Sb_4]^{2-}$ is isoelectronic with $[Bi_4]^{2-}$, which has the planar D_{4h}

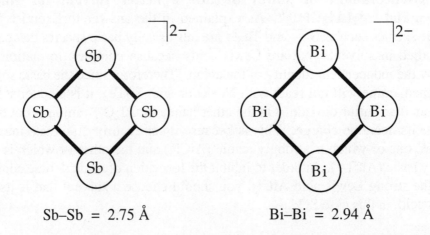

Sb–Sb = 2.75 Å   Bi–Bi = 2.94 Å

structure shown in Structure **37**. Both of these ions have 22 valence electrons ($(5 \times 4) + 2 = 22$). Other *p*-block clusters with four atoms and 22 valence electrons are Se_4^{2+} and Te_4^{2+}. These are also isoelectronic with Sb_4^{2-} and Bi_4^{2-}, and consistent with this, they also have planar D_{4h} structures (see Structure **39**).

(b) $[P_7]^{3-}$? This ion has seven atoms and 38 valence electrons ($(7 \times 5) + 3 = 38$). The compound P_4S_3, shown in Structure **41**, also has seven atoms and 38 valence electrons ($(4 \times 5) + (3 \times 6) = 38$). Therefore, since isoelectronic species frequently have the same structure, the probable structure of the $[P_7]^{3-}$ ion is the structure exhibited by P_4S_3. The compound Sr_3P_{14} does contain discrete $[P_7]^{3-}$ ions with the structure shown below.

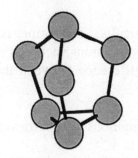

The structure of the $[P_7]^{3-}$
ions in Sr_3P_{14}

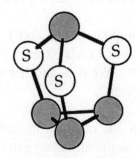

The structure of P_4S_3
(see Structure **41**)

Guide to Solutions Quiz

1   Draw Lewis structures and predict the geometries of NF_4^+, $N_2F_3^+$, NH_2OH, $SPCl_3$, and PF_3Cl_2.

2   How can NMR spectroscopy be used to (a) distinguish between solutions of $Na_5P_3O_{10}$ and $Na_6P_4O_{13}$, and (b) determine the geometry of SF_4.

3   Explain why hydrazine is a weaker base than ammonia.

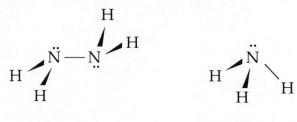

hydrazine   ammonia

4   There are more chemical differences between nitrogen and phosphorus than between any other pair of period 2 and period 3 elements that are in the same group. List as many differences as you can.

5   The boiling point of PF_5 (–75°C) is lower than that of PCl_5 (sublimes at 162°C). In contrast, the boiling point of SbF_5 (150°C) is *higher* than that of $SbCl_5$ (79°C). Explain the difference.

6   At high temperatures, sulfur vapor contains S_2 molecules. The S–S distance, 1.89 Å, is shorter than the S–S distance in S_8, 2.06 Å. Explain.

7   The compound $S_2O_6F_2$ is a powerful oxidizing agent. Draw its structure and account for its oxidizing ability.

8   The compounds SF_6 and TeF_6 are different chemically. The former is inert to addition of Lewis bases, while the latter forms fluoroanions such as TeF_7^- and TeF_8^{2-}. What is another important chemical difference between these two compounds?

9   Review the changes in structure and Brønsted acid/base properties of metal oxides as the metal/oxygen ratio changes from 0.5 to 4.

10   Predict the structures of $S_4N_3^+$ and $(NSF)_4$.

12 The halogens and the noble gases

	18	
16	17	He

O	F	Ne
S	Cl	Ar
Se	Br	Kr
Te	I	Xe
Po	At	Rn

VII VIII

The halogens and noble gases are two groups that contain the most reactive and the least reactive elements, respectively. Nevertheless, there are similarities in the patterns of their reactions and in the structures of their compounds. For example, iodine and xenon form an extensive set of isoelectronic and isostructural species, including IF/XeF^+, IF_4^-/XeF_4, and IO_3^-/XeO_3. The halogens form a very important class of compounds with the d-block metals, which range from metallic to ionic to molecular. Anhydrous metal halides are among the most important starting materials for inorganic syntheses.

S12.1 Recovery of iodine from iodate? Since the IO_3^-/I_2 standard reduction potential is 1.19 V (see Appendix 2), many species can be used as reducing agents, including $SO_2(aq)$ and $Sn^{2+}(aq)$ (the HSO_4^-/H_2SO_3 and Sn^{4+}/Sn^{2+} reduction potentials are 0.158 V and 0.15 V, respectively). The balanced equations would be as follows:

$$2\,IO_3^- + 5\,H_2SO_3 \rightarrow I_2 + 5\,HSO_4^- + H_2O + 3\,H^+$$

$$2\,IO_3^- + 5\,Sn^{2+} + 12\,H^+ \rightarrow I_2 + 5\,Sn^{4+} + 6\,H_2O$$

The reduction of iodate with aqueous sulfur dioxide would be far cheaper than reduction with Sn^{2+}, because sulfur dioxide is much cheaper than tin. One reason this is so is because sulfuric acid, for which SO_2 is an intermediate, is prepared worldwide on an enormous scale.

S12.2 **The structure and point group of ClO_2F?** As you should know by now, a reliable way to predict the structure of a *p*-block compound is to draw its Lewis structure and then apply VSEPR theory. The Lewis structure for ClO_2F is shown below. (Note that chlorine is the central atom, not oxygen or fluorine: the heavier, less electronegative element is always the central atom in interhalogen compounds and in compounds containing two different halogens and oxygen.) Three bonding pairs and one lone pair yield a trigonal pyramidal geometry, also shown below. This structure possesses only a single symmetry element, a mirror plane that bisects the O–Cl–O angle and contains the Cl and F atoms. Therefore, ClO_2F has C_s symmetry.

:Ö:C̈l:Ö: The Lewis structure   The trigonal pyramidal
   :F̈:   of ClO_2F structure of ClO_2F

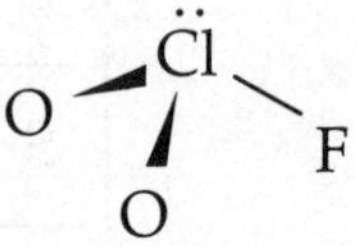

S12.3 **Polyhalides that are analogous to [py–I–py]$^+$?** Examples include I_3^-, IBr_2^-, ICl_2^-, and IF_2^-. In all cases, consider that the central I^+ species has a vacant orbital that interacts with two electron pair donors, py in the case of [py–I–py]$^+$ and I^- in the case of I_3^-. The three centers contribute four electrons to three molecular orbitals, one bonding, one nonbonding, and one antibonding.

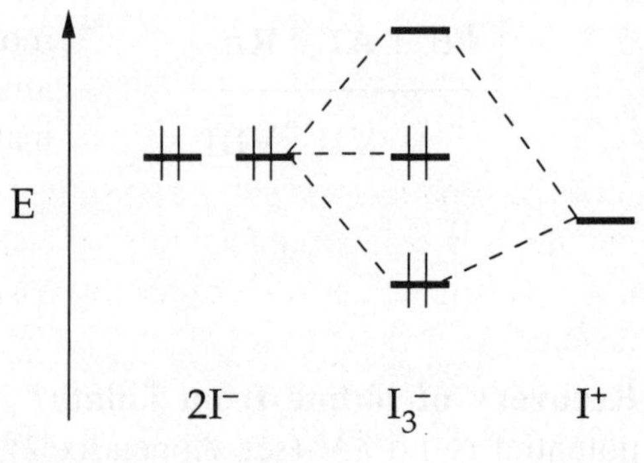

S12.4 Which halogens can oxidize H_2O to O_2? The standard reduction potential for the O_2/H_2O couple is 1.229 V. Any halogen for which the standard reduction potential for the the X_2/X^- couple is more positive than this will oxidize water, since the net potential for the reaction

$$2\,H_2O(aq) + 2\,X_2(aq) \rightarrow O_2(g) + 4\,X^-(aq) + 4\,H^+(aq)$$

would then be positive. The X_2/X^- potentials are 3.053 V for fluorine, 1.358 V for chlorine, 1.087 V for bromine, and 0.535 V for iodine. Therefore, elemental fluorine and chlorine are capable of oxidizing water at low pH. However, only the reaction with F_2 is rapid. The reaction of Cl_2 with H_2O is so slow that commercial production of Cl_2 occurs in the presence of H_2O (see Section 12.1).

S12.5 A balanced equation for the decomposition of xenate ion? You are told that xenate ($HXeO_4^-$) decomposes to perxenate (XeO_6^{4-}), xenon, and oxygen, so the equation you must balance is:

$$HXeO_4^- \rightarrow XeO_6^{4-} + Xe + O_2 \quad \text{(not balanced)}$$

Since the reaction occurs in basic solution, we can use OH^- and H_2O to balance the equation. You notice immediately that there is no species containing hydrogen on the right hand-side of the equation, so obviously H_2O will go on the right and OH^- will go on the left. The balanced equation is:

$$2\,HXeO_4^-(aq) + 2\,OH^-(aq) \rightarrow XeO_6^{4-}(aq) + Xe(g) + O_2(g) + 2\,H_2O(l)$$

Since the products are perxenate (a Xe(VIII) species) and elemental xenon, this reaction is a disproportionation of the Xe(VI) species $HXeO_4^-$. Oxygen is produced from a thermodynamically unstable intermediate of Xe(IV), possibly as follows:

$$HXeO_3(aq) \rightarrow Xe(g) + O_2(g) + OH^-(aq)$$

12.1   General properties of the elements?

	physical state	electronegativity	hardness of halide ion	color
F_2	gas	highest (4.0)	hardest	light yellow
Cl_2	gas	lower	softer	yellow-green
Br_2	liquid	lower	softer	dark red-brown
I_2	solid	lowest	softest	dark violet
He	gas			colorless
Ne	gas			colorless
Ar	gas			colorless
Kr	gas			colorless
Xe	gas			colorless

12.2   Recovery of the halogens from naturally occurring halides? The principal source of fluorine is CaF_2. It is converted to HF by treating it with a strong acid such as sulfuric acid. Liquid HF is electrolyzed to H_2 and F_2 in an anhydrous cell, with KF as the electrolyte:

$$CaF_2 + H_2SO_4 \rightarrow CaSO_4 + 2\,HF$$

$$2\,HF + 2\,KF \rightarrow 2\,K^+HF_2^-$$

$$2\,K^+HF_2^- + \text{electricity} \rightarrow F_2 + H_2 + 2\,KF$$

The principal source of the other halides is sea water (natural brines). Chlorine is liberated by the electrolysis of aqueous NaCl (the chloralkali process):

$$2\,Cl^- + 2\,H_2O + \text{electricity} \rightarrow Cl_2 + H_2 + 2\,OH^-$$

Bromine and iodine are prepared by treating aqueous solutions of the halides with chlorine:

$$2\,X^- + Cl_2 \rightarrow X_2 + 2\,Cl^- \quad (X^- = Br^-, I^-)$$

12.3 **The chloralkali cell?** A drawing of the cell is shown in Figure 12.3. Note that Cl_2 is liberated at the anode and H_2 is liberated at the cathode, according to the following half-reactions:

$$\text{anode:} \quad 2\,Cl^-(aq) \rightarrow Cl_2(g) + 2\,e^-$$

$$\text{cathode:} \quad 2\,H_2O(l) + 2\,e^- \rightarrow 2\,OH^-(aq) + H_2(g)$$

To maintain electroneutrality, Na^+ ions diffuse through the polymeric membrane. Due to the chemical properties of the membrane, anions such as Cl^- and OH^- cannot diffuse through it. If OH^- did diffuse through the membrane, it would react with Cl_2 and spoil the yield of the electrolysis, as shown in the following equation:

$$2\,OH^-(aq) + Cl_2(aq) \rightarrow ClO^-(aq) + Cl^-(aq) + H_2O(l)$$

12.4 **Sketch the vacant σ^* orbital of a halogen molecule?** According to Figure 12.4, the vacant antibonding orbital of a halogen molecule is $2\sigma_u^*$ and is composed primarily of halogen atomic p orbitals (recall that the $ns-np$ gap is larger toward the right hand side of the periodic table, and the larger the gap, the smaller the amount of $s-p$ mixing). Sketches of the filled $2\sigma_g$ bonding orbital and the empty $2\sigma_u^*$ antibonding orbital are shown below:

$$2\sigma_u^* \qquad \text{antibonding}$$

$$2\sigma_g \qquad \text{bonding}$$

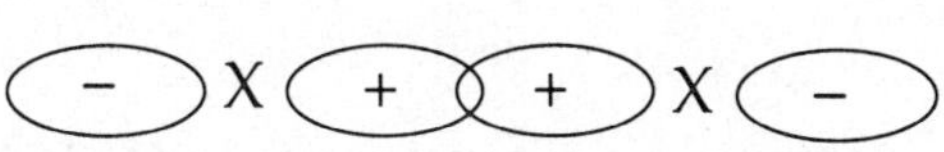

Since the $2\sigma_u^*$ antibonding orbital is the LUMO for a X_2 molecule, it is the orbital that accepts the pair of electrons from a Lewis base B when a dative $B \rightarrow X_2$ bond is formed. From the shape of the LUMO, the B–X–X unit should be linear.

12.5 **NF_3 vs. NH_3?** **(a) Difference in volatility?** Ammonia is one of a number of substances that exhibit very strong hydrogen bonding, as described in Sections 5.2 and 8.4(b). The very strong intermolecular interactions lead to a relatively large enthalpy of vaporization and a relatively high boiling point for NH_3 ($-33°C$). In contrast, the intermolecular forces in liquid NF_3 are relatively weaker dipole–dipole forces, resulting in a smaller enthalpy of vaporization and a low boiling point (–129°C).

(b) Explain the difference in basicity? The strong electron-withdrawing effect of the three fluorine atoms in NF_3 lowers the energy of the nitrogen atom lone pair. This lowering of energy has the effect of reducing the electron donating ability of the nitrogen atom in NF_3, reducing the basicity.

12.6   The analogy between halogens and pseudohalogens?   (a) The reaction of NCCN with NaOH? When a halogen such as chlorine is treated with aqueous base, it undergoes disproportionation to yield chloride ion and hypochlorite ion, as follows:

$$Cl_2(aq) + 2OH^-(aq) \rightarrow Cl^-(aq) + ClO^-(aq) + H_2O(l)$$

The analogous reaction of cyanogen with base is:

$$NCCN(aq) + 2OH^-(aq) \rightarrow CN^-(aq) + NCO^-(aq) + H_2O(l)$$

The linear NCO^- ion is named cyanate.

(b) The reaction of SCN⁻ with MnO₂ in aqueous acid? If MnO_2 is an oxidizing agent, then the probable reaction is oxidation of thiocyanate ion to thiocyanogen, $(SCN)_2$, coupled with reduction of MnO_2 to Mn^{2+} (if you do not recall that Mn^{3+} is unstable to disproportionation, you will discover it when you refer to the Latimer diagram for manganese in Appendix 2). The balanced equation is:

$$2SCN^-(aq) + MnO_2(s) + 4H^+(aq) \rightarrow (SCN)_2(aq) + Mn^{2+}(aq) + 2H_2O(l)$$

(c) The structure of trimethylsilyl cyanide? Just as halides form Si–X single bonds, trimethylsilyl cyanide contains a Si–CN single bond. Its structure is shown in Structure **1**.

12.7   Iodine fluorides?   (a) The structures of [IF₆]⁺ and IF₇? The structures are shown below. The structures actually represent the Lewis structures as well, except that the three lone pairs of electrons on each fluorine atom have been omitted. VSEPR theory predicts that a species with six bonding

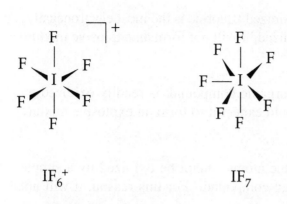

IF_6^+ IF_7

pairs of electrons, like IF_6^+, should be octahedral. Possible structures of species with seven bonding pairs of electrons, like IF_7, was not covered in Section 3.3, but a reasonable and symmetrical structure would be a pentagonal bipyramid.

(b) The preparation of $[IF_6][SbF_6]$? Judging from the structures shown above, it should be possible to abstract a F^- ion from IF_7 to produce IF_6^+. The strong Lewis acid SbF_5 should be used for the fluoride abstraction so that the salt $[IF_6][SbF_6]$ will result:

$$IF_7 + SbF_5 \rightarrow [IF_6][SbF_6]$$

12.8 **The Lewis acidity and basicity of BrF_3?** The acid/base properties of this inorganic solvent result from the following autoionization reaction:

$$2\,BrF_3 \rightarrow BrF_2^+ + BrF_4^-$$

The acidic species is the cation BrF_2^+ and the basic species is the anion BrF_4^-. Adding the powerful Lewis acid SbF_5 will increase the concentration of BrF_2^+, thus increasing the acidity of BrF_3, by the following reaction:

$$BrF_3 + SbF_5 \rightarrow BrF_2^+ + SbF_6^-$$

Adding CsF, which contains the strong Lewis base F^-, will increase the concentration of BrF_4^-, thus increasing the basicity of BrF_3, by the following reaction:

$$BrF_3 + F^- \rightarrow BrF_4^-$$

Adding SF_6 will have no effect on the acidity or basicity of BrF_3 because SF_6 is neither Lewis acidic nor Lewis basic.

12.9 **Which compound is likely to be explosive in contact with BrF_3?** **(a) SbF_5?** All interhalogens, including BrF_3, are strong oxidizing agents. The antimony atom in SbF_5 is already in its highest oxidation state, and the fluorine atom ligands cannot be chemically oxidized (fluorine is the most electronegative element). Since SbF_5 cannot be oxidized, it will not form an explosive mixture with BrF_3.

(b) CH_3OH? Methanol, being an organic compound, is readily oxidized by strong oxidants. Therefore, you should expect it to form an explosive mixture with BrF_3.

(c) F_2? As stated in part (a), fluorine atoms cannot be oxidized by a simple chemical oxidant like an interhalogen compound. For this reason, it will not form an explosive mixture with BrF_3.

(d) S_2Cl_2? In this compound, sulfur is in the +1 oxidation state. Recall that SF_4 and SF_6 are stable compounds. Therefore, you should expect S_2Cl_2 to be oxidized to higher valent sulfur fluorides. The mixture $S_2Cl_2 + BrF_3$ will be an explosion hazard.

12.10   The reaction of $[NR_4][Br_3]$ with I_2? Since Br_3^- has only a moderate formation constant, you can think of it as being chemically equivalent to Br_2 + Br^-. Since IBr is fairly stable, the probable reaction will be:

$$Br_3^- + I_2 \rightarrow 2\,IBr + Br^-$$

The IBr may associate to some extent with Br^- to produce IBr_2^-.

12.11   Explain why CsI_3 is stable but NaI_3 is not? This is an example of an important principle of inorganic chemistry: large cations stabilize large, unstable anions. This trend appears to have a simple electrostatic origin, as described in Section 2.12. A detailed analysis of this particular situation is as follows. The enthalpy change for the reaction:

$$NaI_3(s) \rightarrow NaI(s) + I_2(s)$$

is negative, which is just another way of stating that NaI_3 is not stable. This enthalpy change is composed of four terms, as shown below:

$$
\begin{aligned}
\Delta H \quad = \quad & \text{lattice enthalpy of } NaI_3 \\
+ \quad & I{-}I_2^- \text{ bond enthalpy} \\
- \quad & \text{lattice enthalpy of } I_2 \\
- \quad & \text{lattice enthalpy of } NaI
\end{aligned}
$$

The fourth term is larger than the first term, since I⁻ is smaller than I₃⁻. This difference provides the driving force for the reaction to occur as written. If you substitute Cs^+ for Na^+, the two middle terms will remain constant. Now, the fourth term is still larger than the first term, *but by a significantly smaller amount*. The net result is that ΔH for the cesium system is positive. In the limit where the cation becomes infinitely large, the difference between the first and fourth terms becomes negligible.

12.12 Write Lewis structures and predict shapes? (a) ClO_2? The Lewis structure and the predicted shape of ClO_2 are shown below. The angular shape is a consequence of repulsions between the bonding and nonbonding electrons. With three nonbonding electrons, the O–Cl–O angle in ClO_2 is 118°. With four nonbonding electrons, as in ClO_2^-, the repulsions are greater and the O–Cl–O angle is only 111°.

(b) I_2O_5? The Lewis structure and the predicted shape of I_2O_5 are shown below. The central O atom has two bonding pairs of electrons and two lone pairs, like H_2O, so it should be no surprise that the I–O–I bond angle is less than 180° (it is 139°). Each I atom is trigonal pyramidal, since it has three bonding pairs and one lone pair. If you had difficulty working this exercise, you should review VSEPR theory, covered in Section 3.3.

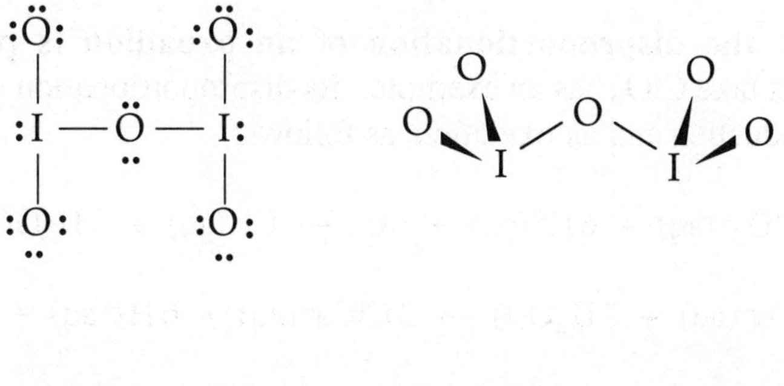

12.13 Perbromic and periodic acid? (a) Formulas and probable acidities? The formulas are $HBrO_4$ and H_5IO_6. The difference lies in iodine's ability to expand its coordination shell, a direct consequence of its large size. Recall from Section 5.4 that the relative strength of an oxoacid can be estimated from the q/p ratio (p is the number of oxo groups and q is the number of OH groups attached to the central atom). A high value of p/q correlates with strong acidity while a low value correlates with weak acidity. For HBrO$_4$, p/q = 3/1, so it is a strong acid. For H$_5$IO$_6$, p/q = 1/5, so it is a weak acid.

(b) Relative stabilities? Periodic acid is thermodynamically more stable with respect to reduction than perbromic acid. Bromine, like its period 4 neighbors arsenic and selenium, is more oxidizing in its highest oxidation state than the members of the group immediately above and below.

12.14 **The pH dependence of reduction potentials of oxoanions? (a) The expected trend?** The trend is that E decreases as the pH increases.

(b) E at pH 0 and pH 7 for ClO$_4^-$? The balanced equation for the reduction of ClO$_4^-$ is:

$$ClO_4^-(aq) + 2H^+(aq) + 2e^- \rightarrow ClO_3^-(aq) + H_2O(l)$$

The value of $E°$ for this reaction is 1.201 V (see Appendix 2). The potential at any [H$^+$], given by the Nernst equation, is

$$E = E° - ((0.059\ V/2))(\log([ClO_3^-]/[ClO_4^-][H^+]^2)$$

At pH 7, [H$^+$] = 10^{-7} M. Assuming that both perchlorate and chlorate ions are present at unit activity, at pH 7 the reduction potential is

$$E = 1.201\ V - (0.0295\ V)(\log 10^{14}) = 1.201\ V - 0.413\ V = 0.788\ V$$

12.15 **Explain why the disproportionation of an oxoanion is promoted by low pH?** Let's take ClO$_3^-$ as an example. Its disproportionation can be broken down into a reduction and an oxidation, as follows:

$$ClO_3^-(aq) + 6H^+(aq) + 6e^- \rightarrow Cl^-(aq) + 3H_2O(l)$$

$$3ClO_3^-(aq) + 3H_2O(l) \rightarrow 3ClO_4^-(aq) + 6H^+(aq) + 6e^-$$

Any effect that changing the pH has on the potential for the reduction reaction will be counteracted by an equal but opposite change on the potential for the oxidation reaction. In other words, the net reaction does not include H$^+$, so the net potential for the disproportionation cannot be pH dependent. Therefore, the promotion of disproportionation reactions of some oxoanions at low pH *cannot* be a thermodynamic promotion. Low pH results in a kinetic promotion:

protonation of an oxo group aids oxygen–halogen bond scission (see Section 12.6). The disproportionation reactions have the same driving force at high pH and at low pH, but they are much faster at low pH.

12.16 Which reacts more quickly in dilute aqueous solution, perchloric acid or periodic acid? Periodic acid is by far the quicker oxidant. Recall that it exists in two forms in aqueous solution, H_5IO_6 (the predominant form) and HIO_4. Even though the concentration of HIO_4 is low, this four-coordinate species can form a complex with a potential reducing agent, providing for an efficient and rapid mechanism by which the redox reaction can occur (see Section 12.6).

12.17 Which oxo anions of chlorine undergo disproportionation in acid? The Frost diagram for chlorine in aqueous acid solution is shown below. It is worth your while to construct this diagram yourself from the potential data in Appendix 2. Then, if you connect the points for Cl^- and ClO_4^- with a line, you will see that the intermediate oxidation state species $HClO$, $HClO_2$, and ClO_3^- lie above that line. Therefore, they are unstable with respect to disproportionation. As discussed in Section 12.6, the redox reactions of halogen oxo anions become progressively faster as the oxidation number of the halogen *decreases*. Therefore, the rates of disproportionation are probably $HClO > HClO_2 > ClO_3^-$. Note that ClO_4^- cannot undergo disproportionation, since there are no species with a higher oxidation number.

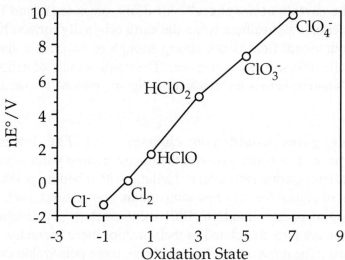

12.18 Which of the following compounds present an explosion hazard? (a) NH_4ClO_4? The key to answering this question is to decide if the perchlorate ion, which is a very strong oxidant, is present along with a species that can be oxidized. If so, the compound *does* represent an explosion hazard.

Ammonium perchlorate is a dangerous compound, since the N atom of the NH_4^+ ion is in its lowest oxidation state and can be oxidized. The explosion hazard of NH_4ClO_4 was mentioned in Section 12.6.

(b) $Mg(ClO_4)_2$? Since Mg(II) cannot be oxidized to a higher oxidation state, magnesium perchlorate is a stable compound and is not an explosion hazard.

(c) $NaClO_4$? The same answer applies here as above for magnesium perchlorate. Sodium has only one common oxidation state. Even the strongest oxidants, such as F_2, FOOF, and ClF_3, cannot oxidize Na(I) to Na(II).

(d) $[Fe(H_2O)_6][ClO_4]_2$? Although the H_2O ligands cannot be oxidized, the metal ion can. This compound presents an explosion hazard, since Fe(II) can be oxidized to Fe(III) by a strong oxidant such as perchlorate ion.

12.19 **Why is He rare in the earth's atmosphere?** All of the original He and H_2 that was present in the earth's atmosphere when the earth originally formed has been lost. Earth's gravitational field is not strong enough to hold these light gases, and they eventually diffuse away into space. The small amount of helium that is present in today's atmosphere is the product of ongoing radioactive decay.

12.20 **Which of the noble gases would you choose?** **(a) The lowest-temperature refrigerant?** The noble gas with the lowest boiling point would best serve as the lowest-temperature refrigerant. Helium, with a boiling point of 4.2 K, is the refrigerant of choice for very low temperature applications, such as cooling superconducting magnets in modern NMR spectrometers. The boiling points of the noble gases are directly related to their size, or more correctly, to their number of electrons. The larger atoms Kr and Xe are more polarizable than He or Ne, so Kr and Xe experience larger van der Waals attractions and have higher boiling points.

(b) An electric discharge light source requiring a safe gas with the lowest ionization potential? Of the "safe" noble gases, i.e. He–Xe, the largest one, Xe, has the lowest ionization potential (see Table 12.1). Radon is even larger than Xe and has a lower ionization potential, but since it is radioactive it is not safe.

(c) The least expensive inert atmosphere? Geographic location might play a role in your answer to this question. Since Ar is so plentiful in the atmosphere relative to the other noble gases, it is generally cheaper than helium, which is rare in the atmosphere. However, a great deal of helium is collected as a byproduct of natural gas production. In places where large amounts of natural gas are produced, such as the United States, helium may be marginally less expensive than argon.

12.21 **Give balanced equations and conditions for syntheses? (a) XeF$_2$?** This compound can be prepared in two different ways. First, a mixture of Xe and F$_2$ that contains an excess of Xe is heated to 400 °C. The excess Xe prevents the formation of XeF$_4$ and XeF$_6$. The second way is to photolyze a mixture of Xe and F$_2$ in a glass reaction vessel. For either method of synthesis, the balanced equation is:

$$\text{Xe(g) + F}_2\text{(g)} \rightarrow \text{XeF}_2\text{(s)}$$

At 400 °C the product is a gas, but at room temperature it is a solid.

(b) XeF$_6$? For the synthesis of this compound you would want to also use a high temperature, but unlike the synthesis of XeF$_2$, you want to have a *large* excess of F$_2$:

$$\text{Xe(g) + 3 F}_2\text{(g)} \rightarrow \text{XeF}_6\text{(s)}$$

As with XeF$_2$, xenon hexafluoride is a solid at room temperature.

(c) XeO$_3$? This compound is endoergic, so it cannot be prepared directly from the elements. However, a sample of XeF$_6$ can be carefully hydrolyzed to form the product:

$$\text{XeF}_6\text{(s) + 3 H}_2\text{O(l)} \rightarrow \text{XeO}_3\text{(s) + 6 HF(g)}$$

If a large excess of water is used, an aqueous solution of XeO$_3$ is formed instead.

12.22 **Isostructural noble gas species? (a) ICl$_4^-$?** The Lewis structure of this anion has four bonding pairs and two lone pairs of electrons around the central iodine atom. Therefore, the anion has a square-planar geometry. The noble gas compound that is isostructural is XeF$_4$, as shown below.

(b) IBr$_2^-$? The Lewis structure of this anion has two bonding pairs and three lone pairs of electrons around the central iodine atom. Therefore, the anion has a linear geometry. The noble gas compound that is isostructural is XeF$_2$, as shown below.

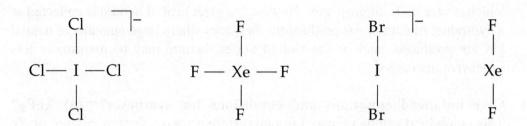

(c) BrO₃⁻? The Lewis structure of this anion has three bonding pairs and one lone pair of electrons around the central bromine atom. Therefore, the anion has a trigonal pyramidal geometry. The noble gas compound that is isostructural is XeO₃, as shown below.

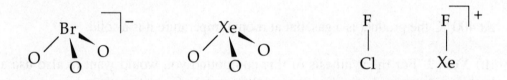

(d) ClF? This diatomic molecule does not have a *molecular* noble gas counterpart (i.e. all noble gas compounds have three or more atoms). However, it is isostructural with the cation XeF⁺, as shown above.

12.23 ClO⁻, Br₂, and XeF⁺? (a) Lewis structures and formal charges? The Lewis structures and formal charges (in parentheses) are shown below:

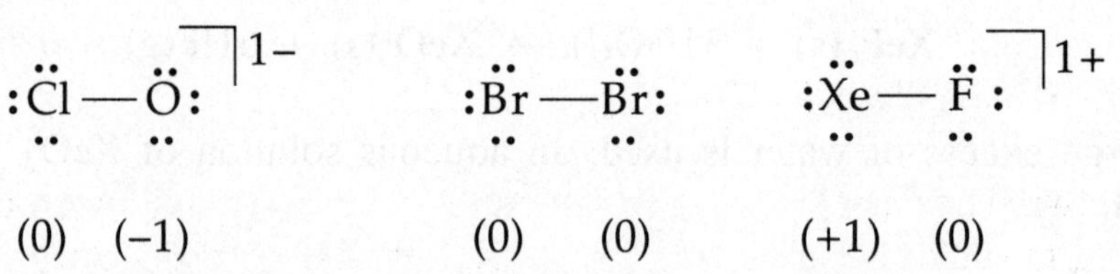

$$:\!\ddot{C}l\!-\!\ddot{O}:\;^{1-}\qquad\qquad:\!\ddot{B}r\!-\!\ddot{B}r:\qquad\qquad:\!\ddot{X}e\!-\!\ddot{F}:\;^{1+}$$

      (0)   (–1)   (0)   (0)   (+1) (0)

(b) Are they isolobal? Yes, the number of orbitals, their shapes, and electron populations are similar, so they are isolobal.

(c) Reactions with nucleophiles? As shown above, ClO⁻ is isoelectronic with diatomic halogen molecules (e.g. Br₂) as well as with diatomic interhalogen molecules (ClF, IF, ICl, IBr, etc.). These neutral molecules are Lewis acids, as discussed in Sections 12.4 and 5.10. Therefore, it is reasonable to assume that the chlorine end of ClO⁻ can act as an acceptor. If so, the disproportionation of

ClO^- might proceed by attack of the oxygen end of ClO^- (the nucleophile) on the chlorine end of another ClO^- anion (the electrophile):

$$ClO^- + ClO^- \rightarrow [ClO\cdots ClO]^{2-} \rightarrow [Cl\cdots OClO]^{2-} \rightarrow Cl^- + ClO_2^-$$

Similarly, XeF^+ can act as an acceptor towards Lewis bases. For example, it reacts with F^- to form XeF_2.

(d) Trends in electrophilicity and basicity? The cationic species XeF^+ would be expected to be the strongest electrophile, followed by Br_2 and then by ClO^-. Hypochlorite ion, in fact, acts a Brønsted base and can act as a Lewis base besides being capable of acting as an electron acceptor (i.e. a Lewis acid).

12.24 XeF_7^-? **(a) Lewis structure?** The Lewis structure is shown at the right.

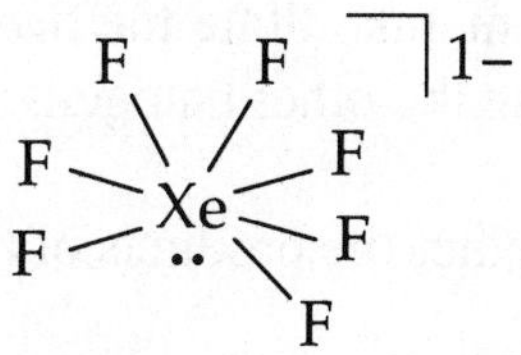

(b) Possible structure? The structure of XeF_5^-, with seven pairs of electrons, is based on a pentagonal bipyramidal array of electron pairs with two stereochemically active lone pairs (see Structure **19**). The ion XeF_8^{2-}, on the otherhand, has nine pairs of electrons, but only eight are stereochemically active (see Structure **20**). It is possible that the structure of XeF_7^-, with eight pairs, would be a pentagonal bipyramid, with one stereochemically inactive lone pair.

Guide to Solutions Quiz

1   Write Lewis structures, draw the structures, and assign point groups to the following molecules or ions: IF_4^-, IF_4^+, XeF_5^-, XeF_5^+, and XeO_3.

2   Write autoionization reactions for BrF_3, ICl, and BrF_5, and draw the structures of the ions produced in the reactions.

3   At first glance, the halogen/pseudohalogen analogy might lead you to think that both of the following oxidation reactions are possible:

$$2\,Cl^- \rightarrow Cl_2 + 2\,e^-$$

$$2\,N_3^- \rightarrow N_6 + 2\,e^-$$

However, the second reaction gives a very different product instead of linear N_6. What is it, and why is linear N_6 unstable?

4   Perbromate is a stronger oxidant than either perchlorate or periodate. Explain this apparent anomaly. Hint: justify your explanation by finding similar "anomalies" in other period 4 p-block elements near bromine.

5   Assign oxidation numbers for the elements in the fluoroxysulfate ion, $SO_3(OF)^-$. Predict this ion's chemical properties.

6   Predict the relative volatilites of H_2SO_4, HSO_3F, and SO_2F_2. Explain your answer.

7   Draw Born–Haber type cycles for the formation of CF_4, CCl_4, and CI_4 from the elements. State the *two* reasons why elemental fluorine is much more reactive than the other halogens.

8   Predict the products of the reaction of XeF_2 with (a) CO, (b) SO_2, and (c) ClO_2.

9   All known compounds of xenon contain this element in positive oxidation states. Suggest a reason why salts such as Na^+Xe^- do not exist. Hint: there are numerous metal oxide salts even though the enthalpy of formation of the oxide ion, O^{2-}, is highly endothermic.

10   Explain why compounds of Xe are thermodynamically more stable than compounds of Kr, even though Kr, which is the smaller of the two, should form stronger bonds to other elements.

13 The electronic spectra of complexes

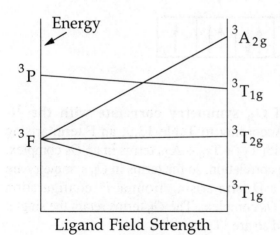

Energy

^3P

^3F

$^3A_{2g}$

$^3T_{1g}$

$^3T_{2g}$

$^3T_{1g}$

Ligand Field Strength

A correlation diagram between a free d^2 metal ion (left side) and the octahedral strong-field terms of a d^2 metal ion (right side). Only the spin triplet terms are shown. There are three spin-allowed transitions between the $^3T_{1g}$ ground state and higher energy spin triplet states, $^3T_{2g} \leftarrow {}^3T_{1g}$, $^3T_{1g} \leftarrow {}^3T_{1g}$, and $^3A_{2g} \leftarrow {}^3T_{1g}$.

S13.1 What terms arise from a p^1d^1 configuration? An atom with a p^1d^1 configuration will have $L = 1 + 2 = 3$ (an electron in a p orbital has $l = 1$, while an electron in a d orbital has $l = 2$). This value of L corresponds to a F term. Since the electrons may be paired ($S = 0$, multiplicity $2S + 1 = 1$) or parallel ($S = 1$, multiplicity $2S + 1 = 3$), both ^1F and ^3F terms are possible.

S13.2 Identifying ground terms? (a) $2p^2$? Two electrons in a p subshell can occupy separate p orbitals and have parallel spins ($S = 1$), so the maximum multiplicity $2S + 1 = 3$. The maximum value of M_L is $1 + 0 = 1$, which corresponds to a P term (according to the Pauli principle, m_l cannot be $+1$ for both electrons if the spins are parallel). Thus, the ground term is ^{3}P (called "a triplet P" term).

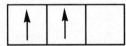

(b) $3d^9$? The largest value of M_S for nine electrons in a d subshell is 1/2 (eight electrons are paired in four of the five orbitals, while the ninth electron has $m_s = \pm 1/2$). Thus, $S = 1/2$ and the multiplicity $2S + 1 = 2$. Notice that the largest value of M_S is the same for one electron in a d subshell. Similarly, the largest value of M_L is 2, which results from the following nine values of m_l: $+2$, $+2$, $+1$, $+1$, 0, 0, -1, -1, -2. Notice that $M_L = 2$ also corresponds to one electron in a d subshell. Thus, the ground term for a d^9 or a d^1 configuration is ^{2}D (called "a doublet D" term).

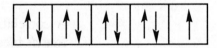

S13.3 What terms in a d^2 complex of O_h symmetry correlate with the ^{3}F and ^{1}D terms of the free ion? According to Table 13.3, an F term arising from a d^n configuration correlates with $T_{1g} + T_{2g} + A_{2g}$ terms in an O_h complex. The multiplicity is unchanged by the correlation, so the terms in O_h symmetry are $^3T_{1g}$, $^3T_{2g}$, and $^3A_{2g}$. Similarly, a D term arising from a d^n configuration correlates with $T_{2g} + E_g$ terms in an O_h complex. The O_h terms retain the singlet character of the ^{1}D free ion term, and so are $^1T_{2g}$ and ^1E_g.

S13.4 Predict the wavenumbers of the first two spin-allowed bands in the spectrum of $[Cr(H_2O)_6]^{3+}$? The two transitions in question are $^4T_2 \leftarrow {}^4A_2$ and $^4T_1 \leftarrow {}^4A_2$ (the first one has a lower energy and hence a lower wavenumber). Since $\Delta_o = 17{,}600$ cm^{-1} and $B = 700$ cm^{-1}, you must find the point $17{,}600/700 = 25.1$ on the x-axis in Figure 13.7. Then, the ratios E/B for the two bands are the y values of the points on the $^4T_2$ and $^4T_1$ lines, 25 and 32, respectively, as shown below. Therefore, the two lowest-energy spin-allowed bands in the spectrum of $[Cr(H_2O)_6]^{3+}$ will be found at $(700$ cm$^{-1})(25) = 17{,}500$ cm^{-1} and $(700$ cm$^{-1})(32) = 22{,}400$ cm^{-1}.

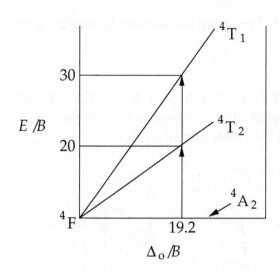

A simplified version of the Tanabe–Sugano diagram for the d^3 configuration of $[Cr(H_2O)_6]^{3+}$.

S13.5 **Assign the bands in the spectrum of $[Cr(NCS)_6]^{3-}$?** This six-coordinate d^3 complex undoubtedly has O_h symmetry, so the general features of its spectrum will resemble the spectrum of $[Cr(NH_3)_6]^{3+}$, shown in Figure 13.1. The very low intensity of the band at 16,000 cm^{-1} is a clue that it is a spin-forbidden transition, probably $^2E_g \leftarrow {}^4A_{2g}$. Spin-allowed but Laporte-forbidden bands typically have $\varepsilon \sim 100$ M^{-1} cm^{-1}, so it is likely that the bands at 17,700 cm^{-1} and 23,800 cm^{-1} are of this type (they correspond to the $^4T_{2g} \leftarrow {}^4A_{2g}$ and $^4T_{1g} \leftarrow {}^4A_{2g}$ transitions, respectively). The band at 32,400 cm^{-1} is probably a charge transfer band, since its intensity is too high to be a ligand field (d–d) band. Since you are provided with the hint that the NCS^- ligands have low-lying π^* orbitals, it is reasonable to conclude that this band corresponds to a MLCT transition. Notice that the two spin-allowed ligand field transitions of $[Cr(NCS)_6]^{3-}$ are at lower energy than those of $[Cr(NH_3)_6]^{3+}$, showing that NCS^- induces a smaller Δ_o on Cr^{3+} than does NH_3. Also notice that $[Cr(NH_3)_6]^{3+}$ lacks an intense MLCT band at ~30,000–40,000 cm^{-1}, showing that NH_3 does not have low-lying empty orbitals.

S13.6 **What evidence from absorption and CD spectra show that $[Co(edta)]^-$ has lower symmetry than $[Co(en)_3]^{3+}$?** The evidence is the greater number of CD bands in the spectrum of $[Co(edta)]^-$ than in the spectrum of $[Co(en)_3]^{3+}$. The CD spectrum of $[Co(en)_3]^{3+}$, shown in Figure 13.14(b), exhibits three bands, near 21,000 cm^{-1}, 24,000 cm^{-1}, and 26,000 cm^{-1}. The spectrum of $[Co(edta)]^-$, shown in Figure 13.16, exhibits five bands, near 17,000 cm^{-1}, 19,000 cm^{-1}, 24,000 cm^{-1}, 26,000 cm^{-1}, and 28,000 cm^{-1}. The greater number of bands for $[Co(edta)]^-$ results from a lifting of degeneracies, which implies that this complex has a lower symmetry than $[Co(en)_3]^{3+}$.

13.1 **Russell–Saunders term symbols?**   (a) $L = 0$, $S = 5/2$?   You should remember that a term, denoted by a capital letter, is related to L in the same way that an orbital, denoted by a lower case letter, is related to l:

If l =	0	1	2	3	4	5	6
orbital =	s	p	d	f	g	h	i

If L =	0	1	2	3	4	5	6
term =	S	P	D	F	G	H	I

The multiplicity of the term, which is always given as a left superscript, can always be determined by using the formula multiplicity $= 2S + 1$:

If S =	0	1/2	1	3/2	2	5/2
multiplicity $2S + 1$ =	1	2	3	4	5	6

The terms and multiplicities listed above are not a complete list to cover all possibilities for all atoms and ions, but they will cover all possible d^n configurations. As far as the situation $L = 0$, $S = 5/2$ is concerned, the term symbol is ^6S. In addition to answering this question, you should try to decide which d^n configurations can give rise to the term. In this case, the only d^n configuration that can give rise to an ^6S term is d^5 (e.g. a gas-phase Mn^{2+} or Fe^{3+} ion).

(b) $L = 3$, $S = 3/2$? According to the relations shown above, this set of angular momentum quantum numbers is described by the term symbol ^4F. The diagram below, which is another way to depict a microstate, shows how the situation $L = 3$, $S = 3/2$ can arise from a d^3 configuration (e.g. a gas-phase Cr^{3+} ion). It can also arise from d^5 and d^7 configurations.

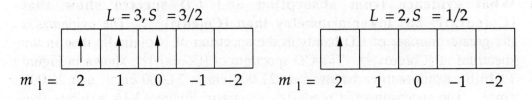

(c) $L = 2$, $S = 1/2$? This set of quantum numbers is described by the term symbol ^2D. The microstate diagram above shows how the situation $L = 2$, $S = 1/2$ can arise from a d^1 configuration (e.g. a gas-phase Ti^{3+} ion). It can also arise from d^3, d^5, d^7, and d^9 configurations.

(d) $L = 1$, $S = 1$? This set of quantum numbers is described by the term symbol ^3P. It can arise from d^2, d^4, d^6, and d^8 configurations.

13.2 **Identify the ground term?** **(a) ^3F, ^3P, ^1P, ^1G?** By definition, the ground term has the lowest energy of all of the terms. Recall Hund's two rules, discussed in Section 13.2: (1) The term with the greatest multiplicity lies lowest in energy; (2) For a given multiplicity, the greater the value of L of a term, the lower the energy. Therefore, the ground term in this case will be a triplet, not a singlet (rule 1). Of the two triplet terms, ^3F lies lower in energy than ^3P: $L = 3$ for ^3F, $L = 1$ for ^3P (rule 2). Therefore, the ground term is ^3F.

(b) ^5D, ^3H, ^3P, ^1G, ^1I? The ground term is ^5D because this term has a higher multiplicity than the other terms.

(c) ^6S, ^4G, ^4P, ^2I? The ground term is ^6S because this term has a higher multiplicity than the other terms.

13.3 **The Russell–Saunders terms for the following configurations and identify the ground term?** **(a) $4s^1$?** You can approach this exercise in the way described in Section 13.2 for the d^2 configuration (see Table 13.1). You first write down all possible microstates for the s^1 configuration, then write down the M_L and M_S values for each microstate, then infer the values of L and S to which the microstates belong. In this case, the procedure is not lengthy, since there are only two possible microstates, (0^+) and (0^-). The only possible values of M_L and M_S are 0 and 1/2, respectively. If M_L can only be 0, then L must be 0, which gives an S term (remember that L can take on all values M_L, (M_L-1), ..., 0, ...$-M_L$). Similarly, if M_S can only be 1/2 or $-1/2$, then S must be 1/2, which gives a multiplicity $2S + 1 = 2$. Therefore, the one and only term that arises from a $4s^1$ configuration is ^2S.

(b) $3p^2$?   This case is more complicated because there are 15 possible microstates:

	$M_S = -1$	$M_S = 0$	$M_S = 1$
$M_L = 2$		$(1^+, 1^-)$	
$M_L = 1$	$(1^-, 0^-)$	$(1^+, 0^-), (1^-, 0^+)$	$(1^+, 0^+)$
$M_L = 0$	$(1^-, -1^-)$	$(1^+, -1^-), (0^+, 0^-), (-1^+, 1^-)$	$(1^+, -1^+)$
$M_L = -1$	$(-1^-, 0^-)$	$(-1^+, 0^-), (-1^-, 0^+)$	$(-1^+, 0^+)$
$M_L = -2$		$(-1^+, -1^-)$	

A term that contains a microstate with $M_L = 2$ must be a D term ($L = 2$). This term can only be a singlet, since if both electrons have $m_l = 1$ they must be spin-paired. Therefore, one of the terms of the $3p^2$ configuration is ^{1}D, which contains five microstates. To account for these, you can cross out $(1^+, 1^-),(-1^+, -1^-)$, and one microstate from each of the other three rows under $M_S = 0$. That leaves ten microstates to be accounted for. The maximum value of M_L of the remaining microstates is 1, so you next consider a P term ($L = 1$). Since M_S can be -1, 0, or 1, this term will be ^{3}P, which contains nine of the ten remaining microstates. The last remaining microstate is one of the original three for which $M_L = 0$ and $M_S = 0$, which is the one and only microstate that belongs to a ^{1}S term ($L = 0$, $S = 0$). Of the three terms ^{1}D, ^{3}P, and ^{1}S that arise from the $3p^2$ configuration, the ground term is ^{3}P, since it has a higher multiplicity than the other two terms (see Hund's rule #1).

13.4   **Ground terms for $3p^5$ and $4d^9$ configurations?**   A neutral chlorine atom has an electron configuration [Ne]$3s^2 3p^5$. An Ag^{2+} ion has an electron configuration [Kr]$4d^9$. As in the answer to Self-test S13.2, you can treat a p^5 or a d^9 configuration just like a p^1 or a d^1 configuration, since having one "hole" in a subshell is in many ways equivalent to having one electron in the subshell. Therefore, for a p^1 or p^5 configuration, the values of L and S are 1 and 1/2, respectively, leading to a ground-state term of ^{2}P. For a d^1 or d^9 configuration, the values of L and S are 2 and 1/2, respectively, leading to a ground term of ^{2}D.

13.5   Identification of ground terms?

species	configuration	L	S	ground term
B^+	$[He]2s^2$	0	0	^1S
Na	$[Ne]3s^1$	0	1/2	^2S
Ti^{2+}	$[Ar]3d^2$	3	1	^3F
Ag^+	$[Kr]3d^{10}$	0	0	^1S

13.6   Calculate the values of B and C for V^{3+}?

The diagram below shows the relative energies of the ^3F, ^1D, and ^3P terms. It can be seen that the 10,642 cm^{-1} energy gap between the ^3F and ^1D terms is $5B + 2C$, while the 12,920 cm^{-1} energy gap between the ^3F and ^3P terms is $15B$. From the two equations

$$5B + 2C = 10,642 \text{ cm}^{-1} \quad \text{and} \quad 15B = 12,920 \text{ cm}^{-1}$$

you can determine that $B = (12,920 \text{ cm}^{-1})/(15) = 861.33 \text{ cm}^{-1}$ and $C = 3167.7$ cm^{-1}.

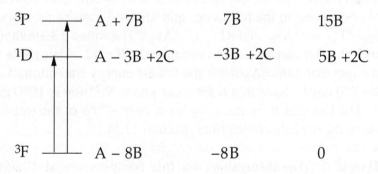

^3P	——————	$A + 7B$	$7B$	$15B$
^1D	——————	$A - 3B + 2C$	$-3B + 2C$	$5B + 2C$
^3F	——————	$A - 8B$	$-8B$	0

Relative Energies

13.7 d^n configurations and ground terms?   (a) Low-spin $[Rh(NH_3)_6]^{3+}$?

There are a number of ways to determine the integer n for a given d-block metal. One straightforward procedure is as follows. Count the number of elements from the left side of the periodic table to the metal in question. This will be the number of d electrons *for a metal atom in a complex* (note that an isolated gas-phase metal atom may have an $s^m d^n$ configuration, but the same neutral metal atom *in a complex* will have a d^{m+n} configuration). Then subtract the positive charge on the metal ion from this number, leaving the integer n. For example, Rh is the ninth element in period 5, so Rh^0 in a complex has a d^9 configuration, Rh^+ has a d^8 configuration, and so on. Using this procedure, the Rh^{3+} ion in the octahedral complex $[Rh(NH_3)_6]^{3+}$ has a d^6 configuration ($9 - 3 = 6$). According to the d^6 Tanabe–Sugano diagram (see Appendix 5), the ground term for a low-spin t_{2g}^6 metal ion is $^1A_{1g}$.

(b) $[Ti(H_2O)_6]^{3+}$? Titanium is the fourth element in period 4, so Ti^0 in a complex has a d^4 configuration. Therefore, the Ti^{3+} ion in the octahedral $[Ti(H_2O)_6]^{3+}$ ion has a d^1 configuration ($4 - 3 = 1$). A correlation diagram for d^1 metal ions is shown in Figure 13.3 (Appendix 5 does not include the d^1 Tanabe–Sugano diagram). According to this diagram, the ground term for a t_{2g}^1 metal ion is $^2T_{2g}$.

(c) High-spin $[Fe(H_2O)_6]^{3+}$? Iron is the eighth element in period 4, so Fe^0 in a complex (such as $Fe(CO)_5$) has a d^8 configuration. Therefore, the Fe^{3+} ion in the octahedral $[Fe(H_2O)_6]^{3+}$ ion has a d^5 configuration ($8 - 3 = 5$). According to the d^5 Tanabe–Sugano diagram, the ground term for a high-spin $t_{2g}^3 e_g^2$ metal ion is $^6A_{1g}$.

13.8 **Estimate Δ_o and B for: (a) $[Ni(H_2O)_6]^{2+}$?** According to the d^8 Tanabe–Sugano diagram (Appendix 5), the absorptions at 8500 cm^{-1}, 13,800 cm^{-1}, and 25,300 cm^{-1} correspond to the following spin-allowed transitions, respectively: $^3T_{2g} \leftarrow {}^3A_{2g}$, $^3T_{1g} \leftarrow {}^3A_{2g}$, and $^3T_{1g} \leftarrow {}^3A_{2g}$. The ratios $13,800/8500 = 1.6$ and $25,300/8500 = 3.0$ can be used to estimate $\Delta_o/B \approx 11$. Using this value of Δ_o/B and the fact that $E/B = \Delta_o/B$ for the lowest-energy transition, $\Delta_o = 8500$ cm^{-1} and $B \approx 770$ cm^{-1}. Note that B for a gas-phase Ni^{2+} ion is 1080 cm^{-1} (see Table 13.2). The fact that B for the complex is only ~70% of the free ion value is an example of the nephalauxetic effect (Section 13.3).

(b) $[Ni(NH_3)_6]^{2+}$? The absorptions for this complex are at 10,750 cm^{-1}, 17,500 cm^{-1}, and 28,200 cm^{-1}. The ratios in this case are $17,500/10,750 = 1.6$ and $28,200/10,750 = 2.6$, and lead to $\Delta_o/B \approx 15$. Thus, $\Delta_o = 10,750$ cm^{-1} and $B \approx 720$ cm^{-1}. It is sensible that B for $[Ni(NH_3)_6]^{2+}$ is smaller than B for $[Ni(H_2O)_6]^{2+}$, since NH_3 is higher in the nephalauxetic series than is H_2O.

13.9 **Ground term and lowest energy transition for a paramagnetic octahedral Fe(II) complex?** A d^6 octahedral Fe(II) complex can be either high-spin ($t_{2g}^4 e_g^2$, $S = 2$) or low-spin (t_{2g}^6, $S = 0$). If the complex has a large paramagnetic susceptibility, it must be high-spin, since a low-spin complex would be diamagnetic. According to the d^6 Tanabe–Sugano diagram (Appendix 5), the ground term for the high-spin case (i.e. to the left of the discontinuity) is $^5T_{2g}$. The only other quintet term is ^5E_g, so the only spin-allowed transition is $^5E_g \leftarrow {}^5T_{2g}$.

13.10 The spectrum of $[Co(NH_3)_6]^{3+}$? If this d^6 complex were high-spin, the only spin-allowed transition possible would be $^5E_g \leftarrow {}^5T_{2g}$ (refer once again to the d^6 Tanabe–Sugano diagram). On the other hand, if it were low-spin, several spin-allowed transitions are possible, including $^1T_{1g} \leftarrow {}^1A_{1g}$, $^1T_{2g} \leftarrow {}^1A_{1g}$, $^1E_g \leftarrow {}^1A_{1g}$, etc. The presence of *two* moderate-intensity bands in the visible/near-UV spectrum of $[Co(NH_3)_6]^{3+}$ suggests that it is low-spin. The first two transitions listed above correspond to these two bands. The very weak band in the red corresponds to a spin-forbidden transition such as $^3T_{2g} \leftarrow {}^1A_{1g}$.

13.11 Why is $[FeF_6]^{3-}$ colorless whereas $[CoF_6]^{3-}$ is colored? The d^5 Fe^{3+} ion in the octahedral hexafluoro complex must be high-spin. According to the d^5 Tanabe–Sugano diagram (Appendix 5), a high-spin complex has no higher energy terms of the same multiplicity as the $^6A_{1g}$ ground term. Therefore, since no spin-allowed transitions are possible, the complex is expected to be colorless (i.e. only very weak spin-forbidden transitions are possible). If this Fe(III) complex were low-spin, spin-allowed transitions such as $^2T_{1g} \leftarrow {}^2T_{2g}$, $^2A_{2g} \leftarrow {}^2T_{2g}$, etc. would render the complex colored. The d^6 Co^{3+} ion in $[CoF_6]^{3-}$ is also high-spin, but in this case a single spin-allowed transition, $^5E_g \leftarrow {}^5T_{2g}$, makes the complex colored and gives it a one-band spectrum.

13.12 Explain why the nephalauxetic effect for CN^- is larger than for NH_3? These two ligands are quite different with respect to the types of bonds they form with metal ions. Ammonia and cyanide ion are both σ-bases, but cyanide is also a π-acid. This difference means that NH_3 can form molecular orbitals only with the metal e_g orbitals, while CN^- can form molecular orbitals with the metal e_g and t_{2g} orbitals. The formation of molecular orbitals is the way that ligands "expand the clouds" of the metal d orbitals.

13.13 The origins of transitions for a complex of Co(III) with ammine and chloro ligands? Let's start with the intense band at relatively high energy with $\varepsilon_{max} = 2 \times 10^4$ M^{-1} cm^{-1}. This is undoubtedly a spin-allowed charge-transfer transition, since it is too intense to be a ligand field (d–d) transition. Furthermore, it is probably a LMCT transition, not a MLCT transition, since the ligands do not have the empty orbitals necessary for a MLCT transition. The two bands with ε_{max} = 60 and 80 M^{-1} cm^{-1} are probably spin-allowed ligand field transitions. Even though the complex is not strictly octahedral, the ligand field bands are still not very intense. The *very* weak peak with ε_{max} = 2 M^{-1} cm^{-1} is most likely a spin-forbidden ligand field transition.

13.14 Describe the transitions of Fe^{3+} impurities in bottle glass? The Fe^{3+} ions in question are d^5 metal ions. If they were low-spin, several spin-allowed ligand field transitions would give the glass a color even when viewed through the wall of the bottle (see the Tanabe–Sugano diagram for d^5 metal ions in Appendix 5). Therefore, the Fe^{3+} ions are high-spin, and as such have no spin-allowed transitions (the ground state of an octahedral high-spin d^5 metal ion is $^6A_{1g}$, and there are no sextet excited states). The faint green color, which is only observed when looking through a *long* pathlength of bottle glass, is due to spin-forbidden ligand field transitions.

13.15 The origins of transitions for $[Cr(H_2O)_6]^{3+}$ and CrO_4^{2-}? The blue-green color of the Cr^{3+} ions in $[Cr(H_2O)_6]^{3+}$ is due to spin-allowed but Laporte-forbidden ligand field transitions. The relatively low molar absorption coefficient, ε, which is a manifestation of the Laporte-forbidden nature of the transitions, is the reason that the intensity of the color is weak. The oxidation state of chromium in dichromate dianion is Cr(VI), which is d^0. Therefore, no ligand field transitions are possible. The intense yellow color is due to LMCT transitions (i.e. electron transfer from the oxide ion ligands to the Cr(VI) metal center). Charge transfer transitions are intense because they are both spin-allowed and Laporte-allowed.

13.16 The orbitals of $[CoCl(NH_3)_5]^{2+}$? The d_{z^2} orbital in this complex is left unchanged by each of the symmetry operations of the C_{4v} point group. It therefore has A_1 symmetry. The Cl atom lone pairs of electrons can form π molecular orbitals with d_{xz} and d_{yz}. These metal atomic orbitals are π-antibonding MOs in $[CoCl(NH_3)_5]^{2+}$ (they are nonbonding in $[Co(NH_3)_6]^{3+}$), and so they will be raised in energy relative to their position in $[Co(NH_3)_6]^{3+}$, in which they were degenerate with d_{xy}. Since Cl^- ion is not as strong a σ base as NH_3, the d_{z2} orbital in $[CoCl(NH_3)_5]^{2+}$ will be at lower energy than in $[Co(NH_3)_6]^{3+}$, in which it was degenerate with d_{x2-y2}. A qualitative d-orbital splitting diagram for both complexes is shown below (L = NH_3).

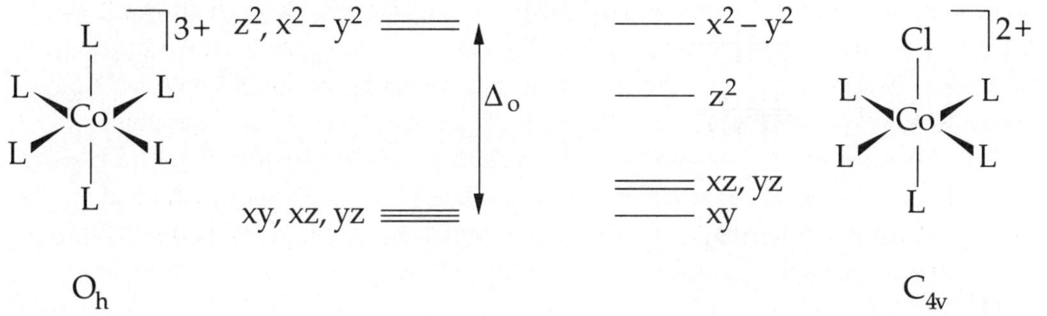

13.17 **Show that the purple color of MnO_4^- cannot arise from a ligand field transition?** As discussed in Section 13.4, *Charge-transfer bands*, ligand field transitions can occur for d-block metal ions with one or more, but fewer than ten, electrons in the metal e and t_2 orbitals. However, the oxidation state of manganese in permanganate anion is Mn(VII), which is d^0. Therefore, no ligand field transitions are possible. The metal e orbitals are the LUMOs, and can act as acceptor orbitals for LMCT transitions. These fully allowed transitions give permanganate its characteristic *intense* purple color.

13.18 **Explain how to estimate Δ_T for MnO_4^-?** You can see that two possible LMCT transitions are possible, $e \leftarrow a_1$ and $t_2 \leftarrow a_1$. These two transitions give rise to the two absorption bands observed at 18,500 cm^{-1} and 32,200 cm^{-1}. The difference in energy between the two transitions, $E(t_2) - E(e) = 13,700$ cm^{-1}, is just equal to Δ_T, as shown in the diagram below:

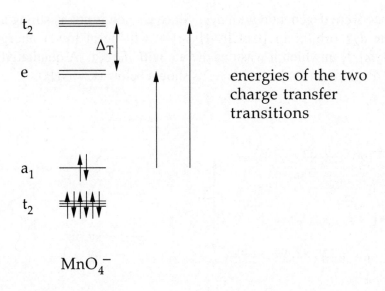

MnO$_4^-$

13.19   The intense absorption bands of Prussian blue?   Both Fe^{2+} and Fe^{3+} are in octahedral environments in this compound. Therefore, the intense transitions cannot be ligand field (*d–d*) transitions, which should be weak for centrosymmetric complexes; they arise from intervalence charge transfers.

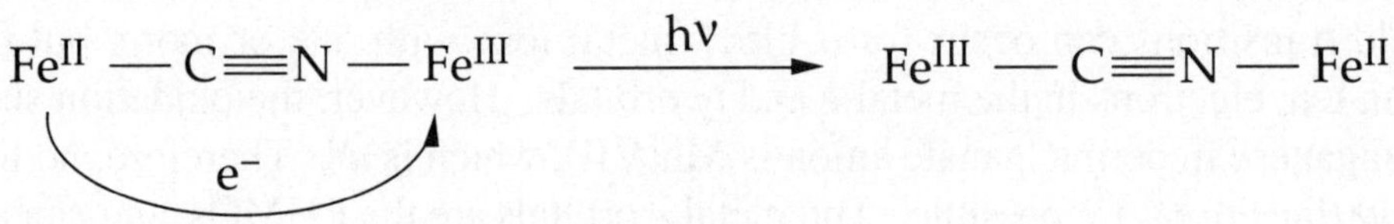

$$Fe^{II} — C\equiv N — Fe^{III} \xrightarrow{\ h\nu\ } Fe^{III} — C\equiv N — Fe^{II}$$

e^-

13.20   Ground terms for Sm^{2+}, Eu^{2+}, Yb^{2+}.   Recall that the ground term is one of the terms with the highest possible spin multiplicity for a particular electron configuration. Specifically, it is the term with the maximum L value and the maximum S value. You should first write down the electron configuration for each ion, then determine the maximum M_S value. This is because the maximum M_S value is the same as the highest possible S value. Once you have done that, you should determine the maximum M_L value for that S value.

Sm^{2+}. Samarium is eight elements past xenon, and since it has a +2 charge, all of the six remaining valence electrons are in the $4f$ subshell. Therefore, the electron configuration is $[Xe]4f^6$. A representation of the configuration is shown below.

$4f$ subshell of Sm^{2+}

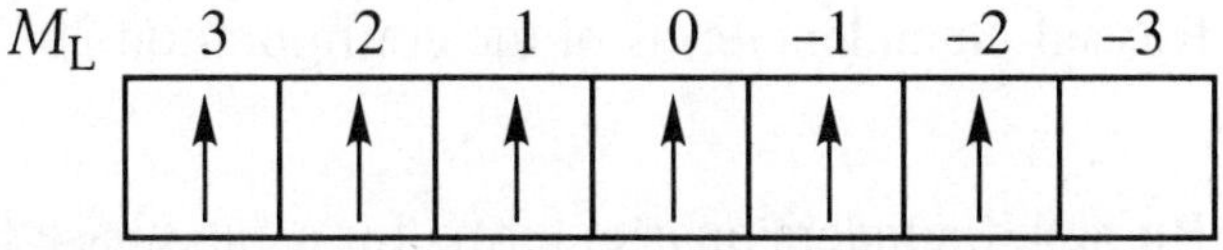

M_L	3	2	1	0	-1	-2	-3
	↑	↑	↑	↑	↑	↑	

Note that this arrangement of the six f electrons gives the maximum possible M_S value and the maximum possible M_L value for this M_S value, both of which are 3 in this case. If $S = 3$, then $2S + 1 = 7$. If $L = 3$, then the term is a F term. Therefore, the ground term of Sm^{2+} is ^7F.

Eu^{2+}. Europium is nine elements past xenon. Therefore, the electron configuration is $[Xe]4f^7$. A representation of the configuration is shown below.

$4f$ subshell of Eu^{2+}

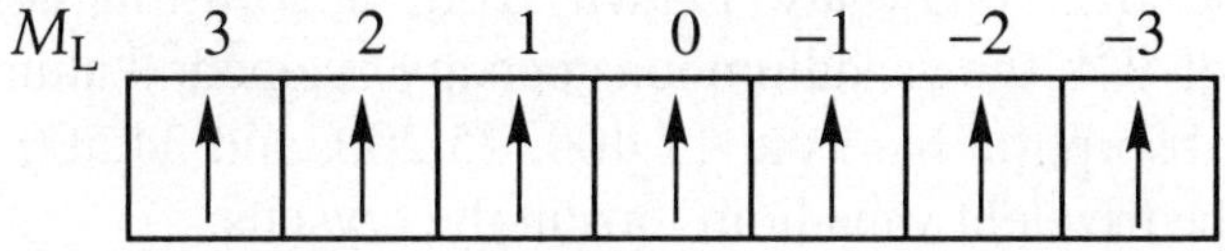

M_L	3	2	1	0	-1	-2	-3
	↑	↑	↑	↑	↑	↑	↑

This arrangement of the seven f electrons gives the maximum possible M_S and M_L values, which are 7/2 and 0, respectively. If $S = 7/2$, then $2S + 1 = 8$. If $L = 0$, then the term is a S term. Therefore, the ground term of Eu^{2+} is ^8S.

Yb^{2+}. Ytterbium is 16 elements past xenon. Therefore, the electron configuration is $[Xe]4f^{14}$. A representation of the configuration is shown below.

$4f$ subshell of Yb^{2+}

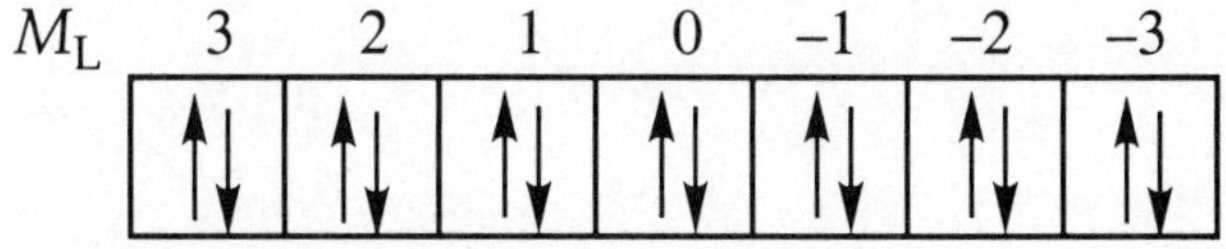

M_L	3	2	1	0	-1	-2	-3
	↑↓	↑↓	↑↓	↑↓	↑↓	↑↓	↑↓

Fourteen f electrons results in only one microstate. M_S and M_L values for this microstate are both 0, and the term is ^1S. It is not only the ground term, it is the only term of this configuration.

Guide to Solutions Quiz

1   Identify the ground terms of the gas-phase species C, Fe^{2+}, Cu^{2+}, and Br.

2   Determine the Russell–Saunders terms of the configuration $3p^1 3d^1$. Identify the ground term.

3   Determine the Russell–Saunders quartet terms for a gas-phase Cr^{3+} ion. Which term is the ground term?

4   The Δ_O values for $[Ru(H_2O)_6]^{2+}$ and $[RuCl_6]^{3-}$ are nearly the same. Is this observation consistent with the positions of H_2O and Cl^- in the spectrochemical series? If not, give an explanation.

5   Both $[Co(NH_3)_6]^{3+}$ and $[Cr(CO)_6]$ are octahedral d^6 complexes. The cobalt complex is bright orange, while the chromium complex is colorless. Explain.

6   Trivalent V^{3+} forms octahedral complexes with two ligands X and Y, $[VX_6]^{3+}$ and $[VY_6]^{3+}$. The complex with X ligands is yellow while the complex with Y ligands is blue-green. Which ligand gives a larger value of Δ_O?

7   When crystals of Al_2O_3 are grown from a solution containing a low concentration of V^{3+}, the vanadium ions occupy octahedral aluminum sites. The crystals have absorption bands at 17,400, 25,200, and 34,500 cm^{-1}. Estimate Δ_O and B for the trivalent vanadium ions in the crystals.

8   Explain why the CrO_4^{2-} ion, which is a d^0 complex, is colored. Do you expect λ_{max} for the transition to be higher or lower than for MnO_4^-?

9   What hyperfine structure should you observe in the EPR spectra of tetranuclear complexes containing one unpaired electron and (a) four equivalent rhodium ions ($I = 1/2$), and (b) four equivalent cobalt ions ($I = 7/2$)?

10 A type of tetranuclear complex of iron is $[Fe_4S_4(SR)_4]^{2-}$, the highly symmetric structure of which is shown in Figure 7.5(b). Is this a mixed-valence complex? If so, what Robin and Day classification does it belong to?

14 Reaction mechanisms of *d*-metal complexes

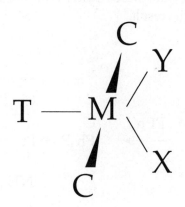

The proposed five-coordinate intermediate in substitution reactions of square-planar complexes. The incoming ligand Y adds to the four-coordinate complex MC_2TX (X is the leaving group; T is *trans* to X; the C ligands are *cis* to X). The trigonal bipyramid that forms has T, X, and Y in the equatorial plane. The electronic properties of T dramatically affect the rate of the reaction (the *trans* effect), while the electronic properties of C do not.

S14.1 **Calculate the second order rate constant for the reaction of *trans*-[PtCl(CH₃)(PEt₃)₂] with NO₂⁻ in MeOH?** We can make use of the equation

$$\log k_2(NO_2^-) = S n_{Pt}(NO_2^-) + C$$

where S is the nucleophilic discrimination factor for this complex, $n_{Pt}(NO_2^-)$ is the nucleophilicity parameter of nitrite ion, and C is the logarithm of the second order rate constant for the substitution for Cl⁻ in this complex by MeOH. S was determined to be 0.41 in the example, and $n_{Pt}(NO_2^-)$ is given as 3.22. C is a constant for a given complex and is -0.61 in this case. Therefore, $k_2(NO_2^-)$ can be determined as follows:

$$\log k_2(NO_2^-) = (0.41)(3.22) - 0.61 = 0.71$$

$$k_2(NO_2^-) = 10^{0.71} = 5.1 \text{ M}^{-1} \text{ s}^{-1}$$

S14.2 **Propose efficient routes to *cis*- and *trans*-[PtCl$_2$(NH$_3$)(PPh$_3$)]?** For the three ligands in question, Cl$^-$, NH$_3$, and PPh$_3$, the *trans* effect series is NH$_3$ < Cl$^-$ < PPh$_3$. This means that a ligand *trans* to Cl$^-$ will be substituted at a faster rate than a ligand *trans* to NH$_3$, and a ligand *trans* to PPh$_3$ will be substituted at a faster rate than a ligand *trans* to Cl$^-$. Since our starting material is [PtCl$_4$]$^{2-}$, two steps involving substitution *of* Cl$^-$ *by* NH$_3$ or PPh$_3$ must be used. If you first add NH$_3$ to [PtCl$_4$]$^{2-}$, you will produce [PtCl$_3$(NH$_3$)]$^-$. Now if you add PPh$_3$, one of the mutually *trans* Cl$^-$ ligands will be substituted faster than the Cl$^-$ ligand *trans* to NH$_3$, and the *cis* isomer will be the result.

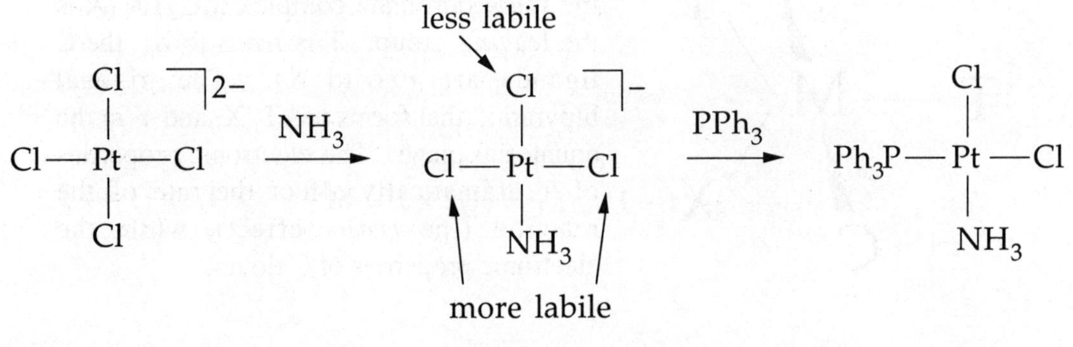

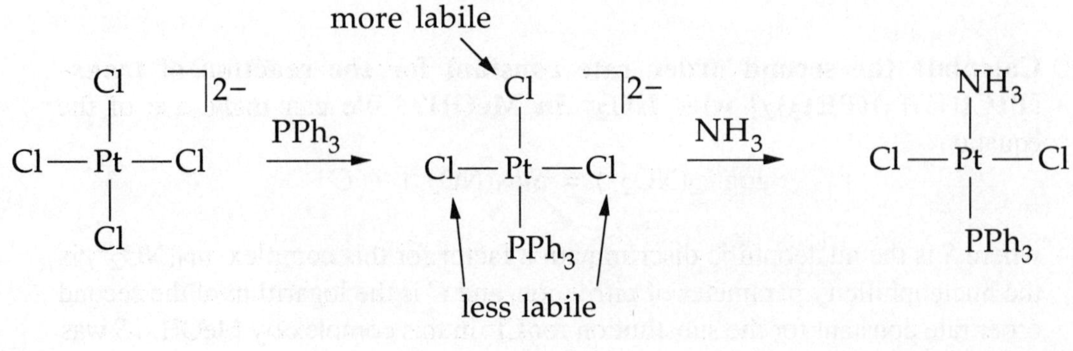

If you first add PPh$_3$ to [PtCl$_4$]$^{2-}$, you will produce [PtCl$_3$(PPh$_3$)]$^-$. Now if you add NH$_3$, the Cl$^-$ ligand _trans_ to PPh$_3$ will be substituted faster than one of the mutually _trans_ Cl$^-$ ligands, and the _trans_ isomer will be the result.

S14.3 **Calculate k for the reaction of [V(H$_2$O)$_6$]$^{2+}$ with Cl$^-$?** As discussed in Section 14.6(a), the Eigen–Wilkins mechanism for substitution in octahedral complexes is:

$$[V(H_2O)_6]^{2+} + Cl^- \underset{}{\overset{K_E}{\rightleftharpoons}} \{[V(H_2O)_6]^{2+}, Cl^-\}$$

$$\{[V(H_2O)_6]^{2+}, Cl^-\} \xrightarrow{k} [VCl(H_2O)_5]^+ + H_2O$$

The observed rate constant, k_{obs}, is given by:

$$k_{obs} = kK_E$$

and in the case of the substitution of H$_2$O by Cl$^-$ in [V(H$_2$O)$_6$]$^{2+}$ is 1.2×10^2 M^{-1} s^{-1}. You will be able to calculate k if you can estimate a proper value for K_E. Inspection of Table 14.6 shows that $K_E = 1$ M^{-1} for the encounter complex formed by F$^-$ or SCN$^-$ and [Ni(H$_2$O)$_6$]$^{2+}$. This value can be used for the reaction in question since (i) the charge and size of Cl$^-$ are similar to those of F$^-$ and SCN$^-$ and (ii) the charge and size of [Ni(H$_2$O)$_6$]$^{2+}$ are similar to those of [V(H$_2$O)$_6$]$^{2+}$. Therefore, $k = k_{obs}/K_E = (1.2 \times 10^2$ M^{-1} s$^{-1})/(1$ M$^{-1}) = 1.2 \times 10^2$ s^{-1}.

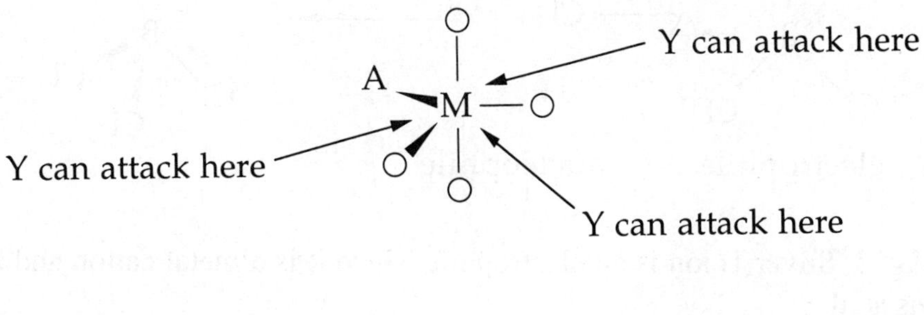

S14.4 **What is the expected _cis-trans_ ratio for [CoA(NH$_3$)$_4$(OH$_2$)]$^{2+}$?** The trigonal bipyramidal intermediate shown in the left path of Figure 14.7 is reproduced below (O = NH$_3$). The three possible positions of attack in the equaorial plane are between the A and NH$_3$ ligands (there are two such positions) and between two NH_3 ligands (there is one such position). If the entering group Y (H_2O) can randomly attack these three positions, twice as much *cis* product as *trans* product will be formed. The *cis–trans* ratio will be 2:1.

S14.5 **What is k if the overall reaction has $E° = 1.00$ V?** If the reaction has $E° = 1.00$ V, the equilibrium constant K will not be the same as in the example. A reaction potential of 1.00 V will lead to a free energy change of $-(1.602 \times 10^{-19}$ J/eV$)(6.023 \times 10^{23}$ mol$^{-1}) = -9.65 \times 10^4$ J mol^{-1}. Knowing this, K can be calculated as follows ($T = 273$ K):

$$K = \exp(-\Delta G/RT) = \exp((9.65 \times 10^4 \text{ J mol}^{-1})/(8.31 \text{ J mol}^{-1} \text{ K}^{-1})(273 \text{ K}))$$

$$K = e^{42.5} = 2.87 \times 10^{18}$$

Setting $f = 1$ gives

$$k = (9.0 \times 48 \times 2.87 \times 10^{18})^{1/2} \text{ M}^{-1} \text{ s}^{-1} = 3.5 \times 10^{10} \text{ M}^{-1} \text{ s}^{-1}$$

14.1 **Classify the following as nucleophiles or electrophiles? (a) NH_3?** Nucleophiles are intrinsically basic species (i.e. Lewis bases), while electrophiles are intrinsically acidic species (i.e. Lewis acids). A Lewis base may be either a good nucleophile or a poor nucleophile, but a base is a nucleophile. Correspondingly, a Lewis acid may be either a good electrophile or a poor electrophile, but it is rarely a nucleophile (a few ligands that are ambiphilic, like SO_2, display Lewis acidity *and* are nucleophiles). Within the framework of these definitions, NH_3 is a nucleophile, since it is a Lewis base.

(b) Cl^-? Chloride ion is a nucleophile, since it is a Lewis base (albeit a weak one). Nucleophiles attack electrophiles (electrophilic centers) such as BCl_3.

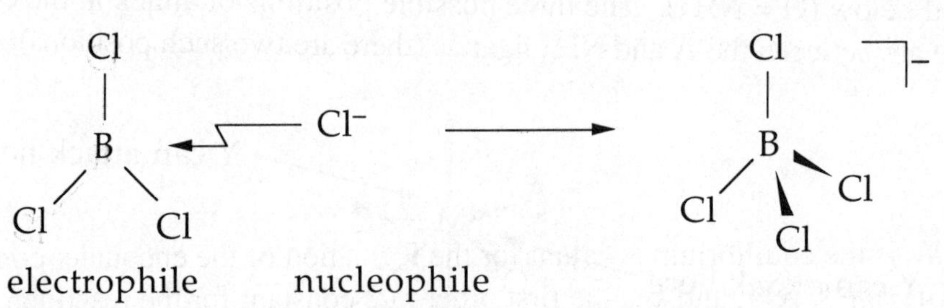

electrophile   nucleophile

(c) Ag^+? Silver(I) ion is an electrophile, since it is a metal cation and hence a Lewis acid.

(d) S^{2-}? Sulfide ion is a nucleophile, since it is a Lewis base.

(e) Al^{3+}? Aluminum(III) ion is an electrophile, since it is a metal cation and hence a Lewis acid.

14.2 **Is the reaction of $Ni(CO)_4$ with phosphanes and phosphites *d* or *a*?** Since the rate of substitution is the same for a variety of different entering ligands L, the activated complex in each case must not include any significant bond making to the entering ligand and the reaction must be *d*. If the rate-determining step included any Ni–L bond making, the rate of substitution would change as the electronic and steric properties of L were changed.

14.3 **Which experiment reveals the stoichiometric mechanism and which reveals the intimate mechanism? (a) $^{36}Cl^-$?** If the *trans* to *cis* isomerization of $[CoCl_2(en)_2]^+$ results in the incorporation of $^{36}Cl^-$ in the complex, then the isomerization reaction proceeded by a mechanism which includes the dissociation of Cl^- from the octahedral complex. This is information that bears on the stoichiometric mechanism, i.e. the sequence of elementary steps by which the isomerization reaction takes place (see Section 14.2).

(b) D replaces H? On the other hand, if you find that the replacement of H by D in the ammine ligands of $[Cr(NCS)_4(NH_3)_2]^-$ reduces the rate of substitution of NCS^- by H_2O, you have learned that N–H bond breaking is involved in the rate-determining step, i.e. in the formation of the activated complex. This is a detail about the energetics of formation of the activated complex, so it bears on the intimate mechanism of substitution (base catalyzed hydrolysis, in this case).

14.4 **How would you determine if the formation of $[MnX(OH_2)_5]^+$ is *d* or *a*?** The rate law for this substitution reaction is:

$$\text{rate} = (kK_E[Mn(OH_2)_6^{2+}][X^-])/(1 + K_E[X^-])$$

where K_E is the equilibrium constant for the formation of the encounter complex $\{[Mn(OH_2)_6]^{2+}, X^-\}$, and k is the first order rate constant for the reaction:

$$\{[Mn(OH_2)_6^{2+}], X^-\} \longrightarrow [Mn(OH_2)_5X]^+ + H_2O$$

The rate law will be the same regardless of whether the transformation of the encounter complex into products is dissociatively or associatively activated. However, you can distinguish d from a by varying the nature of X^-. If k varies as X^- varies, then the reaction is a. If k is relatively constant as X^- varies, then the reaction is d. Note that k cannot be measured directly. It can be found using the expression $k_{obs} = kK_E$ and an estimate of K_E, as described in Section 14.6.

14.5   The nonlability of octahedral complexes of metals with high oxidation numbers or metals of periods 5 and 6? If ligand substitution takes place by a d mechanism, the strength of the metal-leaving group bond is directly related to the substitution rate. Metal centers with high oxidation numbers will have stronger bonds to ligands than metal centers with low oxidation numbers. Furthermore, period 5 and 6 d-block metals have stronger metal ligand bonds (see Section 7.4). Therefore, for reactions that are dissociatively activated, complexes of period 5 and 6 metals are less labile than complexes of period 3 metals, and complexes of metals in high oxidation states are less labile than complexes of metals in low oxidation states (all other things remaining equal).

14.6   Explain in terms of associative activation the fact that a Pt(II) complex of tetraethyldiethylenetriamine is attacked by Cl$^-$ 10^5 times less rapidly than the diethylenetriamine analog? The two complexes are shown below. The ethyl substituted complex presents a greater degree of steric hindrance to an incoming Cl$^-$ ion nucleophile. Since the rate-determining step for associative substitution of X^- by Cl$^-$ is the formation of a Pt–Cl bond, the more hindered complex will react more slowly.

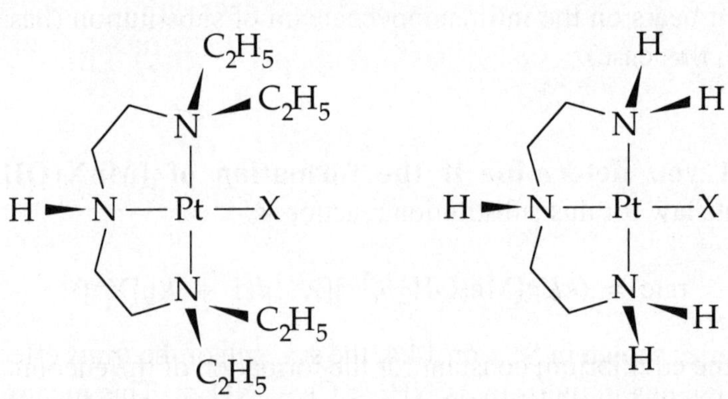

14.7   Substitution of PhCl in [W(CO)$_4$L(PhCl)]?   Since the rate of loss of chlorobenzene, PhCl, from the tungsten complex becomes faster as the cone angle of L increases, this is a case of dissociative activation (i.e., steric crowding in the transition state accelerates the rate).   This is not at odds with the observation that the rate is proportional to the concentration of the entering phosphane at low phosphane concentrations, since K_E for the equilibrium producing the encounter complex, C + Y $\rightleftharpoons$ {CY}, is inversely proportional to [Y] (C is the tungsten complex, Y is the entering phosphane; see Section 14.6, *Rate laws and their interpretation*).   The overall rate is given by:

$$\text{rate} = (kK_E[C][Y])/(1 + K_E[Y])$$

and, in the limit of low [Y], this becomes rate $\sim kK_E[C][Y]$.

14.8   Why substitution reactions of [Ni(CN)$_4$]$^{2-}$ are very fast?   The fact that the five-coordinate complex [Ni(CN)$_5$]$^{3-}$ can be detected does indeed explain why substitution reactions of the four-coordinate complex [Ni(CN)$_4$]$^{2-}$ are fast.   The reason is that, for a detectable amount of [Ni(CN)$_5$]$^{3-}$ to build up in solution, the forward rate constant k_f must be numerically close to or greater than the reverse rate constant k_r:

$$[\text{Ni(CN)}_4]^{2-} + \text{CN}^- \underset{k_f}{\overset{k_r}{\rightleftharpoons}} [\text{Ni(CN)}_5]^{3-}$$

If k_f were much smaller than k_r, the equilibrium constant $K = k_f/k_r$ would be small and the concentration of [Ni(CN)$_5$]$^{3-}$ would be too small to detect. Therefore, since k_f is relatively large, you can infer that rate constants for the association of other nucleophiles are also large, with the result that substitution reactions of [Ni(CN)$_4$]$^{2-}$ are very fast.

14.9   A two-step synthesis for *cis*- and *trans*-[PtCl$_2$(NO$_2$)(NH$_3$)]$^-$?   Starting with [PtCl$_4$]$^{2-}$, you need to perform two separate ligand substitution reactions.   In one, NH$_3$ will replace Cl$^-$ ion.   In the other, NO$_2^-$ ion will replace Cl$^-$ ion.   The question is, which substitution to perform first?   According to the *trans* effect series shown in Section 14.4, the strength of the *trans* effect on Pt(II) for the three ligands in question is NH$_3$ < Cl$^-$ < NO$_2^-$.   This means that a Cl$^-$ ion *trans* to another Cl$^-$ will be substituted faster than a Cl$^-$ ion *trans* to NH$_3$, while a Cl$^-$ ion *trans* to NO$_2^-$ will be substituted faster than a Cl$^-$ ion *trans* to another Cl$^-$ ion.   If you first add NH$_3$ to [PtCl$_4$]$^{2-}$, you will produce

[PtCl$_3$(NH$_3$)]$^-$. Now if you add NO$_2$$^-$, one of the mutually *trans* Cl$^-$ ligands will be substituted faster than the Cl$^-$ ligand *trans* to NH$_3$, and the *cis* isomer will be the result. If you first add NO$_2$$^-$ to [PtCl$_4$]$^{2-}$, you will produce [PtCl$_3$(NO$_2$)]$^{2-}$. Now if you add NH$_3$, the Cl$^-$ ligand *trans* to NO$_2$$^-$ will be substituted faster than one of the mutually *trans* Cl$^-$ ligands, and the *trans* isomer will be the result. These two-step syntheses are shown below:

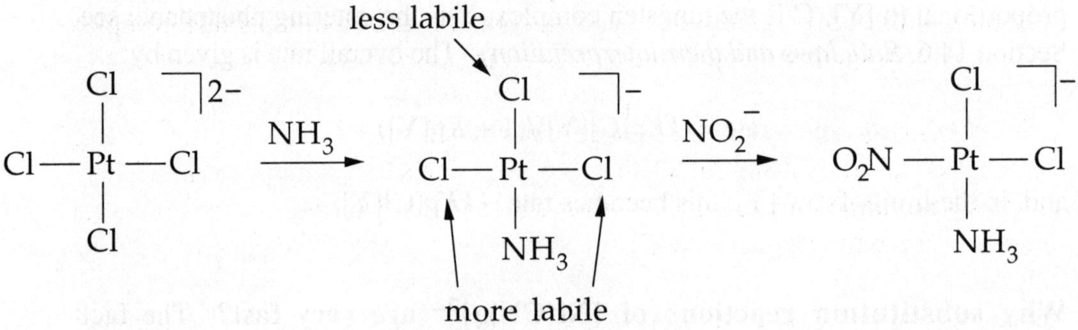

less labile

more labile

cis isomer

more labile

less labile

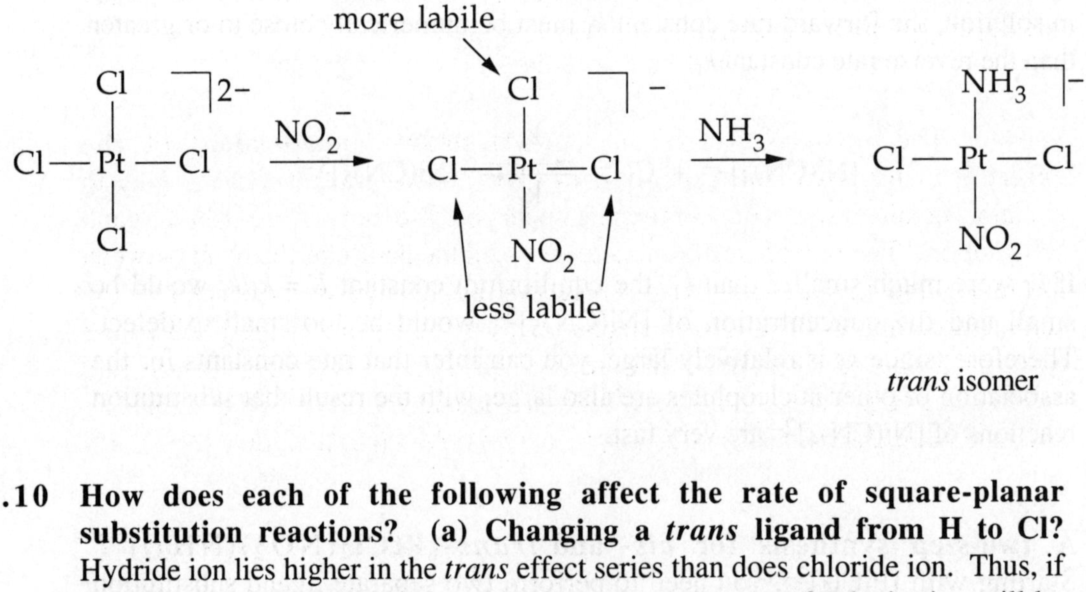

trans isomer

14.10 How does each of the following affect the rate of square-planar substitution reactions? (a) Changing a *trans* ligand from H to Cl? Hydride ion lies higher in the *trans* effect series than does chloride ion. Thus, if the ligand *trans* to H or Cl is the leaving group, its rate of substitution will be decreased if H is changed to Cl. The change in rate can be as large as a factor of 10^4 (see Table 14.4).

(b) Changing the leaving group from Cl$^-$ to I$^-$? The rate at which a ligand is substituted in a square-planar complex is related to its position in the *trans* effect series. If a ligand is high in the series, it is a *good* entering group (i.e. a good nucleophile) and a *poor* leaving group. Since iodide ion is higher in the *trans* effect series than chloride ion, it is a poorer leaving group than chloride ion. Therefore, the iodo complex will undergo I$^-$ substitution more slowly than the chloro complex will undergo Cl$^-$ substitution.

(c) Adding a bulky substituent to a *cis* ligand? This change will hinder the approach of the entering group and will slow the formation of the five-coordinate activated complex. The rate of substitution of a square-planar complex with bulky ligands will be slower than that of a comparable complex with sterically smaller ligands (see also Exercise 14.6).

(d) Increasing the positive charge on the complex? If all other things are kept equal, increasing the positive charge on a square-planar complex will increase the rate of substitution. This is because the entering ligands are either anions or have the negative ends of their dipoles pointing at the metal ion. As explained in Section 14.2, if the charge density of the complex decreases in the activated complex, as would happen when an anionic ligand adds to a cationic complex, the solvent molecules will be *less* ordered around the complex (the opposite of the process called electrostriction). The increased disorder of the solvent makes $\Delta^{\ddagger}S$ less negative (compare the values of $\Delta^{\ddagger}S$ for the Pt(II) and Au(III) complexes in Table 14.5).

14.11   Why is the rate of attack on Co(III) nearly independent of the entering group with the exception of OH$^-$? The general trend is easy to explain: octahedral Co(III) complexes undergo dissociatively activated ligand substitution. The rate of substitution depends on the nature of the bond between the metal and the leaving group, since this bond is partially broken in the activated complex. The rate is independent of the nature of the bond to the entering group, since this bond is formed in a step subsequent to the rate determining step. The anomalously high rate of substitution by OH$^-$ signals an alternate path, that of base hydrolysis, as shown below (see Section 14.9):

$$
\begin{bmatrix} & L & \\ L & | & L \\ & Co & \\ L & | & EH_n \\ & L & \end{bmatrix}^{m+} + \ OH^- \ \rightleftharpoons \ \begin{bmatrix} & L & \\ L & | & L \\ & Co & \\ L & | & EH_{n-1} \\ & L & \end{bmatrix}^{(m-1)+} + \ H_2O
$$

The deprotonated complex $[CoL_5(EH_{n-1})]^{(m-1)+}$ will undergo loss of L faster than the starting complex $[CoL_5(EH_n)]^{m+}$, because the anionic EH_{n-1}^- ligand is a stronger base than the neutral EH_n ligand and can better stabilize the coordinatively unsaturated activated complex. The implication is that a complex without protic ligands will not undergo anomalously fast OH^- ion substitution.

14.12 **Predict the products of the following reactions:** (a) $[Pt(PR_3)_4]^{2+}$ + **$2Cl^-$?** The first Cl^- ion substitution will produce the reaction intermediate $[Pt(PR_3)_3Cl]^+$. This will be attacked by the second equivalent of Cl^- ion. Since phosphanes are higher in the *trans* effect series than chloride ion, a phosphane *trans* to another phosphane will be substituted, giving *cis*-$[PtCl_2(PR_3)_2]$.

(b) $[PtCl_4]^{2-}$ + $2PR_3$? The first PR_3 substitution will produce the reaction intermediate $[PtCl_3(PR_3)]^-$. This will be attacked by the second equivalent of PR_3. Since phosphanes are higher in the *trans* effect series than chloride ion, the Cl^- ion *trans* to PR_3 will be substituted, giving *trans*-$[PtCl_2(PR_3)_2]$.

(c) *cis*-$[Pt(NH_3)_2(py)_2]^{2+}$ + $2Cl^-$? You should assume that NH_3 and pyridine are about equal in the *trans* effect series. The first Cl^- ion substitution will produce one of two reaction intermediates, *cis*-$[PtCl(NH_3)(py)_2]^+$ or *cis*-$[PtCl(NH_3)_2(py)]^+$. Either of these will produce the same product when they are attacked by the second equivalent of Cl^- ion. This is because Cl^- is higher in the *trans* effect series than NH_3 or pyridine. Therefore, the ligand *trans* to the first Cl^- ligand will be substituted faster, and the product will be *trans*-$[PtCl_2(NH_3)(py)]$.

14.13 **Put the following in order of rate of substitution by H_2O:** $[Co(NH_3)_6]^{3+}$, $[Rh(NH_3)_6]^{3+}$, $[Ir(NH_3)_6]^{3+}$, $[Mn(OH_2)_6]^{2+}$, $[Ni(OH_2)_6]^{2+}$? The three ammine complexes will undergo substitution more slowly than the two aqua complexes. This is because of their higher charge and their low-spin d^6 configurations. You should refer to Table 14.8, which lists ligand-field activation energies (LFAE) for various d^n configurations. While low-spin d^6 is not included, note the similarity between d^3 (t_{2g}^3) and low-spin d^6 (t_{2g}^6). A low-spin octahedral d^6 complex has LFAE = $0.4\Delta_o$, and consequently is inert to substitution. Of the three ammine complexes listed above, the iridium complex is the most inert, followed by the rhodium complex, which is more inert than the cobalt complex. This is because Δ_o increases on descending a group in the d-block. Of the two aqua complexes, the Ni complex, with LFAE = $0.2\Delta_o$, undergoes substitution more slowly than the Mn complex, with LFAE = 0. Thus, the order of increasing rate is $[Ir(NH_3)_6]^{3+}$ < $[Rh(NH_3)_6]^{3+}$ < $[Co(NH_3)_6]^{3+}$ < $[Ni(OH_2)_6]^{2+}$ < $[Mn(OH_2)_6]^{2+}$.

14.14 **What is the effect on the rate of dissociatively activated reactions of Rh(III) complexes of each of the following? (a) An increase in the positive charge on the complex?** Since the leaving group (X) is invariably negatively charged or the negative end of a dipole, increasing the positive charge on the complex will retard the rate of M–X bond cleavage. For a dissociatively activated reaction, this change will result in a decreased rate.

(b) Changing the leaving group from NO_3^- to Cl^-? Try to answer this one after inspecting Figure 14.4. For the substitution reaction shown, changing the leaving group from nitrate ion to chloride ion results in a decreased rate. The explanation offered in Section 14.7 is that the Co–Cl bond is stronger than the Co–ONO_2 bond. For a dissociatively activated reaction, a stronger bond to the leaving group will result in a decreased rate.

(c) Changing the entering group from Cl^- to I^-? This change will have little or no effect on the rate. For a dissociatively activated reaction, the bond between the entering group and the metal is formed subsequent to the rate-determining step.

(d) Changing the *cis* ligands from NH_3 to H_2O? These two ligands differ in their σ-basicity. The less basic ligand, H_2O, will decrease the electron density at the metal and will destabilize the coordinatively unsaturated activated complex. Therefore, this change, from NH_3 to the less basic ligand H_2O, will result in a decreased rate.

(e) Changing an ethylenediamine ligand to propylenediamine when the leaving group is Cl^-? Ethylenediamine forms five-membered chelate rings with metal ions while propylenediamine forms six-membered chelate rings. The greater flexibility of the latter will lead to a more stabilized activated complex. Therefore, this change will lead to an increased rate.

14.15 **The mechanism of CO insertion in $RMn(CO)_5$?** This reaction mechanism has a lot in common with the mechanism for a dissociative *D* mechanism of ligand substitution, which is discussed in Section 14.7. For example, you can write the proposed mechanism as follows:

$$RMn(CO)_5 \underset{k_{a'}}{\overset{k_a}{\rightleftharpoons}} (RCO)Mn(CO)_4$$

$$(RCO)Mn(CO)_4 + L \xrightarrow{k_b} (RCO)MnL(CO)_4$$

with the rate given by:

$$\text{rate} = (k_a k_b [RMn(CO)_5][L])/(k_{a'} + k_b[L])$$

When [L] is very high, the denominator becomes approximately equal to $k_b[L]$ and the rate expression simplifies to:

$$\text{rate} = k_a[RMn(CO)_5]$$

Therefore, when [L] is very high, the first-order rate constant k_a can be determined from rate vs. $[RMn(CO)_5]$ data.

14.16 Pressure dependence of PhCl substitution in [W(CO)$_4$(PPh$_3$)(PhCl)]? Since the volume of activation is positive (+11.3 cm^3 mol^{-1}), the activated complex takes up more volume in solution than the reactants, as shown in the drawing below. Therefore, the mechanism of substitution must be dissociative. See Section 14.7(d), *Associative activation*.

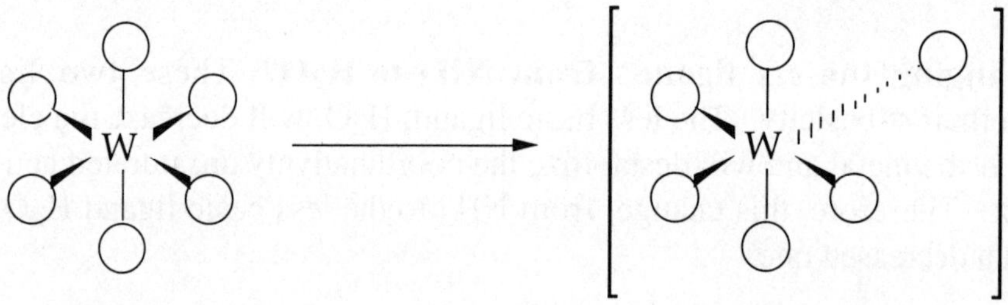

reactant   activated complex

volume = V   volume = V' > V

14.17 What data might distinguish between an inner- and outer-sphere pathway for reduction of [Co(N$_3$)(NH$_3$)$_5$]$^{2+}$ with [V(OH$_2$)$_6$]$^{2+}$? The inner-sphere pathway is shown below (solvent = H$_2$O):

$$[Co(N_3)(NH_3)_5]^{2+} + [V(OH_2)_6]^{2+} \rightarrow \{[Co(N_3)(NH_3)_5]^{2+},[V(OH_2)_6]^{2+}\}$$

$$\{[Co(N_3)(NH_3)_5]^{2+},[V(OH_2)_6]^{2+}\} \rightarrow \{[Co(N_3)(NH_3)_5]^{2+},[V(OH_2)_5]^{2+},H_2O\}$$

$$\{[Co(N_3)(NH_3)_5]^{2+},[V(OH_2)_5]^{2+},H_2O\} \rightarrow [(NH_3)_5Co-N=N=N-V(OH_2)_5]^{4+}$$
$$\text{Co(III)} \qquad \text{V(II)}$$

$$[(NH_3)_5Co-N=N=N-V(OH_2)_5]^{4+} \rightarrow [(NH_3)_5Co-N=N=N-V(OH_2)_5]^{4+}$$
$$\quad\text{Co(III)} \qquad \text{V(II)} \qquad\qquad\qquad \text{Co(II)} \qquad \text{V(III)}$$

$$[(NH_3)_5Co-N=N=N-V(OH_2)_5]^{4+} \rightarrow \rightarrow [Co(OH_2)_6]^{2+} + [V(N_3)(OH_2)_5]^{2+}$$
$$\quad\text{Co(II)} \qquad \text{V(III)}$$

The pathway for outer-sphere electron transfer is shown below:

$$[Co(N_3)(NH_3)_5]^{2+} + [V(OH_2)_6]^{2+} \rightarrow \{[Co(N_3)(NH_3)_5]^{2+}\,[V(OH_2)_6]^{2+}\}$$

$$\{[Co(N_3)(NH_3)_5]^{2+},[V(OH_2)_6]^{2+}\} \rightarrow \{[Co(N_3)(NH_3)_5]^{+},[V(OH_2)_6]^{3+}\}$$

$$\{[Co(N_3)(NH_3)_5]^{+},[V(OH_2)_6]^{3+}\} \rightarrow [Co(OH_2)_6]^{2+} + [V(OH_2)_6]^{3+}$$

In both cases, the cobalt-containing product is the aqua complex, since H_2O is present in abundance and high-spin d^7 complexes of Co(II) are substitution labile. However, something that distinguishes the two pathways is the composition of the vanadium-containing product. If $[V(N_3)(OH_2)_5]^{2+}$ is the product, then the reaction has proceeded *via* an inner-sphere pathway. If $[V(OH_2)_6]^{3+}$ is the product, then the electron transfer reaction is outer-sphere.

14.18 The presence of detectable intermediates in electron transfer reactions? As in Exercise 14.17, the direct transfer of a ligand from the coordination sphere of one redox partner (in this case the oxidizing agent, $[Co(NCS)(NH_3)_5]^{2+}$) to the coordination sphere of the other (in this case the reducing agent, $[Fe(OH_2)_6]^{2+}$), signals an inner-sphere electron transfer reaction. Even if the first formed product $[Fe(NCS)(OH_2)_5]^{2+}$ is short lived and undergoes hydrolysis to $[Fe(OH_2)_6]^{3+}$, its fleeting existence demands that the electron was transfered across a Co–(NCS)–Fe bridge.

14.19 What product and quantum yield do you predict for substitution of $[W(CO)_5(py)]$ in the presence of excess triethylamine? Since the intermediate is believed to be $[W(CO)_5]$, the properties of the entering group (triethylamine vs. triphenylphosphane) should not affect the quantum yield of the reaction, which is a measure of the rate of formation of $[W(CO)_5]$ from the excited state of $[W(CO)_5(py)]$. The product of the photolysis of $[W(CO)_5(py)]$

in the presence of excess triethylamine will be [W(CO)$_5$(NEt$_3$)], and the quantum yield will be 0.4. This photosubstitution is initiated from a ligand-field excited state, not a MLCT excited state. A metal–ligand charge transfer increases the oxidation state of the metal, which would strengthen, not weaken, the bond between the metal and a σ-base like pyridine.

14.20   **Propose a wavelength for transient photoreduction of [CrCl(NH$_3$)$_5$]$^{2+}$?** The intense band at ~250 nm is a LMCT transition (specifically a Cl$^-$-to-Cr^{3+} charge transfer). Irradiation at this wavelength should produce a population of [CrCl(NH$_3$)$_5$]$^{2+}$ ions that contain Cr^{2+} ions and Cl atoms instead of Cr^{3+} ions and Cl$^-$ ions. Irradiation of the complex at wavelengths between 350 and 600 nm will not lead to photoreduction. The bands that are observed between these two wavelengths are ligand-field transitions: the electrons on the metal are rearranged, and the electrons on the Cl$^-$ ion are not involved.

14.21   **The reaction of [Pt(Ph)$_2$(SMe$_2$)$_2$] with phenanthroline?** Since the entropy of activation is a relatively large positive number, you should conclude that the reaction mechanism is dissociative. Phenanthroline (phen) may be too sterically crowded to approach the four-coordinate Pt(II) complex from the top or bottom. Instead, after dissociation of one of the sulfane ligands (SMe$_2$ = dimethylsulfane or dimethylsulfide), phen can attack the three-coordinate intermediate in a fast step. Then, if the plane of the phen ligand is perpendicular to the plane of the Pt(II) complex, the second nitrogen atom of the phen ligand can attack the platinum atom from above in a fast step, forming a five-coordinate intermediate that is typical of square-planar substitution. The five-coordinate [Pt(Ph)$_2$(SMe$_2$)(phen)] complex would then expel the sulfane ligand to form the four-coordinate product, [Pt(Ph)$_2$(phen)]. This proposed mechanism is shown below.

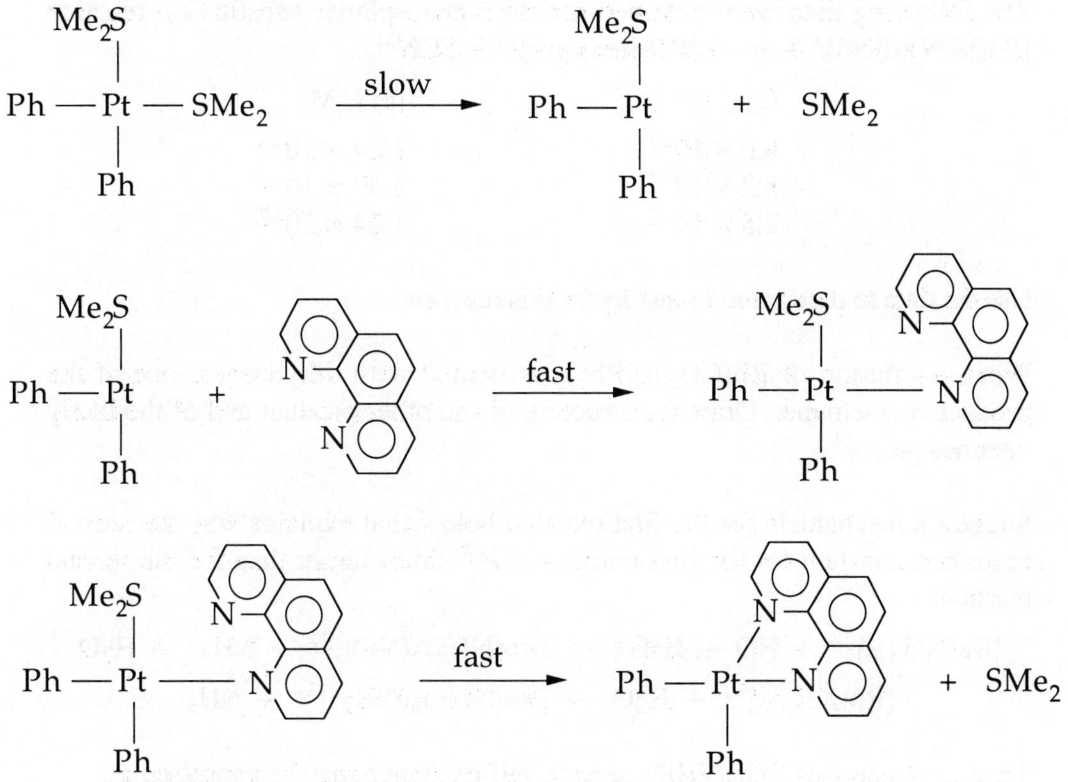

Guide to Solutions Quiz

1   Arrange the following complexes in order of increasing lability: $[Fe(OH_2)_6]^{3+}$, $[Cr(OH_2)_6]^{3+}$, $[Y(OH_2)_6]^{3+}$, $[Al(OH_2)_6]^{3+}$, $[Sr(OH_2)_6]^{2+}$, $[K(OH_2)_6]^+$.

2   Explain why the following reaction is catalyzed by traces of Co^{2+}:

$$[Co(NH_3)_6]^{3+} + 6CN^- \rightarrow [Co(CN)_6]^{3-} + 6NH_3$$

3   Explain why the rate of H_2O exchange in aqueous solution is 10^8 greater for $Ru^{2+}(aq)$ than for $Fe^{2+}(aq)$.

4   Suggest synthetic schemes for the synthesis of the three isomers of $PtBr(NO_2)(NH_3)(py)$.

5   The following data were obtained for the square-planar substitution reaction
         $[Pd(SCN)(dien)]^+ + py \rightarrow [Pd(dien)(py)]^{2+} + SCN^-$:

k_{obs}, s^{-1}	[py], M
6.6×10^{-3}	1.24×10^{-3}
8.2×10^{-3}	2.48×10^{-3}
2.5×10^{-2}	1.24×10^{-2}

         Use the data to determine k_1 and k_2 for this reaction.

6   When a solution of $Rh(CH_3)(PPh_3)_3$ is treated with dihydrogen, one of the
         products is methane. Draw the structure of the other product and of the likely
         intermediate.

7   Suggest a mechanism for the first reaction below that explains why the second
         order rate constant for the first reaction is 10^9 times larger than for the second
         reaction:

         $$[Ru(NH_3)_6]^{3+} + NO + H_3O^+ \rightarrow [Ru(NH_3)_5(NO)]^{3+} + NH_4^+ + H_2O$$

         $$[Ru(NH_3)_6]^{3+} + H_2O \rightarrow [Ru(NH_3)_5(OH_2)]^{3+} + NH_3$$

8   The rate of aqueous V(II)/V(III) electron self-exchange may be expressed as:

         $$rate = k[V(II)][V(III)]$$

         where the rate constant k shows the following dependence on $[H_3O^+]$ (a and b
         are constants):

         $$k = a + b/[H_3O^+]$$

         Explain this observation in terms of two parallel paths.

9   The E° value for the $Fe(CN)_6]^{3-/4-}$ couple is 0.68 V and the self-exchange
         electron transfer rate constant is 7.4×10^2 M^{-1} s^{-1}. The E° value for the
         $[W(CN)_8]^{3-/4-}$ couple is 0.54 V and the self-exchange electron transfer rate
         constant is 7.0×10^4 M^{-1} s^{-1}. Calculate the rate constant for the oxidation of
         $[W(CN)_8]^{4-}$ by $[Fe(CN)_6]^{3-}$. Compare your result with the observed value of
         4.3×10^4 M^{-1} s^{-1}. Is an outer-sphere mechanism a reasonable one for this
         redox reaction?

10   Predict the products of the following photochemical reactions:

         $$W_2(\eta^5\text{-}C_5H_5)_2(CO)_6 + PPh_3 + h\nu \rightarrow$$

         $$W_2(\eta^5\text{-}C_5H_5)_2(CO)_6 + Re_2(CO)_{10} + h\nu \rightarrow$$

15 Main-group organometallic compounds

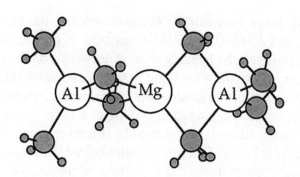

A drawing of $Mg[Al(CH_3)_4]_2$, an organometallic compound containing magnesium, aluminum, and methyl groups. The compound is essentially covalent and is soluble in nonpolar solvents such as cyclopentane and toluene.

S15.1 Classifying and predicting M–C bond forming reactions? In this case, you are given an electropositive metal, magnesium, and the organometallic compound of a less electropositive metal, dimethylmercury. A transmetallation reaction will take place:

$$Mg(s) \;+\; Hg(CH_3)_2(l) \;\rightarrow\; Mg(CH_3)_2(s) \;+\; Hg(l)$$

S15.2 Predicting the products of thermal decomposition? The Pb–C bonds in $Pb(CH_3)_4$ are weak and readily undergo homolysis in much the same way as $Bi(CH_3)_3$, discussed in the example. β-Hydrogen elimination is not possible with methyl groups, since there is no β carbon atom. Therefore, the probable mode of thermal decomposition of tetramethyllead is:

$$Pb(CH_3)_4(l) + heat \rightarrow Pb(s) + 2C_2H_6(g)$$

S15.3 **Propose a structure for $Al_2(i\text{-}Bu)_4Cl_2$?** Since the tendency towards bridge structures is $PR_2^- > X^- > H^- > Ph^- > R^-$, the chloride ligands, and not the isobutyl ligands, will bridge the two aluminum atoms.

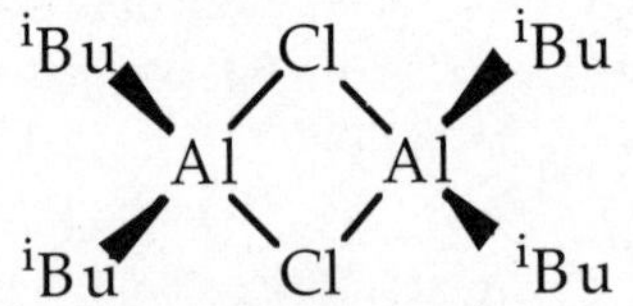

The structure of $Al_2(i\text{-}Bu)_4Cl_2$

S15.4 **Which has the lower bending force constant?** When comparing the two linkages in question, Si–O–Si and C–O–C, your first reaction might be to compare their bond enthalpies. However, the bending motion does not stretch or compress the bond, so to a first approximation bond strength is irrelevant. However, the type of bonding between the three atoms is very relevant. In the case of C–O–C, there are two typical carbon–oxygen σ (single) bonds. In the case of Si–O–Si, on the other hand, there is considerable silicon–oxygen π bonding, but only when the linkage is linear. Therefore, bending, which causes the linkages to become nonlinear, causes less of a reduction in the carbon–oxygen bond order in a C–O–C linkage and more of a reduction in the silicon–oxygen bond order in a Si–O–Si linkage. Hence, the ether Et_2O has a lower bending force constant than the siloxane $(H_3Si)_2O$.

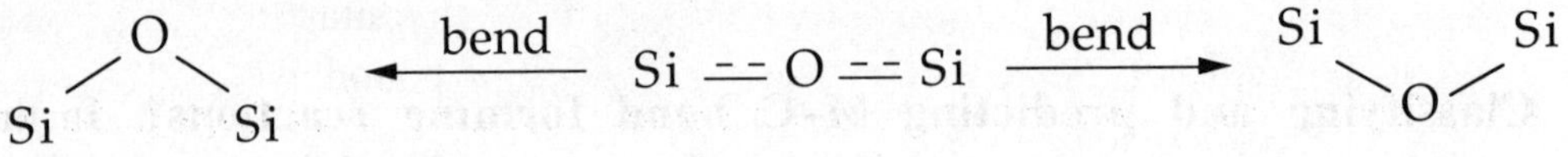

S15.5 **Strategies for the synthesis of aluminum and silicon alkoxides?** Since aluminum–carbon bonds are so polar, they will react directly with alcohols to form alkoxides. Silicon–carbon bonds will not react directly with alcohols. Instead, silicon–halogen bonds will directly react with alcohols.

$$Al_2Me_6 + 6ROH \rightarrow Al_2(OR)_6 + 6CH_4$$

$$SiMe_4 + ROH \rightarrow NR$$

$$SiCl_4 + 4ROH \rightarrow Si(OR)_4 + 4HCl$$

S15.6 **Compare the stable hydrogen compounds of Ge and As?** The trends exhibited by the stable hydrogen compounds of Ge and As follow the trends for their alkyl compounds, which were discussed in the Example. The compound GeH_4 is the only stable hydrogen compound of Ge; GeH_2 is unknown. This parallels the observation that only a few organogermanium(II) compounds are known. In contrast, AsH_3 is the only stable hydrogen compound of As; AsH_5 is unknown. This parallels the fact that $As(CH_3)_5$ is unstable (one of the only pentavalent organoarsenic compounds is $AsPh_5$, which is mentioned in the text). The fact that the trends for hydrogen compounds and alkyl compounds are so similar should make sense to you, since hydrogen and carbon have similar electronegativities (2.20 vs. 2.55, respectively) and the strengths and polarities of element–hydrogen and element–carbon bonds are also quite similar.

15.1 **Organometallic or not organometallic?** **(a) $B(CH_3)_3$?** According to the definition of the authors, a compound is organometallic if it contains at least one carbon–metal bond, and the suffix "metallic" includes the metalloids B, Si, and As. Therefore, trimethylboron is an organometallic compound.

(b) $B(OCH_3)_3$? This compound also contains a boron atom and methyl groups, but it does not have any C–B bonds (only O–B, C–O, and C–H bonds). Therefore, trimethoxyboron (or trimethylborate) is not an organometallic compound.

(c) $(NaCH_3)_4$? The structure of this compound is similar to methyllithium (see Structure **2**). Since it contains C–Na bonds, it is an organometallic compound.

(d) $SiCl_3(CH_3)$? This compound has three Si–Cl bonds and one C–Si bond around the tetrahedral central silicon atom. Since it contains at least one C–Si bond, methyltrichlorosilane is an organometallic compound.

(e) $N(CH_3)_3$? This compound, trimethylamine, does not even contain a metal or metalloid atom, so it cannot be an organometallic compound.

(f) Sodium acetate? This salt does not contain any C–Na bonds. The closest atoms to the Na^+ ions in the lattice would be the carboxylate oxygen atoms. It is not an organometallic compound.

(g) $Na[B(C_6H_5)_4]$? This salt also does not contain any C–Na bonds, but the $B(C_6H_5)_4^-$ anion contains four C–B bonds. Therefore, sodium tetraphenylborate is an organometallic compound. Specifically, it is the sodium salt of an organometallic anion.

15.2 *s- and p-*block and Group 12 organometallics? If you cannot do all of this exercise without consulting reference material, try to do as much as you can. Start by drawing a piece of the periodic table with the relevant elements drawn in (all of the elements from Groups 1, 2, 12, and 13/III, from Si on down in group 14/IV, and from As on down in Group 15/V). Then write out a formula for a representative methyl compound for each group, taking into account the normal oxidation state for elements in that group (+1 for Group 1, +2 for Groups 2 and 12, +3 for Groups 13/III and 15/V, and +4 for Group 14/IV). Then remember that saline methyl compounds, just like saline hydrides, are formed by the most electropositive elements, i.e. those of Groups 1 and 2. The saline methyl compounds are electron deficient (that is why their structures exhibit bridging methyl groups), as are the methyl compounds of Groups 12 and 13/III. Those of group 14/IV are electron precise, i.e. the central atom has an octet with no lone pairs. The organometallic methyl compounds of Group 15/V are electron rich. The central atom in $As(CH_3)_3$ and in similar compounds has an octet, but one of the pairs of electrons is a lone pair. A periodic table showing all of this information is shown below. The final part of this exercise is to recognize trends in $\Delta_f H°$ in the *p-*block. Just as for the *p-*block hydrides (Table 8.6), the standard

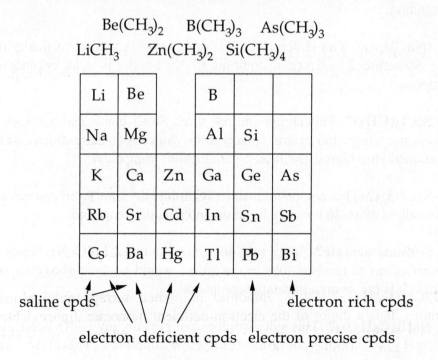

$$Be(CH_3)_2 \quad B(CH_3)_3 \quad As(CH_3)_3$$
$$LiCH_3 \quad Zn(CH_3)_2 \quad Si(CH_3)_4$$

Li	Be		B		
Na	Mg		Al	Si	
K	Ca	Zn	Ga	Ge	As
Rb	Sr	Cd	In	Sn	Sb
Cs	Ba	Hg	Tl	Pb	Bi

saline cpds   electron rich cpds

electron deficient cpds   electron precise cpds

enthalpy of formation of *p*-block methyl compounds becomes less negative or more positive on going down a group, in large part because the C–M bond enthalpies become progressively weaker on going down a group (see Figure 15.3).

15.3 **Formulas/alternative names?** **(a) Trimethylbismuth?** The formula is $Bi(CH_3)_3$. Alternative names are trimethylbismuthine (or trimethylbismuthane), i.e. the methyl derivative of bismuthine (or bismuthane), BiH_3.

(b) Tetraphenylsilane? The formula is $Si(C_6H_5)_4$. An alternative name is tetraphenylsilicon(IV).

(c) Tetraphenylarsonium bromide? The formula is $[As(C_6H_5)_4]Br$. An alternate name is tetraphenylarsenic(1+) bromide.

(d) Potassium tetraphenylborate? The formula is $K[B(C_6H_5)_4]$. An alternative name is potassium tetraphenylboron(1–).

15.4 **Name and classify each of the following compounds?** **(a) $SiH(C_2H_5)_3$?** This compound is named triethylsilane. It is an electron-precise organometallic compound. Its structure, containing a tetrahedral Si atom, is shown below.

(b) $BCl(C_6H_5)_2$? This compound is named diphenylchloroborane. It is an electron-deficient organometallic compound. Its structure, with trigonal planar bonding to the B atom, is also shown below.

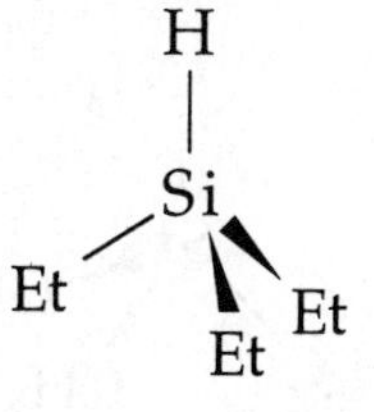

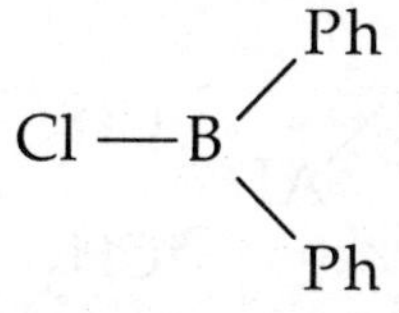

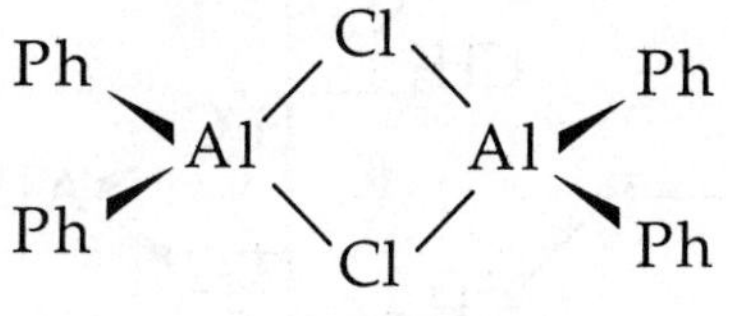

$$SiH(C_2H_5)_3 \qquad BCl(C_6H_5)_2 \qquad Al_2Cl_2(C_6H_5)_4$$

(c) $Al_2Cl_2(C_6H_5)_4$? This compound is named tetraphenyldichlorodialuminum. It is a dimer of the electron-deficient monomer diphenylchloroaluminum. Its structure, a dimer with Cl atom bridges and tetrahedral Al atoms, is shown above.

(d) $Li_4(C_2H_5)_4$? This compound is named ethyllithium (or tetraethyl-tetralithium). It is a tetramer of the electron-deficient monomer ethyllithium. Its structure is shown at the right (one of the ethyl groups has been omitted for clarity).

$$
\begin{array}{c}
Li \!-\!\!-\!\!-\! CH_2CH_3 \\
\diagup \qquad \diagup \; | \\
H_3CH_2C \!-\!\!-\!\!-\! Li \quad | \\
| \qquad\qquad | \quad Li \\
| \qquad\qquad | \diagup \\
Li \!-\!\!-\!\!-\! CH_2CH_3
\end{array}
$$

(e) $RbCH_3$? This compound is named methylrubidium. It is a saline, electron-deficient methyl compound.

15.5 **Sketch the structures of the following?** **(a) Methyllithium?** The structure of this compound, $(LiCH_3)_4$, is based on a cube, with four Li atoms and four C atoms at the eight corners. The structure can also be described as interpenetrating Li_4 and C_4 tetrahedra, and is shown below.

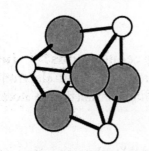

The structure of methyllithium. The larger shaded circles are the Li atoms and the smaller open circles are the C atoms. The H atoms of the methyl groups have been omitted for clarity. The C–Li distances are 2.31 Å (cf. C–B = 1.58 Å in $B(CH_3)_3$, below).

(b) Trimethylboron? This compound contains a trigonal planar B atom with three terminal methyl groups and is shown below.

$$B(CH_3)_3 \qquad\qquad Al_2(CH_3)_6 \qquad\qquad Si(CH_3)_4$$

(c) Hexamethyldialuminum? This compound contains two tetrahedral Al atoms bridged by methyl groups. Each Al atom also has two terminal methyl groups, to give the structure shown above.

(d) Tetramethylsilane? This compound contains a tetrahedral Si atom with four terminal methyl groups and is shown above.

(e) Trimethylarsane? This compound contains a trigonal pyramidal As atom with three terminal methyl groups, similar to the structure of AsH_3. Its structure is shown at the right. The C–As–C bond angles are 96° (the H–As–H bond angles in AsH_3 are 91.8° (Table 8.3)).

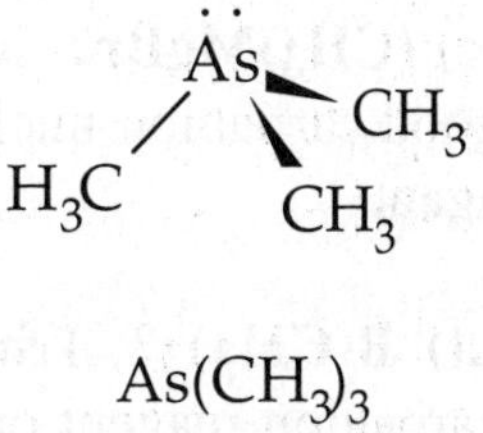

$As(CH_3)_3$

(f) Tetraphenylarsonium? This cation contains a tetrahedral arsenic atom surrounded by four phenyl groups. All of the C–As–C bond angles are 109.5°.

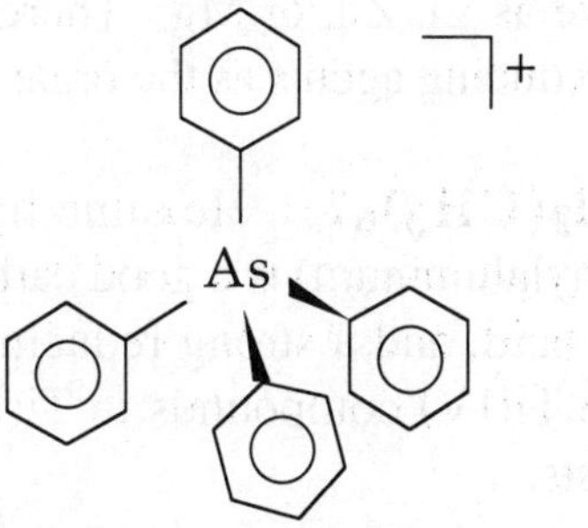

15.6 **The tendency toward association?** For the series of compounds $B(CH_3)_3$, $Al(CH_3)_3$, $Ga(CH_3)_3$, and $In(CH_3)_3$, the tendency to form dimers with bridging methyl groups is greatest for aluminum and decreases dramatically down the group. Trimethylborane is a monomer and shows no tendency to dimerize. The difference between trimethylborane and trimethylaluminum may be due to the small size of boron and the relatively short C–B bonds in the former compound. A structure with bridging methyl groups may be prevented from forming in the case of $B(CH_3)_3$ because of steric hindrance.

15.7 **Which of the following is a (1) good carbanion reagent, (2) mild Lewis acid, (3) mild Lewis base, and/or (4) strong reducing agent?** **(a) $Li_4(CH_3)_4$?** Methyl lithium is a good carbanion nucleophile reagent, a mild Lewis acid (the Li^+ ions will form complexes with Lewis bases such as tetramethylethylenediamine (see Section 15.7 and Structure **8**)), and a strong reducing agent (for example, methyllithium reacts with oxygen to form lithium methoxide, $LiOCH_3$; this is a reduction of oxygen and oxidation of carbon).

(b) $Zn(CH_3)_2$? Dimethylzinc is also a good carbanion reagent, although not as reactive as methyllithium. Recall that zinc alkyls were used quite extensively in organic synthesis until the advent of the more reactive Grignard reagents (see part (c), below). Dimethylzinc is also a mild Lewis acid and a strong reducing agent (recall that it also reacts spontaneously and vigorously with oxygen when it is exposed to air).

(c) (CH₃)MgBr? Methylmagnesium bromide, like all Grignard reagents, is a good carbanion nucleophile reagent, a mild Lewis acid, and a strong reducing agent.

(d) B(CH₃)₃? Trimethylborane is a mild Lewis acid, but it is not a good carbanion reagent or a strong reducing agent. While it will react with oxygen like $Li_4(CH_3)_4$, $Zn(CH_3)_2$, and $(CH_3)MgBr$, the central B atom is not as electropositive as Li, Zn, or Mg. Therefore, its organometallic compounds are not as good reducing agents as the organometallic compounds of these three elements.

(e) Al₂(CH₃)₆? Hexamethyldialuminum (commonly referred to as trimethylaluminum) is a good carbanion nucleophile reagent, a moderately strong Lewis acid, and a strong reducing agent. Aluminum alkyls are in fact used to reduce Ti(IV) compounds to Ti(III) compounds that are olefin polymerization catalysts.

(f) Si(CH₃)₄? Tetramethylsilane, like many other electron precise organometallic compounds, exhibits none of the four properties mentioned above. It has no lone pairs, so it does not exhibit Lewis basicity. It does not have a low-lying vacant orbital or cannot dissociate into a fragment that does, so it does not exhibit Lewis acidity. In fact, its most characteristic property is that it is inert, which is one of the main reasons it can be used as a chemical shift standard for ^1H, ^{13}C, and ^{29}Si NMR spectroscopy (i.e. it can be added to solutions of almost any substance without reaction).

(g) As(CH₃)₃? Trimethylarsene (or trimethylarsine) is a mild Lewis base, since the central As atom has a lone pair of electrons. It is not a good carbanion reagent, a strong reducing agent, or a Lewis acid, since the central As atom is electronegative, not electropositive (compare its electronegativity, 2.18 (Table 1.8), with that of elements whose organometallic compounds are good carbanion reagents, such as Li (0.98), Mg (1.31), Al (1.61), and Si (1.90)).

15.8   Balanced chemical equations? (a) A carbanion reagent with AsCl₃ and with SiPh₂Cl₂? One reagent you might chose is methyllithium, which in ether solution in the presence of $AsCl_3$ would undergo a metathesis reaction:

$$3\,LiCH_3 + AsCl_3 \rightarrow 3\,LiCl(s) + As(CH_3)_3$$

Another reagent might be $(CH_3)MgBr$, which in ether solution in the presence of $SiPh_2Cl_2$ would undergo the following metathesis reaction:

$$2\,(CH_3)MgBr \; + \; SiPh_2Cl_2 \; \rightarrow \; 2\,Mg(Br,\,Cl)_2 \; + \; SiPh_2(CH_3)_2$$

Note that these reactions are spontaneous because Li and Mg are more electropositive than As and Si, respectively.

(b) A Lewis acid with NH_3? A straightforward reaction would be one between trimethylborane and ammonia, which could be carried out without using a solvent, and which would result in the solid complex shown in the balanced equation below:

$$B(CH_3)_3(g) \; + \; NH_3(g) \; \rightarrow \; (CH_3)_3B\!-\!NH_3(s)$$

(c) A Lewis base with $[Hg(CH_3)][BF_4]$? The only Lewis base in Exercise 15.7 is trimethylarsine, which would form a complex with the cationic Lewis acid $Hg(CH_3)^+$ as follows:

$$As(CH_3)_3 \; + \; [Hg(CH_3)][BF_4] \; \rightarrow \; [(CH_3)_3As\!-\!Hg(CH_3)][BF_4]$$

The structure of the $(CH_3)_3As\!-$ $Hg(CH_3)^+$ cation, a complex of the Lewis base $As(CH_3)_3$ and the Lewis acid $Hg(CH_3)^+$. Note that the Hg atom is linear two-coordinate.

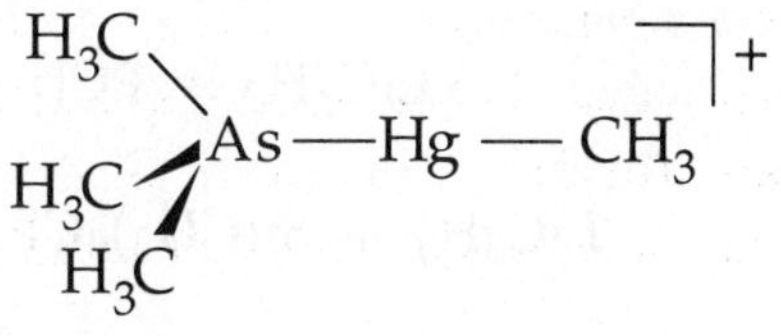

15.9 **Give examples of the following reaction types?** **(a) A metal with an organic halide?** The formation, stability, and reactivity of organometallic compounds is discussed in Sections 15.4–15.6. One of the fundamental ways to prepare an organometallic compound of an electropositive metal is to react the metal with an alkyl or aryl halide. The net reaction can be either (i) or (ii), below, depending on whether the oxidation state of the metal is normally +1 or higher:

$$2\,M \; + \; RX \; \rightarrow \; MR \; + \; MX \qquad\qquad \text{(i)}$$
$$M \; + \; RX \; \rightarrow \; RMX \qquad\qquad \text{(ii)}$$

Specific examples are:

$$2\,Li \; + \; n\text{-BuCl} \; \rightarrow \; n\text{-BuLi} \; + \; LiCl \qquad \text{and} \qquad Mg \; + \; PhBr \; \rightarrow \; PhMgBr$$

(b) Transmetallation? In this type of reaction, one metal takes the place of another, as in:

$$M + M'R \rightarrow M' + MR$$

The most important factor in determining the course of the reaction is the extent to which M is more electropositive than M'. For example, Al will displace Zn from its organometallic compounds, since Al is more electropositive than Zn, but it will not displace lithium from its organometallic compounds, because Al is less electropositive than Li:

$$2Al(s) + 3Zn(C_2H_5)_2 \rightarrow Al_2(C_2H_5)_6 + 3Zn(s)$$

$$Al(s) + LiC_2H_5 \rightarrow NR$$

(c) Double replacement? A double replacement reaction is also called a metathesis reaction. In the context of organometallic chemistry, it is a reaction involving an organometallic MR or MR_n compound and a halide of some element, such as EX_n or R'_nEX. So long as M is more electropositive than E, the following reaction will take place:

$$M-R + E-X \rightarrow M-X + E-R$$

Specific examples are:

$$3\,LiC_3H_7 + PCl_3 \rightarrow 3\,LiCl + P(C_3H_7)_3$$

$$LiC_3H_7 + Si(CH_3)_3Cl \rightarrow LiCl + Si(CH_3)_3(C_3H_7)$$

$$3\,PhMgCl + BCl_3 \rightarrow 3\,MgCl_2 + BPh_3$$

$$PhMgCl + BCl_3 \rightarrow MgCl_2 + BPhCl_2$$

The first example illustrates the way that most alkyl and aryl phosphines are prepared (these are very important ligands for transition metals, as you will see later). The second example demonstrates the utility of double replacement reactions for the preparation of *p*-block organometallics containing more than one type of alkyl group. The third and fourth examples show that, in many cases, controlling the stoichiometry of the reaction can affect the composition of the final product.

15.10 **Which compound is likely to be the stronger reducing agent?   (a)**
$Na[C_{10}H_8]$ or $Na[C_{14}H_{10}]$? Sodium naphthalide and sodium anthracenide
are examples of organometallic salts with a delocalized anion. If you remember
the structures of naphthalene and anthracene, shown below, you can appreciate

naphthalene  anthracene

the fact that naphthalene has a smaller π system and hence a more negative
reduction potential than anthracene (see Table 15.3; remember that the electron
given up by sodium and taken up by the organic molecule is added to the lowest
unoccupied π^* orbital). For this reason, the anion radical of naphthalene will
give up its extra electron more readily than the anion radical of anthracene.
Thus, sodium naphthalide is the stronger reducing agent.

(b) $Na[C_{10}H_8]$ or $Na_2[C_{10}H_8]$? In this case the organic molecule is the
same, but the charge is different. That is, in $Na[C_{10}H_8]$ the organic species is a
radical anion while in $Na_2[C_{10}H_8]$ it is a dianion. In almost every case, the first
reduction potential of a neutral molecule will be less negative than the second
reduction potential (the few exceptions are species that undergo a single two–
electron reduction, a rare occurrence). For this reason, the dianion will give up
an electron more readily than the radical anion. (Compare this situation to the
first and second electron affinities of an O atom, listed in Table 1.7.) Thus,
$Na_2[C_{10}H_8]$ is the stronger reducing agent. As far as the synthesis of
$Na_2[C_{10}H_8]$ is concerned, the relative reduction potentials require that excess
sodium, and not excess naphthalene, be present. Otherwise, the conpro-
portionation reaction below would occur:

$$Na_2[C_{10}H_8] + C_{10}H_8 \rightarrow 2\,Na[C_{10}H_8]$$

15.11 **Balanced chemical equations and reaction type?   (a) Ca with**
$Hg(CH_3)_2$? Whenever a metal, in this case Ca, is mixed with an
organometallic compound, in this case dimethylmercury, a transmetallation can
potentially take place. Remember that a more electropositive element will replace
a less electropositive element and, since Ca is more electropositive than Hg, the
following reaction will take place.

$$Ca(s) + Hg(CH_3)_2 \rightarrow Ca(CH_3)_2 + Hg(l)$$

(b) Hg with Zn(C$_2$H$_5$)$_2$? Once again a transmetallation is the potential reaction, since a metal and an organometallic compound are mixed together. However, since Hg is less electropositive than Zn, no reaction will take place:

$$Hg(l) + Zn(C_2H_5)_2 \rightarrow NR$$

(c) LiCH$_3$ with SiPh$_3$Cl in ether? Whenever an organometallic compound and a halide of some element are mixed together, a metathesis reaction is the possible occurrence (see the answer to Exercise 15.9(c), above). The metathesis will occur if the organometallic metal atom (Li in this case) is more electropositive than the element of the halide compound (Si in this case). Since Li is far more electropositive than Si, the metathesis reaction will occur, as shown below:

$$LiCH_3 + SiPh_3Cl \rightarrow LiCl + SiPh_3(CH_3)$$

(d) Si(CH$_3$)$_4$ with ZnCl$_2$ in ether? As in part (c), above, you have mixed an organometallic compound with a halide. However, in this case the organometallic compound contains Si, which is less electropositive than the element in the halide, namely Zn. Therefore, no reaction will occur:

$$Si(CH_3)_4 + ZnCl_2 \rightarrow NR$$

(e) SiH(CH$_3$)$_3$ with C$_2$H$_4$? In this case, none of the reaction types of organometallic compounds you have studied is evident. Since a metal is not present, you cannot have a transmetallation. Since a halide is not present, you cannot have a metathesis reaction. However, before you conclude that no reaction will occur, consider that the presence of C–M bonds in a compound might not change the reactivity of other portions of the molecule. In other words, even though you cannot think of a reaction that will cleave the C–Si bonds, you must consider reactions that might cleave the H–Si bond. Recall that reagents containing Si–H bonds can add across double bonds in reactions called hydrosilylations (see the end of Section 8.12). The platinic acid is added to the reaction medium as a catalyst, so the net reaction involves only the silane and the olefin:

$$SiH(CH_3)_3 + C_2H_4 \rightarrow Si(C_2H_5)(CH_3)_3$$

15.12 Give balanced chemical equations? (a) **Direct synthesis of Si(CH$_3$)$_2$Cl$_2$?** The reaction of two equivalents of CH$_3$Cl with elemental silicon can be carried out at high temperatures in the presence of a catalyst (Cu is generally used):

$$Si(s) + 2CH_3Cl(g) \rightarrow Si(CH_3)_2Cl_2(g)$$

(b) Redistribution of Si(CH$_3$)$_2$Cl$_2$? The exact ratio of products is difficult to predict, because subtle aspects of bonding and steric interactions affect the relative thermodynamic stabilities of the various species Si(CH$_3$)$_x$Cl$_{4-x}$ ($x = 0 - 4$). A nonbalanced equation is:

$$Si(CH_3)_2Cl_2 \rightarrow Si(CH_3)_4 + Si(CH_3)_3Cl + Si(CH_3)Cl_3 + SiCl_4$$

15.13 The synthesis of poly(dimethylsiloxane)? Two different kinds of siloxanes are generally used, a cyclic compound like [(CH$_3$)$_2$SiO]$_4$, with two methyl groups and two oxygen atoms per silicon atom, and a noncyclic compound such as ((CH$_3$)$_3$Si)$_2$O, with three methyl groups and only one oxygen atom per silicon atom. These compounds are prepared as follows:

$$Si(s) + 2\,CH_3Cl \rightarrow Si(CH_3)_2Cl_2 \quad \text{(formation of C–Si bonds)}$$

$$4\,Si(CH_3)_2Cl_2 + 4\,H_2O \rightarrow [(CH_3)_2SiO]_4 + 8\,HCl \quad \text{(hydrolysis rxn)}$$

$$2\,Si(CH_3)_2Cl_2 \rightarrow Si(CH_3)_3Cl + Si(CH_3)Cl_3 \quad \text{(redistribution rxn)}$$

$$2\,Si(CH_3)_3Cl + H_2O \rightarrow ((CH_3)_3Si)_2O + 2\,HCl \quad \text{(hydrolysis rxn)}$$

Then a mixture of [(CH$_3$)$_2$SiO]$_4$ and ((CH$_3$)$_3$Si)$_2$O is polymerized using a sulfuric acid catalyst:

$$n\,[(CH_3)_2SiO]_4 + ((CH_3)_3Si)_2O \rightarrow (CH_3)_3SiO[Si(CH_3)_2O]_{4n}Si(CH_3)_3$$

15.14 Trends in the oxidation states for Group 13/III and Group 15/V organometallics? For the elements of Groups 13 and 14, the general trends exhibited by all of their compounds are also observed for their organometallic derivatives. That is, the maximum oxidation states of +3 and +4, respectively, are the only important oxidation states at the top of the group, and oxidation states +1 and +2, respectively, are only important at the bottom of each group. In other words, the only organometallic derivatives of boron, aluminum, and silicon are BR$_3$, AlR$_3$, and SiR$_4$; no compounds of composition BR, AlR, or SiR$_2$ are known. At the other extreme, stable organometallic derivatives of thallium have formulas TlR and TlR$_3$, and tin and lead form both divalent and tetravalent organometallic compounds having formulas SnR$_2$ and PbR$_2$, and SnR$_4$ and PbR$_4$, respectively. The elements of Group 15, As, Sb, and Bi, exhibit a parallel but slightly different trend. Oxidation state +5 is found in some organometallic compounds of As and Sb, such as [AsPh$_4$]Br and SbPh$_5$, but this oxidation state is not as important as +3, even for arsenic.

15.15 **Describe the periodic trends?** **(a) Metal–carbon bond enthalpies?** These exhibit a uniform decrease going down a group (see Figure 15.3), similar to the trend exhibited by the element–hydrogen bond energies for p-block elements (see Figure 8.11). Recall that the weak bonds formed by the heavier element near the bottom of each p-block group is attributed to poor overlap with the diffuse s and p orbitals of these very large atoms.

(b) Lewis acidity? With one very important exception, Lewis acidity decreases down a group. As discussed in part (a), above, bond enthalpies decrease down each group, including the enthalpy of the bond between the Lewis acid and the base. The exception is the period 2 element in each group, which is so small that steric hindrance can prevent a strong complex from forming between a period 2 organometallic Lewis acid and a base. For example, aluminum alkyls are the strongest Lewis acids among group 13/III organometallics. Boron alkyls are considerably weaker acids (see below).

Lewis acidity: BR$_3$ < AlR$_3$ > GaR$_3$ > InR$_3$ > TlR$_3$

(c) Lewis basicity for Groups 1, 2, 12, 13/III, and 14/IV? The lighter elements in Groups 1 and 2 form organometallic compounds that have a higher degree of covalency than their heavier congeners. Therefore, the Lewis basicity (i.e. carbanionic character) of organometallic compounds of these elements *increases* down these groups. In contrast, Lewis basicity of group 12 organometallics *decreases* down the group from zinc to mercury. The organometallic compounds of Group 13/III and 14/IV elements do not display Lewis basicity.

15.16 **Summarize the trend in each of the following?** **(a) The relative ease of pyrolysis of Si(CH$_3$)$_4$ and Pb(CH$_3$)$_4$?** Due to the relative bond strengths C–Sn < C–Si, tetramethyllead will undergo thermal decomposition (pyrolysis) more rapidly at 300°C than will tetramethylsilane.

Rate of thermal decomposition: Pb(CH$_3$)$_4$ > Si(CH$_3$)$_4$

(b) The relative Lewis acidity of Li$_4$(CH$_3$)$_4$, B(CH$_3$)$_3$, Si(CH$_3$)$_4$, and Si(CH$_3$)Cl$_3$? Tetramethylsilane exhibits no Lewis acidity, so it is by default the weakest Lewis acid of these four compounds. The presence of the three chlorine ligands in $Si(CH_3)Cl_3$ renders the silicon center more acidic than $Si(CH_3)_4$, but it is still a relatively weak Lewis acid. Considering the other two, trimethylboron is commonly referred to as a weak acid, but it is stronger than methyllithium. Recall that the synthesis of the tetraphenylborate anion may be viewed as the transfer of the strong base Ph^- from the weak Lewis acid Li^+ to the stronger acid BPh_3:

$$BPh_3 + LiPh \rightarrow Li[BPh_4]$$

Therefore, the order of acidity is $Si(CH_3)_4 < Si(CH_3)Cl_3 < Li_4(CH_3)_4 < B(CH_3)_3$.

(c) The relative Lewis basicity of $Si(CH_3)_4$ and $As(CH_3)_3$? For a substance to be basic, it must have either one or more lone pairs of electrons or one or more loosely held bonding pairs of electrons (generally these are π electrons). Tetramethylsilane has neither of these, so it is not basic at all. On the other hand, trimethylarsine has a lone pair of electrons on the central As atom, rendering it a mild Lewis base.

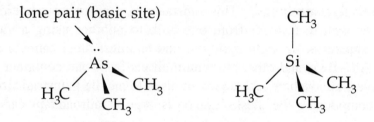

(d) The tendency of $Li_4(CH_3)_4$ and $Hg(CH_3)_2$ to displace halide from $GeCl_4$? A halide displacement by an organometallic compound is an example of a metathesis reaction. You should recall that in this type of metathesis reaction, the more electropositive element will wind up with the halide, and the less electropositive element will wind up with the organic group. Methyllithium *will* displace chloride from tetrachlorogermane, because lithium is more electropositive than germanium, and dimethylmercury *will not* displace chloride from tetrachlorogermane, because mercury is less electropositive than germanium:

$$Li_4(CH_3)_4 + GeCl_4 \rightarrow 4\,LiCl + Ge(CH_3)_4$$

$$Hg(CH_3)_2 + GeCl_4 \rightarrow NR$$

15.17 **What air-free technique is appropriate?** **(a) Li$_4$(CH$_3$)$_4$ in ether?** Methyllithium is sensitive to both oxygen and water, so it cannot be handled in open flasks. Generally, Schlenk apparatus, air-tight syringes, and stainless steel cannula are used to handle air-sensitive solutions (see Box 15.1). Alternatively, solutions can be handled in an inert-atmosphere glovebox, but this is usually not as convenient since a glovebox is a relatively elaborate piece of equipment and requires constant maintenance. Gloveboxes are ideal for handling nonvolatile air-sensitive solids.

(b) Trimethylboron? Since this substance is a pyrophoric gas at room temperature, it is most conveniently handled using a vacuum line (pyrophoric: capable of igniting spontaneously when exposed to air). Of all the air-free techniques mentioned in the text, a vacuum line offers the best protection from the atmosphere. In general, however, it cannot be used to handle solutions, since the solvent and solute will usually have different volatilities (in some cases the solute will be nonvolatile).

$$B(CH_3)_3 \quad mp -162°C, bp -20°C$$

(c) Triisobutylaluminum? This substance is a pyrophoric liquid and is not volatile enough to transfer from one bulb to another using a vacuum line. Schlenk apparatus, air-tight syringes, and stainless steel cannula are used to transfer Al(i–Bu)$_3$ and other aluminum alkyls from one container to another. Similar inert-atmosphere techniques are used to handle industrial sized amounts of these compounds: the "flasks" can be as large as railroad tank cars.

$$Al_2(i\text{-}Bu)_6 \quad mp 4°C, bp 73°C \text{ at } 5 \text{ Torr}$$

(d) Solid AsPh$_3$? Triphenylarsine is not oxygen- or water-sensitive, so it can be handled in an open flask. If it is to be added to an air-sensitive reaction mixture, it can be weighed out in the air and transfered to a Schlenk flask, which is then purged with an inert gas. At this point, an air-sensitive liquid compound or solution can be added using cannula.

$$AsPh_3 \quad mp 61°C, bp 233°C \text{ at } 14 \text{ Torr}$$

(e) Liquid (CH$_3$)$_3$SiOSi(CH$_3$)$_3$? Hexamethyldisiloxane is a liquid that is not oxygen- or water-sensitive. It can be handled in an open flask. Since it is relatively volatile (its boiling point is 101°C), it can be added to an air-sensitive reaction mixture by any of the air-free techniques, including a vacuum line. To help develop a "feel" for relative volatilities, consider the following:

$Si(CH_3)_4$	b.p. 27°C	CH_4	b.p. −164°C
$(CH_3)_3SiOSi(CH_3)_3$	b.p. 101°C	H_3COCH_3	b.p. −23°C

15.18 **The electronic structure of $Si_6(CH_3)_{12}$?** As discussed in Section 15.13, the near-UV absorption band is indicative of a relatively small HOMO–LUMO gap, which is a manifestation of a delocalized low-lying vacant orbital. Recall that *unsaturated* hydrocarbons such as naphthalene, which also have a small HOMO–LUMO gap, are reduced by sodium to form anion radicals (see Exercise 15.10). You should recognize that this behavior of cyclic silanes is in sharp contrast with that of their carbon analogues, cyclohexane and substituted cyclohexanes, which do not have near-UV absorptions. The ability of $Si_6(CH_3)_{12}$ to form a stable anion such as $[Si_6(CH_3)_{12}]^-$ upon treatment with sodium metal is also in sharp contrast with the behavior of cyclohexanes, which do not react with sodium.

15.19 **The synthesis of $R_2Si=SiR_2$?** As discussed in Section 15.13, disilenes can be formed by the photochemical cleavage of two of the Si–Si single bonds in a cyclic trisilane:

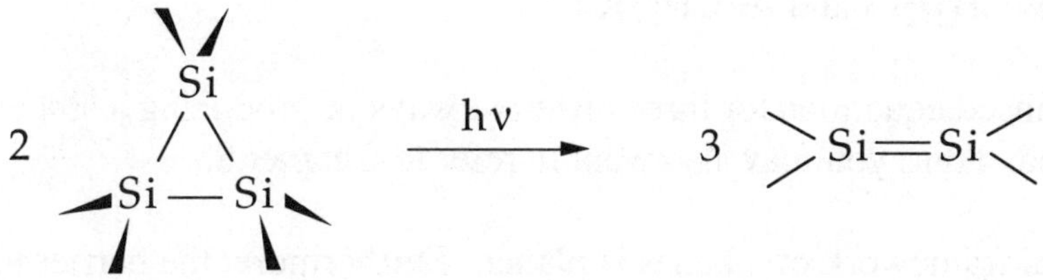

It will be of paramount importance to use R groups that are very bulky so that the disilene does not dimerize or polymerize. So, now you must develop a synthesis for a cyclic trisilane with bulky substituents starting with SiR_2Cl_2. A convenient and general route for the formation of element–element single bonds of the *p*-block elements is reduction of a halide of that element with an alkali metal or with another strong reducing agent. For example, $(CH_3)_3SiSi(CH_3)_3$ can be formed by reacting $Si(CH_3)_3Cl$ with sodium as follows:

$$2\,Si(CH_3)_3Cl + 2\,Na(s) \rightarrow (CH_3)_3SiSi(CH_3)_3 + 2\,NaCl(s)$$

This provides the necessary analogy for a route to our cyclic compound:

$$3SiR_2Cl_2 + 6Na(s) \rightarrow cyclo\text{-}Si_3R_6 + 6NaCl(s)$$

Guide to Solutions Quiz

1   Describe three important general methods for preparing main-group organometallic compounds.

2   Give the name and the chemical formula for: (a) a covalent organometallic compound containing methyl groups that has fewer than eight valence electrons around the central atom; (b) a covalent organometallic compound containing phenyl groups that has more than eight valence electrons around the central atom; (c) an ionic organometallic compound containing methyl groups.

3   Write a synthetic scheme, using balanced equations, for the synthesis of KCH_3, $Mg(CH_3)I$, $Li[AlEt_4]$, $Si(CH_3)_2Cl_2$, and $AsHPh_2$.

4   What is the problem with obtaining highly pure $Si(CH_3)_2Cl_2$, especially when Lewis acid impurities are present?

5   Give a likely example of a transmetallation involving a mercury-containing organometallic compound. Do not use any examples in the text.

6   Compare and contrast the structures of the substances with the empirical formulas $B(CH_3)_2Cl$ and $Al(CH_3)_2Cl$.

7   Write balanced equations for three different ways of producing a silicon–carbon single bond. Hint: you may also want to refer to Chapter 8.

8   The central framework of alkenes is planar. Furthermore, the barrier to rotation about the carbon–carbon double bond is very high (i.e. *cis–trans* isomerization is slow). In contrast, disilaethenes, digermaethenes, and distannaethenes are nonplanar. Explain the difference.

9   Starting with $AsCl_3$ and reagents of your choice, how would you prepare the important ligand $Ph_2AsCH_2CH_2AsPh_2$?

10 Compare and contrast the methods for preparing the $As(CH_3)_4^+$ and $AsPh_4^+$ cations. Why are different methods necessary?

16 *d*- and *f*-block organometallic compounds

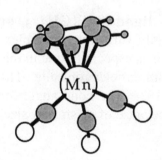

The structure of tricarbonyl(η^5-cyclopentadienyl)-manganese(I) (the oxygen atoms of the carbonyl ligands are not shaded). This molecule contains two important types of ligands frequently found in organometallic compounds, π-acids (the CO ligands) and unsaturated organic molecules or molecular fragments that donate their π electrons to the metal center (the η^5-C_5H_5 ligand).

S16.1 Is Mo(CO)$_7$ likely to exist? A Mo atom (Group 6) has six valence electrons, and each CO ligand is a two-electron donor. Therefore, the total number of valence electrons on the Mo atom in this compound would be 6 + 7(2) = 20. Since organometallic compounds with more than 18 valence electrons on the central metal are never stable, Mo(CO)$_7$ is *not* likely to exist. The compound Mo(CO)$_6$, with exactly 18 valence electrons, is very stable. Note that throughout Chapter 16, the authors do not always write the formulas for organometallic complexes in square brackets. According to the rules of inorganic nomenclature, *all* complexes should be written in square brackets. Nevertheless, the authors are following a common, informal set of rules: neutral organometallic complexes are written without square brackets, while cationic or anionic organometallic complexes are written with them.

S16.2 **Assign the oxidation number of Co in $Co(\eta^5\text{-}C_5H_5)(CO)_2$?** The CO ligands are neutral two-electron donors, so their oxidation number is 0. The pentahaptocyclopentadienyl ligand is assigned an oxidation number -1. The complex as a whole is neutral, so the sum of ligand oxidation numbers ($-1 + 0 + 0 = -1$) must be equal in magnitude but opposite in sign to the metal oxidation number. Therefore, the oxidation number of Co in this compound is $+1$.

S16.3 **Does $Ni_2Cp_2(CO)_2$ contain bridging or terminal CO ligands or both?** The CO stretching bands at 1857 cm^{-1} and 1897 cm^{-1} are both lower in frequency than typical terminal CO ligands (for terminal CO ligands, $v(CO) >$ 1900 cm^{-1} (see Figure 16.2)). Therefore, it seems likely that it only contains bridging CO ligands. The presence of two bands suggests that the bridging CO ligands are probably *not* collinear, since only one band would be observed if they were (see Figure 16.5).

S16.4 **If $Mo(CO)_3L_3$ is desired, which of the ligands $P(CH_3)_3$ or $P(t\text{-}Bu)_3$ would be preferred?** Since this is a highly substituted complex, the effects of steric crowding must be considered. This is especially true in this case, since the two ligands in question should be very similar electronically. The cone angle for $P(CH_3)_3$, given in Table 16.6, is $118°$. The cone angle for $P(t\text{-}Bu)_3$, also given in the table, is $182°$. Therefore, because of its smaller size, PMe_3 would be preferred.

S16.5 **Propose a synthesis for $Mn(CO)_4(PPh_3)(COCH_3)$?** Consider the reactions of carbonyl complexes discussed in Section 16.6. If you use $Mn_2(CO)_{10}$ as the source of manganese, you can reductively cleave the Mn–Mn bond with sodium, forming $Na[Mn(CO)_5]$:

$$Mn_2(CO)_{10} + 2Na \rightarrow 2Na[Mn(CO)_5]$$
oxidation number: 0   0   +1 –1

The anionic carbonyl complex is a relatively good nucleophile. When it is treated with CH_3I, it displaces I^- to form $Mn(CH_3)(CO)_5$ (sometimes written as $CH_3Mn(CO)_5$):

$$Na[Mn(CO)_5] + CH_3I \rightarrow Mn(CH_3)(CO)_5 + NaI$$

Many alkyl-substituted metal carbonyls undergo a migratory insertion reaction when treated with basic ligands. The alkyl group (methyl in this case) migrates from the Mn atom to an adjacent C atom of a CO ligand, leaving an open coordination site for the entering group (PPh₃ in this case) to attack:

$$Mn(CH_3)(CO)_5 + PPh_3 \rightarrow Mn(CO)_4(PPh_3)(COCH_3)$$

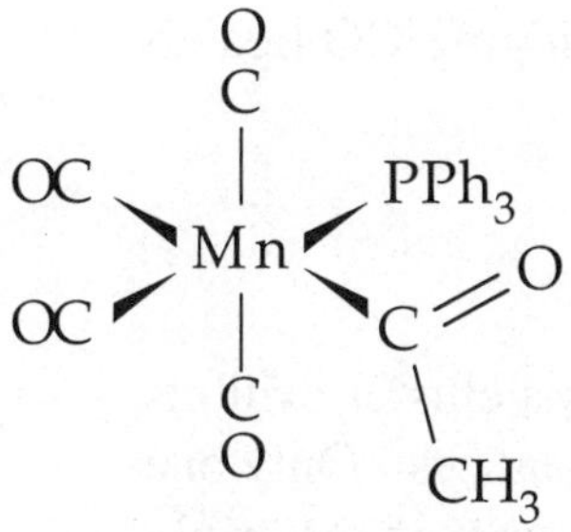

The structure of Mn(CO)₄(PPh₃)(COCH₃)

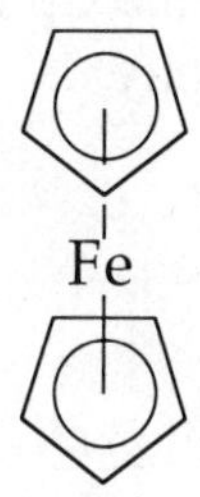

The structure of ferrocene

S16.6 Will oxidation of FeCp₂ to [FeCp₂]⁺ produce a substantial change in M–C bond length?

Neutral ferrocene (shown above) contains 18 valence electrons (8 from Fe and 10 from the two Cp ligands). The MO diagram for ferrocene is shown in Figure 16.11. Eighteen electrons will fill it up to the second a_1' orbital. Since this orbital is the HOMO, oxidation of ferrocene will result in removal of an electron from it, leaving the 17-electron ferricinium cation, [FeCp₂]⁺. If this orbital were strongly bonding, removal of an electron would result in weaker Fe–C bonds. If this orbital were strongly antibonding, removal of an electron would result in stronger Fe–C bonds. However, this orbital is essentially nonbonding (see Figure 16.11). Therefore, oxidation of FeCp₂ to [FeCp₂]⁺ will not produce a substantial change in the Fe–C bond order or the Fe–C bond length.

S16.7 Propose a structure for Fe₄Cp₄(CO)₄?

There are four relevant pieces of information given. First of all, the fact that the compound is highly colored suggests that it contains metal–metal bonds. Second, the composition can be used to determine the cluster valence electron count, which can be used to predict which polyhedral structure is likely:

Fe₄	4 × 8 e⁻ =	32 valence e⁻
Cp₄	4 × 5 =	20
(CO)₄	4 × 2 =	8
Total =		60 valence e⁻ ⇒ a tetrahedral cluster

Third, the presence of only one line in the ^{1}H NMR spectrum suggests that the Cp ligands are all equivalent and are η^5. Finally, the single CO stretch at 1640 cm^{-1} suggests that the CO ligands are triply bridging (see Figure 16.2) and that they form a relatively high symmetry array (otherwise there would be more bands; see Table 16.5). A likely structure for Fe$_4$Cp$_4$(CO)$_4$ is shown below, with only one of the Cp ligands and only one of the triply bridging CO ligands shown for simplicity.

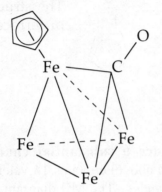

The tetrahedral Fe$_4$ core of this cluster exhibits six equivalent Fe–Fe bond distances. Only one of the η^5-C$_5$H$_5$ ligands and one of the triply bridging CO ligands is shown. Each Fe atom has a Cp ligand, and each of the four triangular faces of the Fe$_4$ tetrahedron is capped by a triply bridging CO ligand.

16.1   **Name and draw the structures of the following?   (a) Fe(CO)$_5$?** The names of metal carbonyls first specify the number of CO ligands, then the number and type of metal atom(s), and finally the oxidation number of the metal(s). Therefore, the name of Fe(CO)$_5$ is pentacarbonyliron(0). As discussed in Section 16.5, the structures of metal carbonyls generally have simple, symmetrical shapes that correspond to the CO ligands taking up positions that place them as far apart from one another as possible. In this case the analogy is between an Fe atom with five CO ligands and an atom with five bonding pairs of valence electrons, like the P atom in PF$_5$. Thus, the structure of Fe(CO)$_5$ is a trigonal bipyramid, like PF$_5$, as shown below.

(b) Ni(CO)$_4$? The name of this complex is tetracarbonylnickel(0). In this case the analogy is between a Ni atom with four CO ligands and an atom with four bonding pairs of valence electrons, like the C atom in CH$_4$. Thus, the structure of Ni(CO)$_4$ is tetrahedral, as shown below.

(c) Mo(CO)$_6$? The name of this complex is hexacarbonylmolybdenum(0). The analogy here is between a Mo atom with six CO ligands and an atom with six bonding pairs of valence electrons, like the S atom in SF$_6$. Thus, the structure of Mo(CO)$_6$ is octahedral, as shown below.

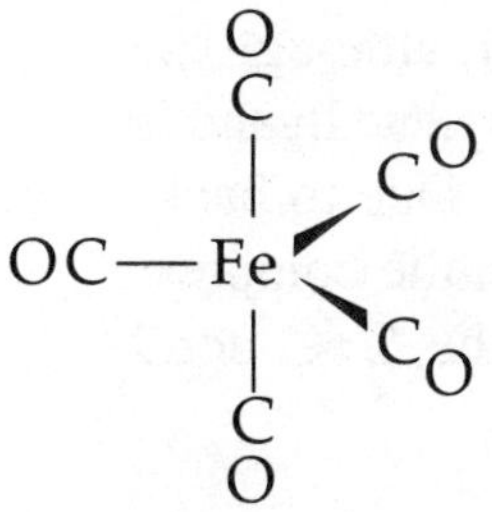

Fe(CO)$_5$ 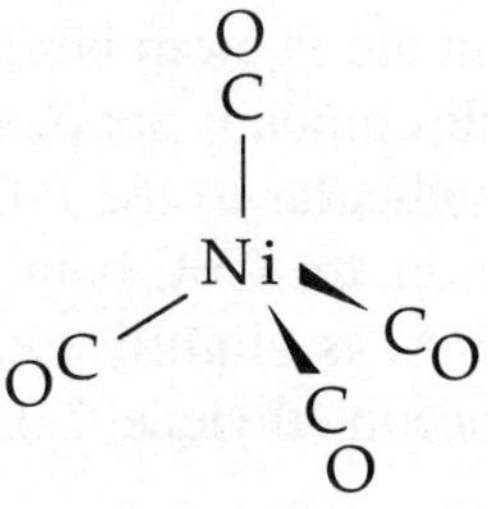   Ni(CO)$_4$    Mo(CO)$_6$

(d) Mn$_2$(CO)$_{10}$? The name of this *dinuclear* (i.e. two metal atoms) complex is decacarbonyldimanganese(0). Its structure is based on two square-pyramidal Mn(CO)$_5$ fragments joined by a Mn–Mn bond. The Mn(CO)$_5$ fragments are staggered with respect to each other. The geometry around each Mn atom is octahedral, although the complex itself has D_{4d} symmetry, not O_h symmetry. Its structure is shown below.

(e) V(CO)$_6$? The name of this complex is hexacarbonylvanadium(0). Its structure is the same as that of Mo(CO)$_6$, octahedral, as shown below.

(f) [PtCl$_3$(C$_2$H$_4$)]$^-$? The name of this complex, which has two ligand types, is trichloro(ethylene)platinate(II) or trichloro(ethylene)platinate(1−). You should review the nomenclature rules for metal complexes. Complexes are named with their ligands in alphabetical order. The prefixes di-, tri-, tetra-, etc. do not count as far as the alphabetical order is concerned. Thus, *c* before *e* in trichloro-(*ethylene*)platinate(II). The structure of this complex is square-planar, the usual structure for period 4 and 5 d^8 metal ions, and is shown below.

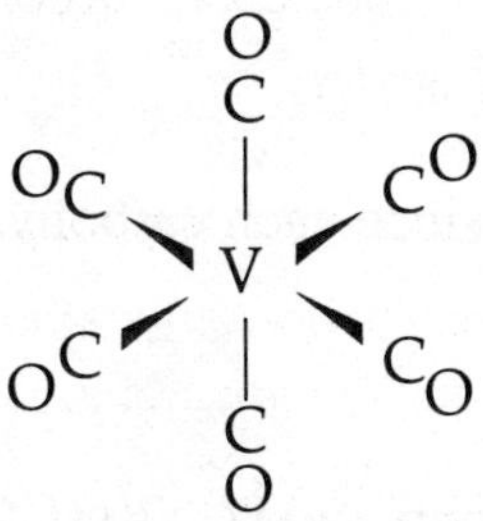

V(CO)$_6$ 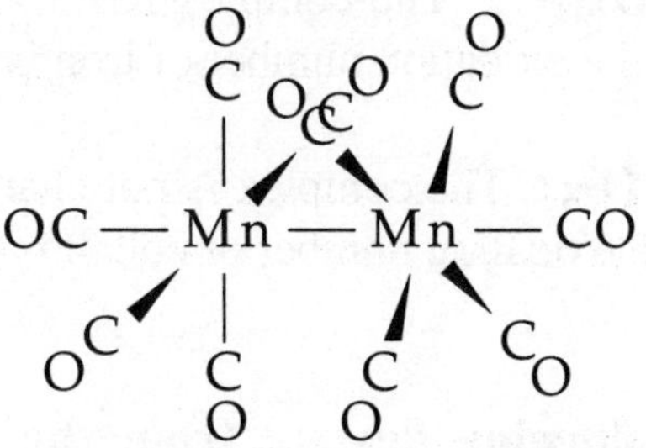   Mn$_2$(CO)$_{10}$

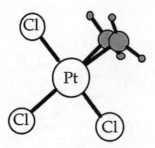

The structure of $[PtCl_3(C_2H_4)]^-$. The geometry around the Pt atom is square-planar, although the complex anion is not planar (the ethylene ligand is perpendicular to the $PtCl_3$ plane). Due to back-donation, the C=C bond distance in the complex, 1.375 Å, is slightly longer than the C=C bond distance in ethylene, 1.337 Å.

16.2   **Sketch the interactions of 1,4-butadiene with a metal atom?**

dihapto   tetrahapto

16.3   **Assign oxidation numbers in the following?   (a) $[Fe(\eta^5\text{-}C_5H_5)_2][BF_4]$?**   The tetrafluoroborate anion has a –1 charge, so the iron complex must have a +1 charge. Since each cyclopentadienyl ligand is generally considered to be a –1 anion, the oxidation number of iron is +3.

(b) $Fe(CO)_5$? The complex is not charged and, since each carbonyl ligand is neutral, the oxidation number of iron is 0.

(c) $[Fe(CO)_4]^{2-}$? The complex has a –2 charge. Since each carbonyl ligand is neutral, the oxidation number of iron is –2.

(d) $Co_2(CO)_8$? The complex is not charged and, since each carbonyl ligand is neutral, the oxidation number of cobalt is 0.

16.4   **Do the following deviate from the 18-electron rule? $Fe(CO)_5$?** An Fe atom (Group 8) has eight valence electrons, and each CO ligand is a two-electron donor. Therefore, the total number of valence electrons on the Fe atom in this compound is $8 + 5(2) = 18$.

Ni(CO)$_4$? A Ni atom (Group 10) has ten valence electrons, and each CO ligand is a two-electron donor. Therefore, the total number of valence electrons on the Ni atom in this compound is $10 + 4(2) = 18$.

Mo(CO)$_6$? A Mo atom (Group 6) has six valence electrons, and each CO ligand is a two-electron donor. Therefore, the total number of valence electrons on the Mo atom in this compound is $6 + 6(2) = 18$.

Mn$_2$(CO)$_{10}$? A Mn atom (Group 7) has seven valence electrons, and each CO is a two-electron donor. Polynuclear metal carbonyls typically have metal–metal bonds, and this compound is no exception. The Mn–Mn bond allows each Mn atom to share an additional valence electron. The total number of valence electrons for Mn in this compound is $7 + 5(2) + 1 = 18$.

V(CO)$_6$? A V atom (Group 5) has five valence electrons, and each CO ligand is a two-electron donor. Therefore, the total number of valence electrons on the V atom in this compound is $5 + 6(2) = 17$. This octahedral complex deviates from the 18-electron rule. It is much more reactive than the first three metal carbonyls, undergoing rapid ligand substitution reactions. For example, rate(V)/rate(Cr) for the following reaction is ~10^{10}.

$$M(CO)_6 + PR_3 \rightarrow M(CO)_5(PR_3) + CO$$

V(CO)$_6$ is readily reduced to [V(CO)$_6$]$^-$, an 18-electron complex anion. It is also *very* highly colored, in sharp contrast to Fe(CO)$_5$, Ni(CO)$_4$, and Mo(CO)$_6$, which are colorless or very faintly colored.

[PtCl$_3$(C$_2$H$_4$)]$^-$? The Pt atom (Group 10) has ten valence electrons, each Cl atom is a one-electron donor, ethylene is a two-electron donor, and one electron must be added for the –1 charge of the complex. Therefore, the total number of valence electrons on the Pt atom in this complex is $10 + 3(1) + 2 + 1 = 16$. This complex deviates from the 18-electron rule, as do many four-coordinate period 4 and 5 d^8 complexes (see Section 16.1). As a consequence of being two electrons short of 18, this complex undergoes ligand substitution by an associative mechanism (i.e. with the formation of a five-coordinate 18-electron intermediate).

[Fe(η^5-C$_5$H$_5$)$_2$]$^+$? An Fe atom has eight valence electrons, each η^5-cyclopentadienyl ligand is a five-electron donor, and one electron must be subtracted for the +1 charge of the complex. Therefore, the total number of valence electrons on the Fe atom in this complex is $8 + 2(5) - 1 = 17$. Like $V(CO)_6$, the deviation from the 18-electron rule causes this compound to be readily reduced by one electron (i.e. it is a strong oxidant).

$[Fe(CO)_4]^{2-}$? An Fe atom has eight valence electrons, each carbonyl ligand is a two-electron donor, and two electrons must be added for the 2– charge on the complex. Therefore, the total number of valence electrons on the Fe atom in this complex is $8 + 4(2) + 2 = 18$.

$Co_2(CO)_8$? Each Co atom (Group 9) has nine valence electrons, and each carbonyl is a two-electron donor. The Co–Co bond allows each Co atom to share an additional valence electron. Therefore, since the total number of valence electrons of the *two* Co atoms in this compound is $2(9) + 8(2) + 2 = 36$, each Co atom obeys the 18-electron rule. In the solid-state, this compound has two bridging carbonyl ligands and six terminal carbonyl ligands, but in solution it has only terminal carbonyls, as shown below:

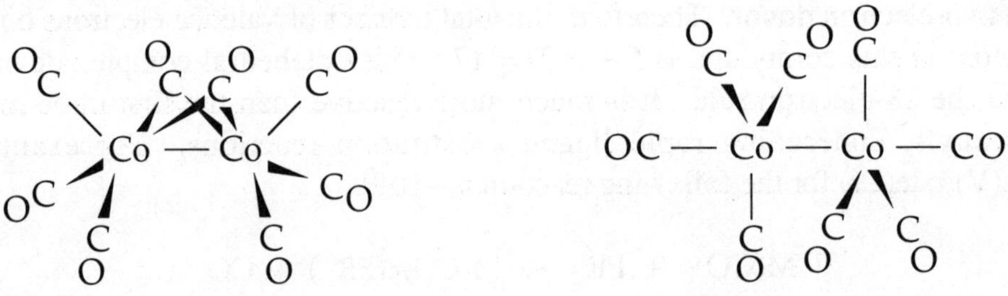

$Co_2(CO)_8$ in the solid–state $Co_2(CO)_8$ in solution

16.5 **The common methods for the preparation of simple metal carbonyls?** As discussed in Section 16.4, the two principal methods are (1) direct combination of CO with a finely divided metal and (2) reduction of a metal salt in the presence of CO under pressure. Two examples are shown below, the preparation of hexacarbonylmolybdenum(0) and octacarbonyldicobalt(0). Other examples are given in the text.

(1) $Mo(s) + 6CO(g) \rightarrow Mo(CO)_6(s)$ (high temp. and pressure required)

(2) $2\,CoCO_3(s) + 2\,H_2(g) + 8\,CO(g) \rightarrow Co_2(CO)_8(s) + 2\,CO_2 + 2\,H_2O$

The reason that the second method is preferred is kinetic, not thermodynamic. The atomization energy (i.e. sublimation energy) of most metals is simply too high for the first method to proceed at a practical rate.

16.6 **A sequence of reactions for the preparation of Fe(CO)$_3$(diphos)?** The most general way to prepare ligand substituted metal carbonyl complexes is to treat the parent binary metal carbonyl, in this case Fe(CO)$_5$, with the ligand of choice, in this case diphos (1,2-bis(diphenylphosphinoethane)). The two-step reaction sequence is:

$$Fe(s) + 5\,CO(g) \rightarrow Fe(CO)_5(l) \quad \text{(high temp. and pressure required)}$$

$$Fe(CO)_5(l) + diphos(s) \rightarrow Fe(CO)_3(diphos)(s) + 2\,CO(g)$$

The second step would require a slightly elevated temperature. A convenient way to achieve this would be to perform the ligand substitution in a refluxing solvent such as THF.

16.7 **IR spectra of metal tricarbonyl complexes?** In general, the lower the symmetry of a M(CO)$_n$ fragment, the greater the number of CO stretching bands in the IR spectrum. Therefore, given complexes with M(CO)$_3$ fragments that have either C_{3v}, D_{3h}, or C_s symmetry, the complex that has C_s symmetry will have the greatest number of bands. When you consult Table 16.5, you will see that a M(CO)$_3$ fragment with D_{3h} symmetry will give rise to one band, one with C_{3v} symmetry will give rise to two bands, and one with C_s symmetry will give rise to three bands.

16.8 **Ni$_3$(C$_5$H$_5$)$_3$(CO)$_2$?** **(a) Propose a structure based on IR data?** The structure on the right is consistent with the IR spectroscopic data. This D_{3h} complex has all three η^5-C$_5$H$_5$ ligands in identical environments. Furthermore, it has two collinear bridging CO ligands, which fits the single CO stretching band at a relatively low frequency. Although it was not given, you should expect this complex to be highly colored, due to the presence of metal–metal bonds.

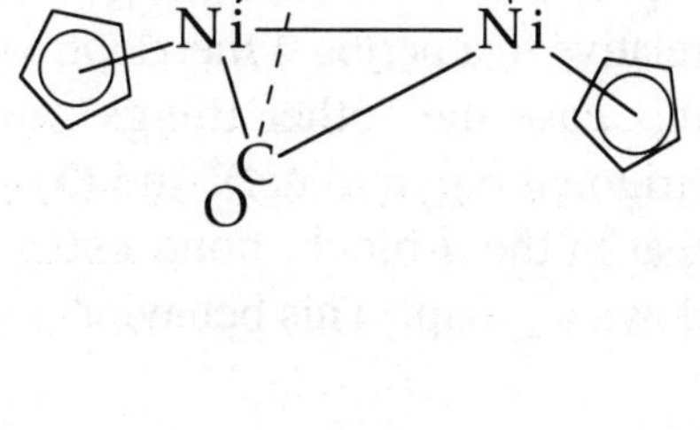

(b) Does each Ni atom obey the 18-electron rule? The three Ni atoms are in identical environments, so you only have to determine the number of valence electrons for one of them:

Ni	10 valence e$^-$
η^5–C_5H_5	5
2 Ni–Ni	2
1/3(2 CO)	4/3
	——
Total	18 1/3

Therefore, the Ni atoms in this trinuclear complex do not obey the 18-electron rule. Deviations from the rule are common for cyclopentadienyl complexes to the right of the d-block. For example, the stable complex $(\eta^5\text{-}C_5H_5)_2Co$ is a 19-electron compound.

16.9 **Which of the two complexes $W(CO)_6$ or $IrCl(CO)(PPh_3)_2$ should undergo the faster exchange with ^{13}CO?** The 18-electron tungsten complex undergoes ligand substitution by a dissociative mechanism. The rate-determining step involves cleavage of a relatively strong W–CO bond. In contrast, the 16-electron iridium complex undergoes ligand substitution by an associative mechanism, which does not involve Ir–CO bond cleavage in the activated complex. Accordingly, $IrCl(CO)(PPh_3)_2$ undergoes faster exchange with ^{13}CO than does $W(CO)_6$.

16.10 **Which complex should be more basic toward a proton? (a) $[Fe(CO)_4]^{2-}$ or $[Co(CO)_4]^-$?** The dianionic complex $[Fe(CO)_4]^{2-}$ should be the more basic. The trend involved is the greater affinity for a cation that a species with a higher negative charge will have, all other things being equal. In this case the "other things" are (1) same set of ligands, (2) same structure (tetrahedral), and (3) same electron configuration (d^{10}). For a more detailed explanation, see *Metal basicity* in Section 16.6.

(b) $[Mn(CO)_5]^-$ or $[Re(CO)_5]^-$? The rhenium complex is the more basic. The trend involved is the greater M–H bond enthalpy for a period 6 metal ion relative to a period 4 metal ion in the same group, all other things being equal. In this case the "other things" are (1) same set of ligands, (2) same structure (trigonal bipyramidal), and (3) same metal oxidation number (–1). Remember that in the d-block, bond enthalpies such as M–M, M–H, and M–R *increase* down a group. This behavior is opposite to that exhibited by p-block elements.

16.11 A chemical equation, structure of the product, and reason for the course of the reaction? **(a) MeLi + W(CO)$_6$?** The methyl carbanion, CH_3^-, is a strong nucleophile. In this reaction, it will attack the electrophilic carbon atom of one of the six carbonyl ligands, producing an acetyl ligand:

$$MeLi + W(CO)_6 \rightarrow Li^+[W(CO)_5(COCH_3)]^-$$

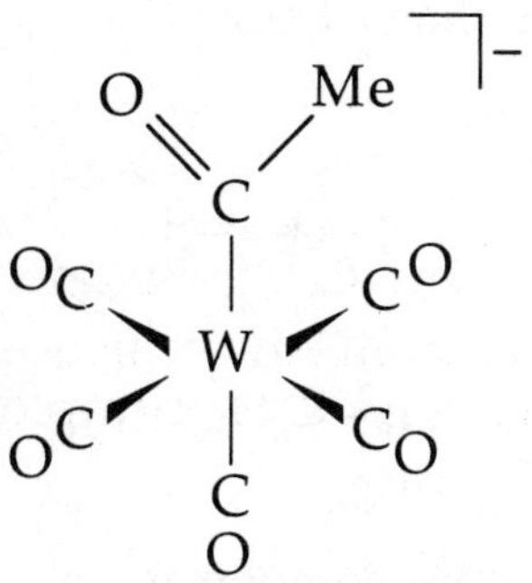

$$[W(CO)_5(COMe)]^- \qquad\qquad [Co_2(CO)_7(COAlBr_3)]$$

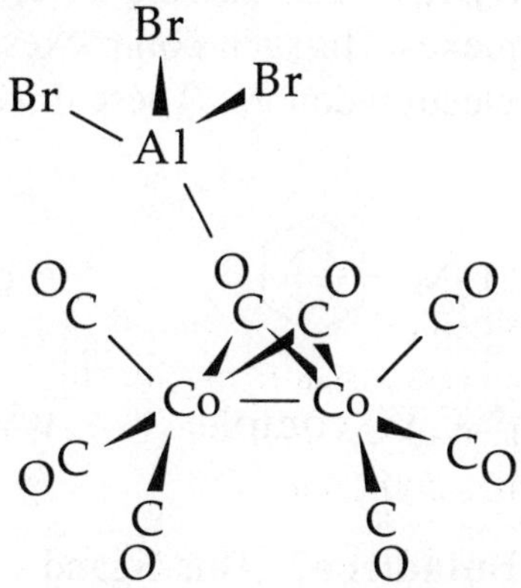

(b) Co$_2$(CO)$_8$ + AlBr$_3$? Aluminum tribromide is a powerful electrophile, and it will form a complex with an oxygen atom of a bridging carbonyl ligand (the bridging carbonyl oxygen atoms are more basic than terminal carbonyl oxygen atoms, since a bridging carbonyl receives electron density from two metal centers instead of only one). The balanced equation is as follows, and the structure of the product is shown above:

$$Co_2(CO)_8 + AlBr_3 \rightarrow Co_2(CO)_7(COAlBr_3)$$

16.12 What hapticities are possible for the following ligands? **(a) C$_2$H$_4$?** Ethylene coordinates to *d*-block metals in only one way, using its π electrons to form a metal–ethylene σ bond (there may also be a significant amount of back donation, if ethylene is substituted with electron withdrawing groups; see Section 16.7). Therefore, C$_2$H$_4$ is always η^2, as shown below.

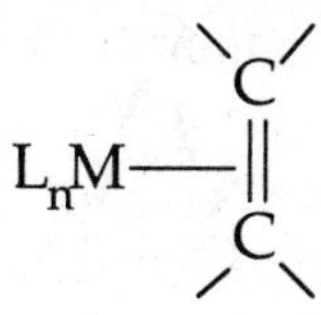

C$_2$H$_4$ complex   η^5– Cp complex   η^3–Cp complex   η^1–Cp complex

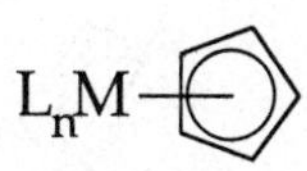

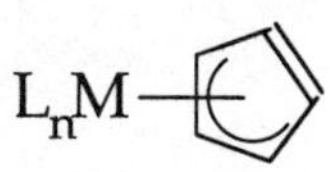

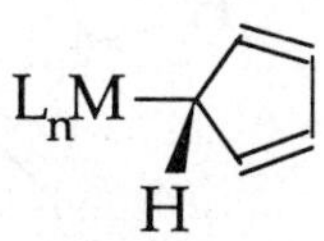

(b) Cyclopentadienyl? This is a very versatile ligand that can be η^5 (a five-electron donor), η^3 (a three-electron donor similar to simple allyl ligands; see Structure **8**), or η^1 (a one-electron donor similar to simple alkyl and aryl ligands; see Structure **7**). These three bonding modes are shown above (see Section 16.8).

(c) C_6H_6? This is also a versatile ligand, which can form η^6, η^4, and η^2 complexes. In such complexes, the ligands are, respectively, six-, four-, and two-electron donors. These three bonding modes are shown below.

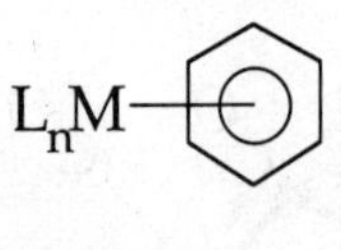

η^6–C_6H_6 complex 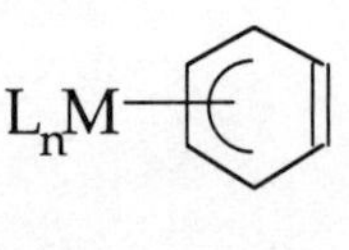 η^4–C_6H_6 complex 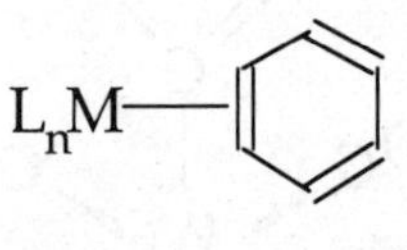 η^2–C_6H_6 complex

(d) Butadiene? This ligand can form both η^4 and η^2 complexes, in which they are four- and two-electron donors, respectively. A drawing of an η^4-butadiene complex is shown in Section 16.7(g). An η^2-butadiene complex would resemble an η^2-ethylene complex, except that one of the two C=C double bonds would remain uncoordinated.

(e) Cyclooctatetraene? This ligand contains four C=C double bonds, any combination of which can coordinate to a *d*-block (or *f*-block) metal. Thus *cyclo*-C_8H_8 can be η^8 (an eight-electron donor; see Table 16.1), η^6 (a six-electron donor), η^4 (a four-electron donor; see Structures **50**, **51**, **52**, and **54** and Figure 16.13(b)), and η^2 (a two-electron donor, in which it would resemble an η^2-ethylene complex, except that three of the four C=C double bonds would remain uncoordinated).

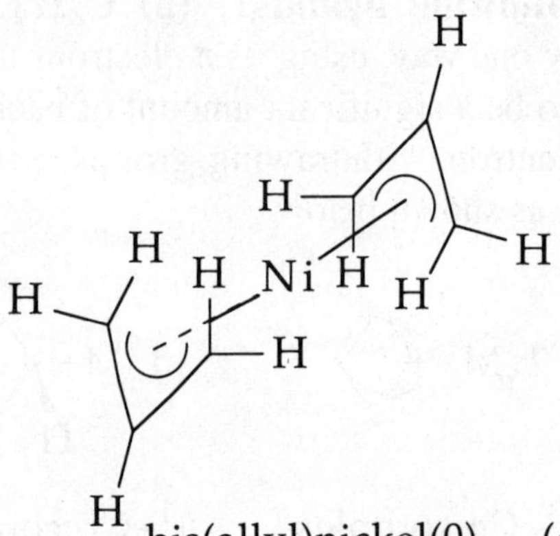

bis(allyl)nickel(0) 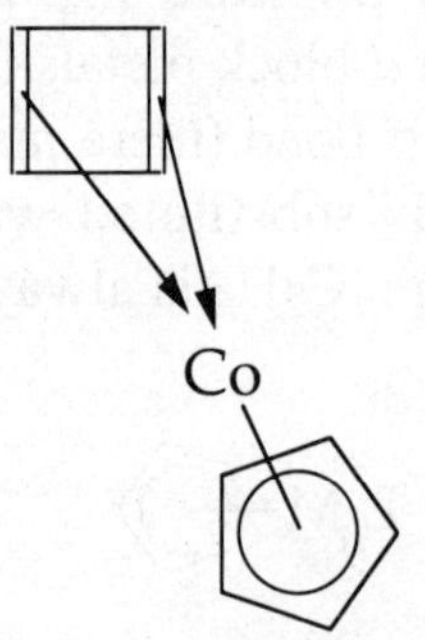   (cyclobutadiene)CpCo 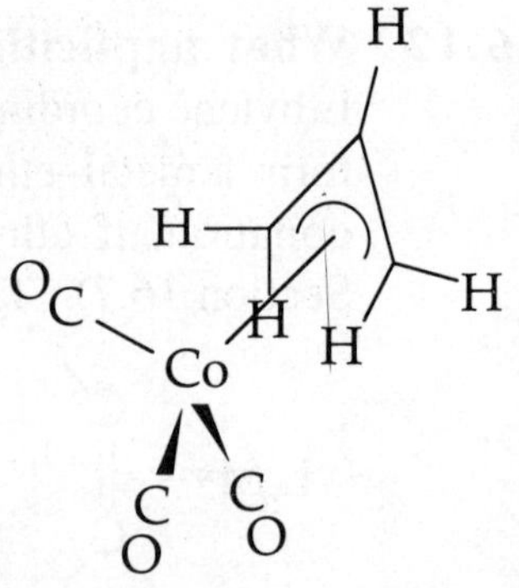   (allyl)tricarbonylcobalt(0)

16.13 Draw structures and give the electron count of the following?
(a) $Ni(\eta^3-C_3H_5)_2$? Bis(allyl)nickel(0) has the structure shown above. Each allyl ligand, which is planar, is a three-electron donor, so the number of valance electrons around the Ni atom (Group 10) is $10 + 2(3) = 16$. Sixteen-electron complexes are very common for Group 9 and Group 10 elements, especially for Rh^+, Ir^+, Ni^{2+}, Pd^{2+}, and Pt^{2+} (all of which are d^8).

(b) $Co(\eta^4-C_4H_4)(\eta^5-C_5H_5)$? This complex has the structure shown above. Since the $\eta^4-C_4H_4$ ligand is a four-electron donor and the $\eta^5-C_5H_5$ ligand is a five-electron donor, the number of valence electrons around the Co atom (group 9) is $9 + 4 + 5 = 18$.

(c) $Co(\eta^3-C_3H_5)(CO)_3$? This complex has the structure shown above. The electron count for the Co atom is:

Co	9 valence e^-
$\eta^3-C_3H_5$	3
3CO	6
	———
Total	18 e^-

16.14 The *d*-block of the periodic table? (a) **Which elements form neutral 18-electron Cp_2M compounds?** Since η^5-Cp is a five-electron donor, the only metals that form such compounds have 8 valence electrons. These are the Group 8 metals, Fe, Ru, and Os.

(b) **Which period 4 elements form dimeric binary carbonyls?** The binary metal carbonyls for period 4 elements are $V(CO)_6$, $Cr(CO)_6$, $Mn_2(CO)_{10}$, $Fe(CO)_5$, $Co_2(CO)_8$, and $Ni(CO)_4$ (see Table 16.2). The Mn and Co complexes are the only dimeric ones.

(c) **Which period 4 elements form neutral carbonyls with six, five, and four carbonyl ligands?** From the list of neutral binary carbonyls given in part (b), above, the elements in question are V, Cr, Fe, and Ni, respectively.

(d) **Which elements have the greatest tendency to follow the 18-electron rule?** The elements in the middle of the *d*-block have the greatest tendency. These are the chromium group (Cr, Mo, and W), the manganese group (Mn, Tc, and Re), and the iron group (Fe, Ru, and Os).

16.15 **Indicate the number of CO ligands in the following complexes?**
(a) $W(\eta^6\text{-}C_6H_6)(CO)_n$? A W atom (Group 6) has six valence electrons, an $\eta^6\text{-}C_6H_6$ ligand is a six-electron donor, and each carbonyl ligand is a two-electron donor. Therefore, to satisfy the 18-electron rule, $n = 3$:

$$18 = 6 + 6 + 2n$$

(b) $Rh(\eta^5\text{-}Cp)(CO)_n$? A Rh atom (Group 9) has nine valence electrons, an $\eta^5\text{-}C_5H_5$ ligand is a five-electron donor, and each carbonyl ligand is a two-electron donor. Therefore, to satisfy the 18-electron rule, $n = 2$:

$$18 = 9 + 5 + 2n$$

(c) $Ru_3(CO)_n$? This complex could be linear (i.e. Ru–Ru–Ru), with two Ru–Ru bonds, or triangular, with three Ru–Ru bonds. If it were linear, there would be no way to satisfy the 18-electron rule for the two Ru atoms (Group 8) at the end, regardless of the value of n:

$$\text{for all } n, \ 18 \neq 8 + 1 \text{ (for Ru–Ru bond)} + 2n$$

Therefore, this must be a triangular cluster, with two Ru–Ru bonds per Ru atom. Therefore, $m = 4$ and $n = 3m = 12$, so the complex must be $Ru_3(CO)_{12}$:

$$18 = 8 + 2 \text{ (for two Ru–Ru bonds)} + 2m$$

16.16 **Give plausible equations for the synthesis of M–C, M–H, and M–M' bonds using metal carbonyl anions?** Metal carbonyl anions are Lewis bases, and as such can displace weaker bases from their complexes with Lewis acids such as H^+, R^+, and L_nM^+. Examples are shown in the following equations:

$$(\eta^5\text{-}Cp)W(CO)_3^- + HCl \rightarrow (\eta^5\text{-}Cp)W(CO)_3H + Cl^-$$

$$(\eta^5\text{-}Cp)Fe(CO)_2^- + C_2H_5Cl \rightarrow (\eta^5\text{-}Cp)Fe(CO)_2(C_2H_5) + Cl^-$$

$$(\eta^5\text{-}Cp)W(CO)_3^- + (\eta^5\text{-}Cp)Fe(CO)_2Cl \rightarrow$$
$$(\eta^5\text{-}Cp)(CO)_3W\text{-}Fe(\eta^5\text{-}Cp)(CO)_2 + Cl^-$$

16.17 **Propose a synthesis for $MnH(CO)_5$?** The most general way to prepare metal carbonyl hydrides is to protonate metal carbonyl anions. In this case, the single protonation of $Mn(CO)_5^-$ would yield the desired product. The anion could be produced by the reductive cleavage of $Mn_2(CO)_{10}$ with Na. Since sodium is a solid that is not soluble in typical solvents such as THF, a liquid amalgam of sodium with mercury is generally used. The mixture of two liquid phases, a THF solution of $Mn_2(CO)_{10}$ and Na(Hg), reacts more rapidly than a mixture of a THF solution and solid Na:

$$Mn_2(CO)_{10}(soln) + 2\,Na(Hg)(l) \rightarrow 2\,Na[Mn(CO)_5](soln) + Hg(l)$$

$$Na[Mn(CO)_5](s) + H_3PO_4(l) \rightarrow MnH(CO)_5(s) + NaH_2PO_4(s)$$

The two solid products of the second reaction can be easily separated by sublimation, since $MnH(CO)_5$, like many metal carbonyls and metal carbonyl hydrides, is a volatile compound. The reason that phosphoric acid is preferred is that it is not oxidizing (metal carbonyl anions are notoriously prone to oxidation).

16.18 **The product obtained when $Mo(CO)_6$ is treated with PhLi and then with $CH_3OSO_2CF_3$?** The product of the first reaction contains a $-C(=O)Ph$ ligand formed by the nucleophilic attack of Ph^- on one of the carbonyl C atoms:

$$Mo(CO)_6 + PhLi \rightarrow Li[Mo(CO)_5(COPh)]$$

$$[Mo(CO)_5(COPh)]^- \qquad\qquad Mo(CO)_5(C(OMe)Ph)$$

The most basic site on this anion is the acyl oxygen atom, and it is the site of attack by the methylating agent $CH_3OSO_2CF_3$:

$$Li[Mo(CO)_5(COPh)] + CH_3OSO_2CF_3 \rightarrow$$

$$Mo(CO)_5(C(OCH_3)Ph) + LiSO_3CF_3$$

The final product is a carbene (alkylidene) complex. Since the carbene C atom has an oxygen-containing substituent, it is an example of a Fischer carbene (see Section 16.7). The two reaction products are shown above.

16.19 **Explain the difference in IR spectra of the following?**
(a) Mo(PF₃)₃(CO)₃ vs. Mo(PMe₃)₃(CO)₃? The two CO bands of the trimethylphosphane complex are 100 cm⁻¹ or more lower in frequency than the corresponding bands of the trifluorophosphane complex. This is because PMe_3 is primarily a σ-donor ligand while PF_3 is primarily a π-acid ligand (PF_3 is the ligand that most resembles CO electronically). The CO ligands in $Mo(PF_3)_3(CO)_3$ have to compete with the PF_3 ligands for electron density from the Mo atom for backdonation. Therefore, less electron density is transferred from the Mo atom to the CO ligands in $Mo(PF_3)_3(CO)_3$ than in $Mo(PMe_3)_3(CO)_3$. This makes the Mo–C bonds in $Mo(PF_3)_3(CO)_3$ weaker than those in $Mo(PMe_3)_3(CO)_3$, but it also makes the C–O bonds in $Mo(PF_3)_3(CO)_3$ stronger than those in $Mo(PMe_3)_3(CO)_3$, and stronger C–O bonds will have higher CO stretching frequencies.

(b) MnCp(CO)₃ vs. MnCp*(CO)₃? The two CO bands of the Cp* complex are slightly lower in frequency than the corresponding bands of the Cp complex. Recall that the Cp* ligand is η⁵-C₅Me₅. It is a stronger donor ligand than Cp, due to the inductive effect of the five methyl groups. Therefore, the Mn atom in the Cp* complex has a greater electron density than in the Cp complex, and hence there is a greater degree of backdonation to the CO ligands in the Cp* complex than in the Cp complex. As explained in part (a), above, more back-donation leads to lower stretching frequencies.

16.20 **Which is more stable, RhCp₂ or RuCp₂?** Refer to Figure 16.11, the MO diagram for metallocenes. The number of valence electrons for the two complexes differs by one; RhCp₂ has 19 electrons while RuCp₂ has 18 electrons. The MOs of ruthenocene are filled up to the very weakly bonding a_1' orbital (see Figure 16.12). Its electron configuration, starting with the e_2' orbitals, is $(e_2')^4(a_1')^2$. Rhodocene, with its extra electron, has a $(e_2')^4(a_1')^2(e_1'')^1$ configuration. Since the e_1'' orbitals are antibonding, the Rh–C bonds in rhodocene will be longer than the Ru–C bonds in ruthenocene (refer to Table 16.10 and compare the Fe–C and Co–C bond distances in FeCp₂ and CoCp₂, which are 2.06 Å and 2.12 Å, respectively, and which are isoelectronic with RuCp₂ and Rh(Cp₂)). Therefore, ruthenocene is the more stable of these two complexes.

16.21 A workable reaction for the conversion of $Fe(\eta^5\text{-}C_5H_5)_2$ to $Fe(\eta^5\text{-}C_5H_5)(\eta^5\text{-}C_5H_4COCH_3)$? The Cp rings in cyclopentadienyl complexes behave like simple aromatic compounds such as benzene, and so are subject to typical reactions of aromatic compounds such as Freidel–Crafts alkylation and acylation. If you treat ferrocene with acetyl chloride and some aluminum(III) chloride as a catalyst, you will obtain the desired compound:

$$Fe(\eta^5\text{-}C_5H_5)_2 + CH_3COCl \xrightarrow{AlCl_3} Fe(\eta^5\text{-}C_5H_5)(\eta^5\text{-}C_5H_4COCH_3) + HCl$$

16.22 The a_1' molecular orbital of a D_{5h} Cp_2M complex? The symmetry-adapted orbitals of the two eclipsed C_5H_5 rings in a metallocene are shown in Figure 16.12. As shown in that figure and in Appendix 3, the d_{z2} orbital on the metal has a_1' symmetry and so can form MOs with this symmetry-adapted orbital. Appendix 3 shows that the s orbital on the metal also has a_1' symmetry, so it can form MOs with the d_{z2} orbital as well as the symmetry-adapted orbital. Therefore, three a_1' MOs will be formed, since three orbitals of a_1' symmetry are available on the metal and the ligands. Figure 16.11 shows the energies of the three a_1' MOs: one is the most stable orbital of a metallocene involving metal–ligand bonding; one is relatively nonbonding (it is the HOMO in $FeCp_2$); and one is a relatively high-lying antibonding orbital.

16.23 Provide an explanation for the different sites of protonation of $NiCp_2$ and $FeCp_2$? Protonation of $FeCp_2$ at iron does not change its number of valence electrons: Both $FeCp_2$ and $[FeCp_2H]^+$ are 18-electron complexes:

$FeCp_2$		$[FeCp_2H]^+$	
Fe	8 valence e^-	Fe	8 valence e^-
2 Cp	10 e^-	2 Cp	10 e^-
	——	H	1 e^-
		+1 charge	–1 e^-
			——
Total	18 e^-		18 e^-

By the same token, since $NiCp_2$ is a 20-electron complex, the hypothetical metal-protonated species $[NiCp_2H]^+$ would also be a 20-electron complex. On the other hand, protonation of $NiCp_2$ at a Cp carbon atom produces the 18-electron complex $[NiCp(\eta^4\text{-}C_5H_6)]^+$. Therefore, the reason that the Ni complex is protonated at a carbon atom is that a more stable (i.e. 18-electron) product is formed:

$[NiCp(\eta^4\text{-}C_5H_6)]^+$

Ni	10 valence e^-
Cp	5 e^-
$\eta^4\text{-}C_5H_6$	4 e^-
+1 charge	-1 e^-
Total	18 e^-

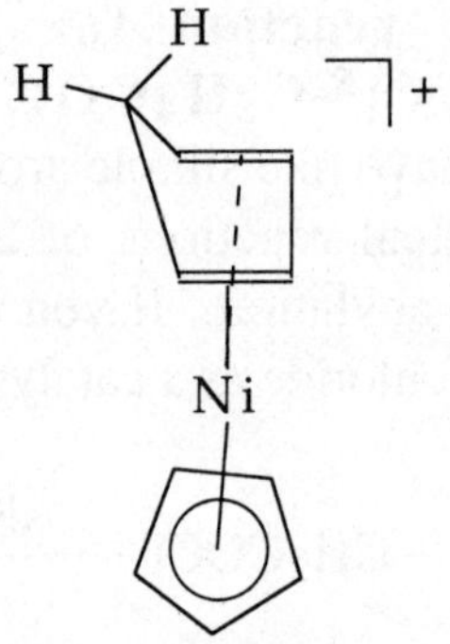

16.24 **Write a plausible mechanism for the following reactions:** **(a)** $[Mn(CO)_5(CF_2)]^+ + H_2O \rightarrow [Mn(CO)_6]^+ + 2HF$? The $=CF_2$ carbene ligand is similar electronically to a Fischer carbene (see Section 16.7). The electronegative F atoms render the C atom subject to nucleophilic attack, in this case by a water molecule. The $[(CO)_5Mn(CF_2(OH_2))]^+$ complex can then eliminate two equivalents of HF, as shown in the mechanism below:

$$[(CO)_5Mn=CF_2]^+ + H_2O \longrightarrow [(CO)_5Mn-C-O]^+ \xrightarrow{-HF}$$

$$[(CO)_5Mn-C=O]^+ \xrightarrow{-HF} [(CO)_5Mn-C\equiv O]^+ = [Mn(CO)_6]^+$$

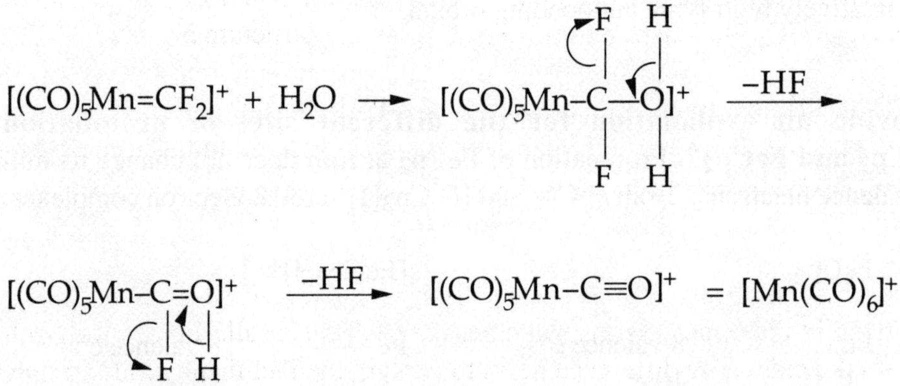

(b) $Rh(C_2H_5)(CO)(PR_3)_2 \rightarrow RhH(CO)(PR_3)_2 + C_2H_4$? This is an example of a β-hydrogen elimination reaction, discussed at the beginning of Section 16.7. This reaction is believed to proceed through a cyclic intermediate involving a (3c,2e) M–H–C interaction, as shown in the mechanism below:

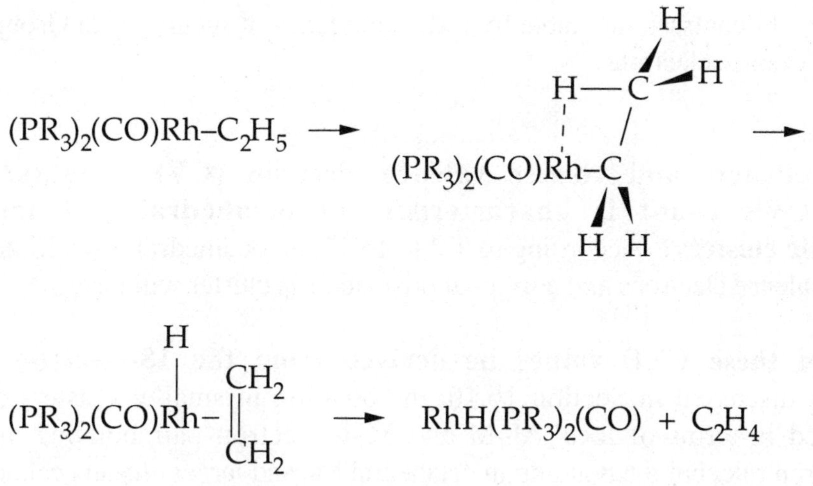

$$(PR_3)_2(CO)Rh-C_2H_5 \longrightarrow$$

(structure: a carbon with H—C bearing H, H at top, bonded to (PR$_3$)$_2$(CO)Rh—C with H H below) $\longrightarrow$

(structure: (PR$_3$)$_2$(CO)Rh— with H above and CH$_2$=CH$_2$ coordinated) $\longrightarrow$ $RhH(PR_3)_2(CO) + C_2H_4$

16.25   **Contrast groups 6–8 with Groups 3 and 4 with respect to:**   **(a) Stability of complexes of the π-C$_5$H$_5$ ligand?**   The cyclopentadienyl ligand forms very stable complexes with metals from Groups 6–8. It can also form stable complexes with metals from Groups 3 and 4 (e.g. TiCp$_2$Cl$_2$), but in many cases an unwanted side reaction is activation of one of the C–H bonds of the ligand. An example of this behavior was discussed in Section 16.9: attempts to prepare TiCp$_2$ led instead to the dimer shown in Structure **55**.

(b) Hydridic or protonic character of M–H bonds?   Group 3 and group 4 hydride complexes are truly hydridic. The M–H bond in these complexes is polarized as follows:

$$M^{d+}-H^{d-}$$

In contrast, hydride complexes of Groups 6–8 have M–H bonds that are polarized in the opposite sense, and are not hydridic at all. In fact, they behave like weak acids. The difference helps to explain the fact that hydride complexes of the early *d*-block metals react with Brønsted acids as weak as alcohols, while hydride complexes of metals from Groups 6–8 do not:

$$TiCp_2H(CH_3) + ROH \rightarrow TiCp_2(OR)(CH_3) + H_2$$

$$MnH(CO)_5 + ROH \rightarrow NR$$

(c) Adherence to the 18-electron rule?   Organometallic hydride complexes of the early *d*-block metals rarely adhere to the 18-electron rule. A good example is the titanium complex in part (b), above; TiCp$_2$H(CH$_3$) has 16 valence electrons. Another example is Sc(η^5-C$_5$Me$_5$)$_2$H, which is a 14-electron complex. In contrast, all stable hydride complexes of metals from Groups 6–8 have 18 valence electrons.

16.26 **Metal clusters and cluster valence electron (CVE) counts? (a) What CVE count is characteristic of octahedral and trigonal prismatic clusters?** According to Table 16.11, an octahedral M_6 will have 86 cluster valence electrons and a trigonal prismatic M_6 cluster will have 90.

(b) Can these CVE values be derived from the 18-electron rule? No. As discussed in Section 16.10, the bonding in smaller clusters can be explained in terms of local M–M and M–L electron pair bonding and the 18-electron rule, but the bonding in octahedral M_6 and larger clusters cannot.

(c) Determine the probable geometry of $[Fe_6(C)(CO)_{16}]^{2-}$ and $[Co_6(C)(CO)_{15}]^{2-}$? The iron complex has 86 CVEs, while the cobalt complex has 90 (see the calculations below). Therefore, the iron complex probably contains an octahedral Fe_6 array, while the cobalt complex probably contains a trigonal prismatic Co_6 array.

6 Fe	$6 \times 8e^- = 48\ e^-$	6 Co	$6 \times 9\ e^- = 54\ e^-$
C	$4\ e^-$	C	$4\ e^-$
16 CO	$16 \times 2e^- = 32\ e^-$	15 CO	$15 \times 2e^- = 30\ e^-$
2– charge	$2\ e^-$	2– charge	$2\ e^-$
Totals	$86\ e^-$		$90\ e^-$

16.27 **Choose the groups that might replace the following? (a) The CH group in $Co_3(CO)_9CH$: OCH_3, $N(CH_3)_2$, or $SiCH_3$?** Isolobal groups have the same number and shape valence orbitals *and* the same number of electrons in those orbitals. The CH group has three sp^3 hybrid orbitals that each contain a single electron. The $SiCH_3$ group has three similar orbitals similarly occupied, so it is isolobal with CH and would probably replace it in $Co_3(CO)_9CH$ to form $Co_3(CO)_9SiCH_3$. In contrast, the OCH_3 and $N(CH_3)_2$ groups are not isolobal with CH. Instead, they have, respectively, three sp^3 orbitals that each contain a pair of electrons and two sp^3 orbitals that each contain a pair of electrons.

(b) One of the $Mn(CO)_5$ groups in $(OC)_5MnMn(CO)_5$: I, CH_2, or CCH_3? The $Mn(CO)_5$ group has a single σ type orbital that contains a single electron. An iodine atom is isolobal with it, since it also has a singly occupied σ orbital. Therefore, you can expect the compound $MnI(CO)_5$ to be reasonably stable. In contrast, the CH_2 and CCH_3 are not isolobal with $Mn(CO)_5$. The CH_2 group has either a doubly occupied σ orbital and an empty p orbital or a singly occupied σ orbital and a singly occupied p orbital. The CCH_3 group has three singly occupied σ orbitals (note that it is isolobal with $SiCH_3$).

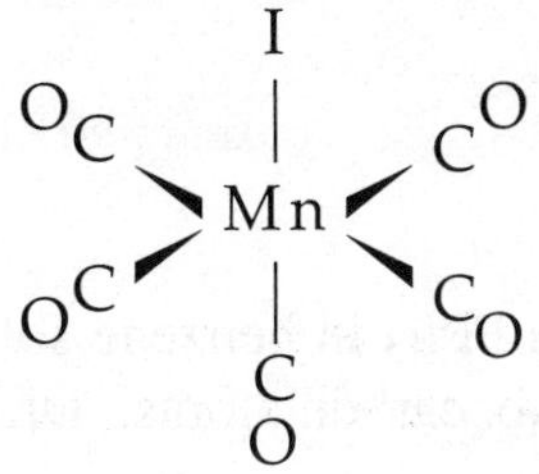

$MnI(CO)_5$ is a stable complex

16.28   **Which cluster undergoes faster associative exchange with ^{13}CO, $Co_4(CO)_{12}$ or $Ir_4(CO)_{12}$?**   If the rate-determining step in the substitution is cleavage of one of the metal–metal bonds in the cluster, the cobalt complex will exhibit faster exchange. This is because metal–metal bond strengths *increase* down a group in the *d*-block. Therefore, Co–Co bonds are weaker than Ir–Ir bonds, all other things like geometry and ligand types kept the same.

Guide to Solutions Quiz

1   Metals in low oxidation states are generally strong reducing agents. Give an example of a complex with a metal in a zero oxidation state that is a good oxidizing agent.

2   What is n in (a) $RhH(CO)_n$, (b) $W(C_6H_6)(CO)_n$, (c) $Ni(CO)_n(PPh_3)_2$, and (d) $Tc_2(CO)_n$?

3   Assuming that it obeys the 18-electron rule, what is the likely structure of the compound with the empirical formula $Fe(CO)_4$?

4   Identify the $3d$ metal M that forms the following 18-electron complexes: (a) $M(\eta^4\text{-}1,3\text{-butadiene})(CO)_2$; (b) $M(\eta^4\text{-}1,3\text{-butadiene})(CO)_3$; (c) $M(CO)_3(NO)_2$, (d) $M(\eta^4\text{-cyclobutadiene})(C_6H_6)$; (e) $M(\eta^3\text{-}C_3H_5)(\eta^5\text{-}C_5H_5)$.

5   How could you determine the correct structure of $Fe(CO)_3(PPh_3)_2$ by infrared spectroscopy?

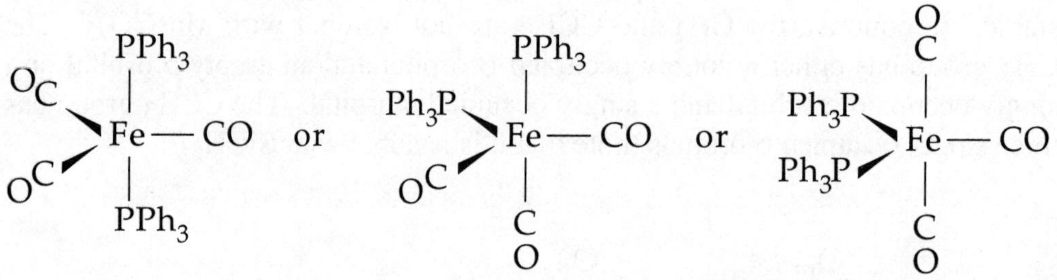

6   The compound $Re(CH_3)(CO)_5$ reacts with PPh_3 in benzene solution. Besides the phosphine ligand, the product still has six carbon atoms. Draw the structure of the product.

7   Draw structures for complexes **I**, **II**, and **III**:

$$Fe_2(C_5H_5)_2(CO)_4 \xrightarrow{\text{Na(Hg)}} \textbf{I} \xrightarrow{\text{Br}_2} \textbf{II} \xrightarrow{\text{LiAlH}_4} \textbf{III}$$

8   The compound $Mo_2H_2(CO)_8$ has D_{2h} symmetry and obeys the 18-electron rule. Does this complex contain a metal–metal bond? If it does, determine whether it is a single, double, triple, or quadruple bond.

9   A metal acyl complex (i.e. a complex containing a M–C(O)R group) is easier to protonate than either a metal carbonyl or an organic ketone. Suggest an explanation.

10   Use isolobal analogies to determine n in (a) $Co_3(CO)_9CH_n$, (b) $Fe_2(CO)_8CH_n$, and (c) $Co_3FeH(CO)_n$.

17 Catalysis

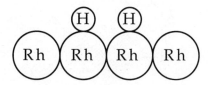

There are many chemical parallels between homogeneous and heterogeneous catalysis, despite the different terminology used to describe them and the different physical methods used to study them. For example, soluble rhodium complexes and rhodium metal both react with hydrogen. In the case of the soluble complex, we call the reaction an oxidative addition; in the case of rhodium metal, we call the reaction a chemisorption. In both cases, however, reactive Rh–H bonds are formed.

S17.1 **The effect of added phosphine on the catalytic activity of RhH(CO)(PPh₃)₃?** The compound $RhH(CO)(PPh_3)_3$ is an 18-electron complex that must lose a phosphine ligand before it can enter the catalytic cycle:

$$RhH(CO)(PPh_3)_3 \rightleftharpoons RhH(CO)(PPh_3)_2 + PPh_3$$
$$\text{18 valence e}^- \qquad\qquad \text{16 valence e}^- \qquad \text{2e}^- \text{ donor}$$

The coordinatively unsaturated complex $RhH(CO)(PPh_3)_2$ can add the alkene that is to be hydroformylated. Added phosphine will shift the above equilibrium to the left, resulting in a lower concentration of the catalytically active 16-electron complex. Thus, you can predict that the rate of hydroformylation will be decreased by added phosphine.

273

S17.2 **γ-Alumina heated to 900°C, cooled, and exposed to pyridine vapor?** Heating alumina to 900°C results in complete dehydroxylation. Therefore, only Lewis acid sites are present in γ-alumina heated to 900°C. The IR spectrum of a sample of dehydroxylated γ-alumina exposed to pyridine will exhibit bands near $1465 \ cm^{-1}$. It will not exhibit bands near $1540 \ cm^{-1}$, since these are due to pyridine that is hydrogen bonded to the surface, and no surface OH groups are present in dehydroxylated alumina.

S17.3 **Would a pure silica analog of ZSM-5 be an active catalyst for benzene alkylation?** A pure silica analog of ZSM-5 would contain Si–OH groups, which are only moderate Brønsted acids, and not the strongly acidic Al–OH_2 groups found in aluminosilicates (see Section 17.5(b), *Surface acidic and basic sites*). Only a very strong Brønsted acid such as an Al–OH_2 group in an aluminosilicate can protonate an alkene to form the carbocations that are necessary intermediates in benzene alkylation. Therefore, a pure silica analog of ZSM-5 would not be an active catalyst for benzene alkylation.

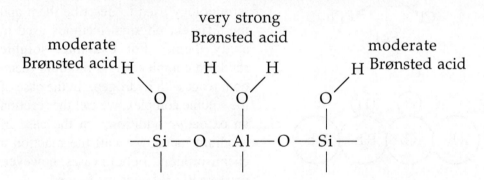

17.1 **Which of the following constitute catalysis?** **(a) H_2 and C_2H_4 in contact with Pt?** This is an example of genuine catalysis. The formation of ethane ($H_2 + C_2H_4 \rightarrow C_2H_6$) has a very high activation barrier. The presence of platinum causes the reaction to proceed at a useful rate. Furthermore, the platinum can be recovered unchanged after many turnovers, so it fits the two criteria of a catalyst, a substance that increases the rate of a reaction but is not itself consumed.

(b) H_2 plus O_2 plus an electrical arc? This gas mixture will be completely converted into H_2O once the electrical arc is struck. Nevertheless, it does not constitute an example of catalysis. The arc provides activation energy to initiate the reaction. Once started, the heat liberated by the reaction provides activation energy to sustain the reaction. The activation energy of the reaction has *not* been lowered by an added substance, so catalysis has not occurred.

(c) The production of Li₃N and its reaction with H₂O? This too is not an example of catalysis. Lithium and nitrogen are both consumed in the formation of Li₃N, which occurs at an appreciable rate even at room temperature. Water and Li₃N react rapidly to produce NH₃ and LiOH at room temperature. As in the formation of Li₃N, both substances are consumed. Since both reactions have intrinsically low activation energies, a catalyst is not necessary.

17.2   **Define the following terms?**   **(a) Turnover frequency?** This is defined differently for homogeneous and heterogeneous catalysis. In both cases, however, it is really the same thing, the *amount* of product formed per unit time per unit *amount* of catalyst. In homogeneous catalysis, the turnover frequency is the rate of formation of product, given in mol L^{-1} s^{-1}, divided by the concentration of catalyst used, in mol L^{-1}. This gives the turnover frequency in units of s^{-1}. In heterogeneous catalysis, the turnover frequency is typically the amount of product formed per unit time, given in mol s^{-1}, divided by the number of moles of catalyst present. In this case, the turnover frequency also has units of s^{-1}. Since one mole of a finely divided heterogeneous catalyst is more active than one mole of the same catalyst with a small surface area, the turnover frequency is sometimes expressed as the amount of product formed per unit time divided by the surface area of the catalyst. This gives the turnover frequency in units of mol s^{-1} cm^{-2}. Often one finds the turnover frequency for commercial heterogeneous catalysts expressed in rate per gram of catalyst.

(b) Selectivity? This is a measure of how much of the desired product is formed relative to undesired byproducts. Unlike enzymes (see Chapter 19), man-made catalysts rarely are 100% selective. The catalytic chemist usually has to deal with the often difficult problem of separating the various products. The separations, by distillation, fractional crystallization, or chromatography, are generally expensive and are always time-consuming. Furthermore, the byproducts represent a waste of raw materials. Recall that an expensive rhodium hydroformylation catalyst is sometimes used industrially instead of a relatively inexpensive cobalt catalyst because the rhodium catalyst is more selective (see Section 17.4).

(c) Catalyst? The first line of Chapter 17 reads "A catalyst is a substance that increases the rate of a reaction but is not itself consumed." This does not mean that the "catalyst" that is added to the reaction mixture is left unchanged during the course of the reaction. Frequently, the substance that is added is a *catalyst precursor* that is transformed under the reaction conditions into the active catalytic species.

(d) Catalytic cycle? This is a sequence of chemical reactions, each involving the catalyst, that transform the reactants into products. It is called a cycle because the actual catalytic species involved in the first step is regenerated during the last step. Note that the concept of a "first" and "last" step may lose its meaning once the cycle is started.

(e) Catalyst support? In cases where a heterogeneous catalyst does not remain a finely divided pure substance with a large surface area under the reaction conditions, it must be dispersed on a support material, which is generally a ceramic like γ-alumina or silica gel. In some cases, the support is relatively inert and only serves to maintain the integrity of the small catalyst particles. In other cases, the support interacts strongly with the catalyst and may affect the rate and selectivity of the reaction.

17.3 **Classify the following as homogeneous or heterogeneous catalysis? (a) The increased rate of SO_2 oxidation in the presence of NO?** The balanced equation for SO_2 oxidation is:

$$2\,SO_2(g) + O_2(g) \rightarrow 2\,SO_3(g)$$

All of these substances, as well as the catalyst, NO, are gases. Since they are all present in the same phase, this is an example of homogeneous catalysis (see Section 17.1).

(b) The hydrogenation of oil using a finely divided Ni catalyst? In this case, the balanced equation is:

$$RHC{=}CHR' + H_2 \rightarrow RH_2C{-}CH_2R'$$

The reactants and the products are all present in the liquid phase (the hydrogen is dissolved in the liquid oil), but the catalyst is a solid. Therefore, this is an example of heterogeneous catalysis.

(c) The conversion of D-glucose to a D,L mixture by HCl? The catalyst for the racemization of D-glucose is HCl (really H_3O^+), which is present in the same aqueous phase as the D-glucose. Therefore, since the substrate and the catalyst are both in the same phase, this is homogeneous catalysis.

$$
\begin{array}{ccc}
\text{CHO} & & \text{CHO} \\
\mid & & \mid \\
\text{H} \!-\! \text{C} \!-\! \text{OH} & & \text{H} \!-\! \text{C} \!-\! \text{OH} \\
\mid & & \mid \\
\text{HO} \!-\! \text{C} \!-\! \text{H} & \underset{\qquad\qquad}{\overset{\text{H}_3\text{O}^+}{\rightleftharpoons}} & \text{HO} \!-\! \text{C} \!-\! \text{H} \\
\mid & & \mid \\
\text{H} \!-\! \text{C} \!-\! \text{OH} & & \text{H} \!-\! \text{C} \!-\! \text{OH} \\
\mid & & \mid \\
\text{H} \!-\! \text{C} \!-\! \text{OH} & & \text{HO} \!-\! \text{C} \!-\! \text{H} \\
\mid & & \mid \\
\text{CH}_2\text{OH} & & \text{CH}_2\text{OH}
\end{array}
$$

D–Glucose   L–Glucose

17.4 **Which of the following processes would be worth investigating? (a) The splitting of H_2O into H_2 and O_2?** A catalyst does not affect the free energy (ΔG) of a reaction, only the activation free energy ($\Delta G^{\ddagger}$). A thermodynamically unfavorable reaction (one with a positive ΔG) will not result in a useful amount of products unless energy in the form of light or electric current is added to the reaction mixture. For example, if $\Delta G \approx +17$ kJ mol^{-1}, $K_{eq} \approx 10^{-3}$. Therefore, it would not be a worthwhile endeavor to try to develop a catalyst to split water into hydrogen and oxygen because water is a thermodynamically stable compound (i.e., it is stable with respect to its elements; $\Delta_f G°(\text{H}_2\text{O}) = -237$ kJ mol^{-1}).

(b) The decomposition of CO_2 into C and O_2? As in part (a), you would be trying to catalyzė the decomposition of a very stable compound into its constituent elements. It would be a waste of time.

$$\text{CO}_2(\text{g}) \rightarrow \text{C}(\text{s}) + \text{O}_2(\text{g}) \qquad \Delta G° = +394 \text{ kJ mol}^{-1}$$

(c) The combination of N_2 with H_2 to produce NH_3? This would be a very worthwhile reaction to try to catalyze efficiently at 80 °C. Ammonia is a stable compound with respect to nitrogen and hydrogen. Furthermore, it is a compound that is important in commerce, since it is used in many types of fertilizers. $\Delta G°$ is only -16.5 kJ mol^{-1} for the reaction $\text{N}_2 + 3\,\text{H}_2 \rightarrow 2\text{NH}_3$ (see Table 8.6). The high temperatures usually required for ammonia synthesis (~400°C) make ΔG less negative and result in a smaller yield. An efficient *low temperature* synthesis of ammonia from nitrogen and hydrogen that could be carried out on a large scale would probably make you and your industrialist rich.

(d) The hydrogenation of double bonds in vegetable oil? The reaction RHC=CHR + H_2 → $RH_2C–CH_2R$ has a negative $\Delta G°$ (see Section 17.4(a)). Therefore, the hydrogenation of vegetable oil would be a candidate for catalyst development. However, there are many homogeneous and heterogeneous catalysts for olefin hydrogenation, including those that operate at low temperatures (i.e. ~80 °C). For this reason, the process can be readily set up with existing technology.

17.5   **Why does the addition of PPh_3 to $RhCl(PPh_3)_3$ reduce the hydrogenation turnover frequency?** The catalytic cycle for homogeneous hydrogenation of alkenes by Wilkinson's catalyst, $RhCl(PPh_3)_3$, is shown in Cycle 2. Let us focus on the dominant path. The catalytic species that enters the cycle is not $RhCl(PPh_3)_3$ but $RhCl(PPh_3)_2(Sol)$ (Sol = a solvent molecule), formed by the following equilibrium:

$$RhCl(PPh_3)_3 + Sol \rightleftharpoons RhCl(PPh_3)_2(Sol) + PPh_3$$

Therefore, added PPh_3 will shift this equilibrium to the left, resulting in a lower concentration of the active catalytic species. Note that this is the only reaction that added PPh_3 can affect, since no other step involves free PPh_3.

17.6   **Explain the trend in rates of H_2 absorption by various olefins catalyzed by $RhCl(PPh_3)_3$?** The data show that hydrogenation is faster for hexene than for *cis*-4-methyl-2-pentene (a factor of 2910/990 = 3). It is also faster for cyclohexene than for 1-methylcyclohexene (a factor of 3160/60 = 53). In both cases, the alkene that is hydrogenated more slowly has a greater degree of substitution and so is sterically more demanding: hexene is a monosubstituted alkene while *cis*-4-methyl-2-pentene is a disubstituted alkene; cyclohexene is a disubstituted alkene while 1-methylcyclohexene is a trisubstituted alkene. According to the catalytic cycle for hydrogenation shown in Cycle 17.2, different alkenes could affect the equilibrium (b) $\rightleftharpoons$ (c) or the reaction (c) → (d). A sterically more demanding alkene could result in (i) a smaller equilibrium concentration of (c) ($RhClH_2L_2(alkene)$), or (ii) a slower rate of conversion of (c) to (d) ($RhClHL_2(alkyl)(Sol)$). Since the reaction (c) → (d) is the rate determining step for hydrogenation by $RhCl(PPh_3)_3$, either effect would lower the rate of hydrogenation.

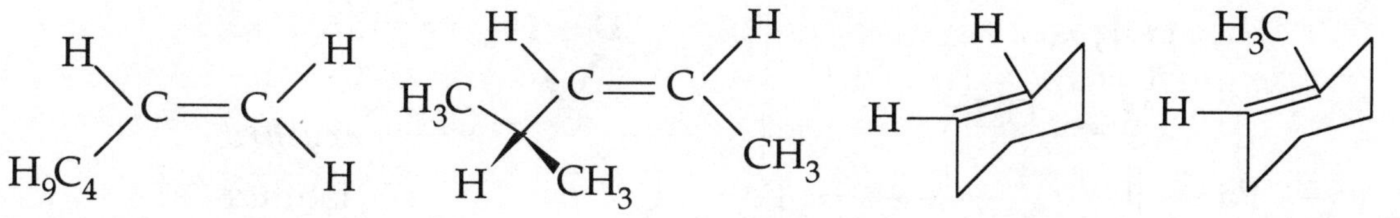

2910 L mol⁻¹s⁻¹   990 L mol⁻¹s⁻¹   3160 L mol⁻¹s⁻¹   60 L mol⁻¹s⁻¹

17.7 Hydroformylation catalysis with and without added P(n-Bu)$_3$?
Since compound (c) is observed under catalytic conditions, its formation must be faster than its transformation into products. If this were not true, a spectroscopically observable amount of it would not build up. Therefore, the transformation of (c) into CoH(CO)$_4$ must be the rate-determining step in the absence of added P(n-Bu)$_3$. In the presence of added P(n-Bu)$_3$, neither compound (c) nor its phosphine-substituted equivalent Co(C(=O)C$_4$H$_9$)-(CO)$_3$(P(n-Bu)$_3$) is observed, requiring that their formation must be slower than their transformation into products. Thus, in the presence of added P(n-Bu)$_3$, the formation of either (a), (c), or their phosphine-substituted equivalents is the rate-determining step.

17.8 The two likely mechanisms for the Wacker process? The two β-hydroxyethyl complexes that result from attack by free OH⁻ or coordinated OH⁻ on coordinated *trans*-1,2-C$_2$H$_2$D$_2$ are shown below. In the case of attack by coordinated OH⁻, two views of the same product are shown, differing by a free rotation of 180° about the C–C bond. You can see that the two products are stereochemical isomers, *erythro* and *threo*. If only the *erythro* product were formed, for example, you could safely infer that the reaction proceeded by the attack on the olefin by free OH⁻.

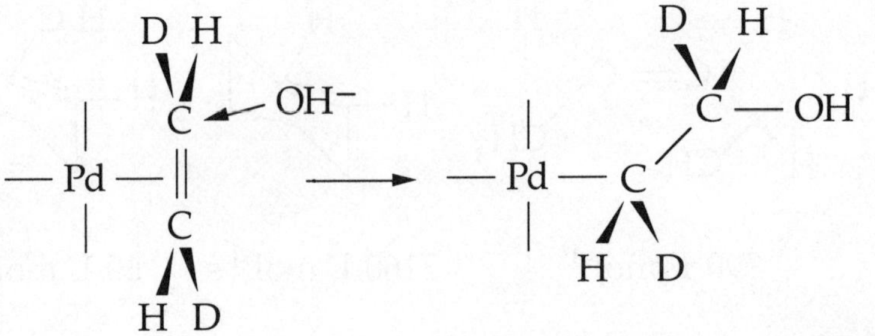

D H
C ← OH⁻
|
—Pd—‖
|
C
H D

D H
C—OH   *erythro*
| isomer
—Pd—C
|
H D

HO   D H
\   C
—Pd—‖
| C
H D

HO   H
\   C
—Pd—C   D
|
H D

H D
C—OH
|
≡ —Pd—C
|
H D

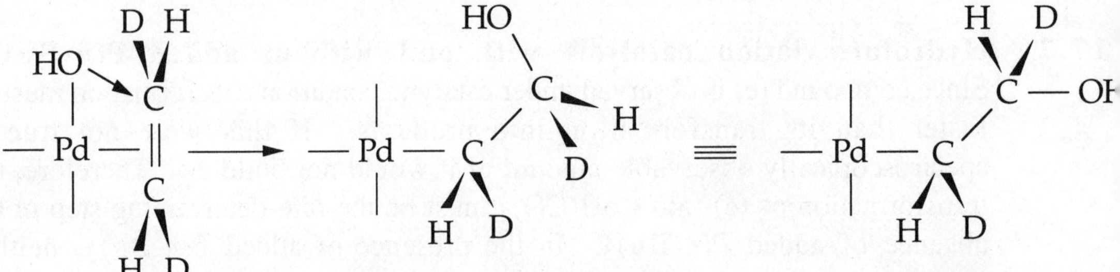

threo isomer

17.9   **A plausible mechanism for the catalysis of formic acid decomposition by IrClH₂(PPh₃)₃?** Formic acid can be decomposed by the following sequence of reactions: oxidative addition of the O–H bond, β-hydride elimination, reductive elimination of H₂, and CO₂ dissociation, as shown below for the generic *coordinatively unsaturated* metal complex M^0 (the reductive elimination of H₂ and CO₂ dissociation are combined into a single final step).

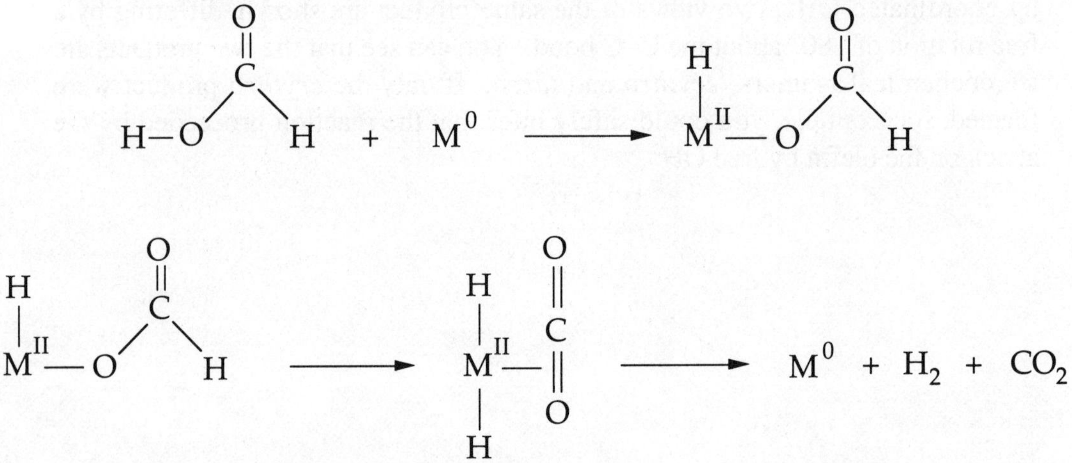

$$H–O \quad \overset{\overset{O}{\|}}{C} \quad H \; + \; M^0 \longrightarrow \overset{H}{\underset{}{M^{II}}}–O \quad \overset{\overset{O}{\|}}{C} \quad H$$

$$\overset{H}{\underset{}{M^{II}}}–O \quad \overset{\overset{O}{\|}}{C} \quad H \longrightarrow \overset{H}{\underset{H}{M^{II}}}–\overset{\overset{O}{\underset{\|}{C}}}{\underset{O}{}} \longrightarrow M^0 \; + \; H_2 \; + \; CO_2$$

The iridium complex $IrClH_2(PPh_3)_3$ contains Ir(III), is six-coordinate, and has 18 valence electrons. As such, it cannot coordinate to formic acid without prior dissociation of a phosphine ligand, and it cannot undergo oxidative addition of formic acid without prior reductive elimination of two other anionic ligands, such as the two hydride ligands.

A plausible mechanism for catalysis by $IrClH_2(PPh_3)_3$ is as follows:

$$IrClH_2(PPh_3)_3 \rightarrow IrClH_2(PPh_3)_2 + PPh_3$$

$$IrClH_2(PPh_3)_2 + HCOOH \rightarrow IrClH_2(PPh_3)_2(HCOOH)$$

$$IrClH_2(PPh_3)_2(HCOOH) \rightarrow IrCl(PPh_3)_2(HCOOH) + H_2$$

$$IrCl(PPh_3)_2(HCOOH) \rightarrow IrClH(HCO_2)(PPh_3)_2$$

$$IrClH(HCO_2)(PPh_3)_2 \rightarrow IrClH_2(CO_2)(PPh_3)_2$$

$$IrClH_2(CO_2)(PPh_3)_2 \rightarrow IrClH_2(PPh_3)_2 + CO_2$$

17.10 Why is titanium ineffective whereas iron is effective as a catalyst for ammonia synthesis? The problem with titanium is clearly stated in the exercise. Titanium reacts with N_2 to form a *stable* nitride. If the stability of a catalytic intermediate is too great, it will build up and will not be transformed into products. Refer to Figure 17.1, which shows schematic representations of the energetics in a catalytic cycle. The formation of titanium nitride would be represented by the deep trough (the dashed line). An efficient catalytic cycle must have a relatively smooth free energy vs. extent of reaction graph, free of deep troughs and high peaks.

17.11 Aluminosilicate surface acidity? (a) Give an explanation for the enhancement of acidity by the presence of Al^{3+} in a silica lattice? The substitution of Al(III) for Si(IV) in a silica lattice would lead to a charge imbalance (there would be one too few positive charges per Al/Si substitution). To compensate for the charge imbalance, an extra H^+ ion per Al/Si substitution is associated with the aluminosilicate lattice, as shown in the diagram accompanying the solution to Self-test S17.3. The Al–OH_2 groups are much more acidic than Si–OH groups.

(b) Name three other ions that might enhance the acidity of silica gel? A number of other +3 cations would lead to the same charge imbalance produced by substituting Al(III) for Si(IV). These include Sc(III), Cr(III), Fe(III), and Ga(III).

17.12 Indicate the difference between each of the following? (a) The Lewis acidity of γ-alumina that has been heated to 900°C versus γ-alumina that has been heated to 100 °C? As discussed in Example 17.2, γ-alumina that has not been heated to above 150 °C is completely hydroxylated. This means that all of the surface Al(III) ions are four-coordinate, and as such they will not exhibit appreciable Lewis acidity. On the other hand, γ-alumina is completely dehydroxylated by heating to 900 °C. The surface of dehydroxylated γ-alumina contains three-coordinate, strongly acidic Al(III) sites, as shown in Structure **7**.

(b) The Brønsted acidity of silica gel versus γ-alumina? While aluminosilicates are strong Brønsted acids (see the solution to Exercise 17.11), silica gel and γ-alumina are not. The Si–OH groups found on the surface of silica gel are moderate Brønsted acids (about as acidic as acetic acid; see the section on *Heterogeneous acid–base reactions* in Chapter 5). The Al–OH groups found on the surface of γ-alumina are weak Brønsted acids. They are weaker than Si–OH groups because of the lower charge of Al(III) relative to Si(IV). Recall that for aqua ions the charge on the metal was directly related to the Brønsted acidity of the ion (see Section 5.3). For example, $Fe(H_2O)_6^{3+}$ is six orders of magnitude more acidic than $Fe(H_2O)_6^{2+}$.

(c) The structural regularity of the channels and voids in silica gel versus that for ZSM-5? As can be seen from a schematic diagram of silica gel, such as the one shown in Figure 17.5, the structure of silica gel is quite irregular: channels and voids of a variety of sizes are evident. In contrast, the structure of Theta-1 zeolite, which is shown in Figure 17.6, has channels and voids of uniform size. Theta-1 is a metastable, crystalline material with a regular structure. Silica gel is a metastable amorphous solid.

17.13 Why is the platinum–rhodium in automobile catalytic converters dispersed on the surface of a ceramic rather than used in the form of thin foil? As discussed in Section 17.5, special measures are required in heterogeneous catalysis to ensure that the reactants achieve contact with catalytic sites. For a given amount of catalyst, the greater the surface area the greater the number of catalytic sites. A thin foil of platinum–rhodium will not have as much surface area as an equal amount of small particles finely dispersed on the surface of a ceramic support. The diagram below shows this for a "foil" of catalyst that is 1000 times larger in two dimensions than in the third dimension (i.e. the thickness is 1; note that the diagram is not to scale). If the same amount of catalyst is broken into cubes that are 1 unit on a side, the surface area of the catalyst is increased by nearly a factor of three.

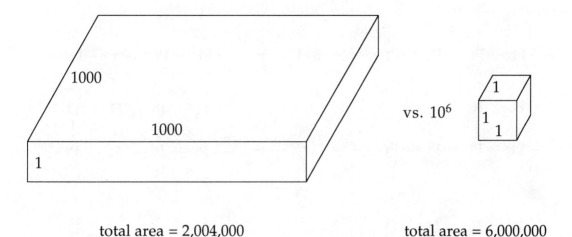

total area = 2,004,000   total area = 6,000,000

17.14 **Describe the concept of shape-selective catalysts?** Certain catalysts such as zeolites have channels and voids of specific sizes and shapes (see Section 10.14). It has been found that the product distribution can be changed by switching from one zeolite to another with different size channels and voids. There are two explanations, as described in Section 17.7. The first is that products of certain sizes diffuse through the zeolite lattice more quickly than others. The remaining products, with longer residence times, can isomerize to products that diffuse away quickly. According to this view, the relative rate of diffusion of a product *once it is formed* is the key property. The currently favored view is that the sizes and shapes of the channels and voids allow only certain products to be formed in the first place.

17.15 **Devise a plausible mechanism to explain the deuteration of 3,3-dimethylpentane?** There are *two* observations that must be explained. The first is that ethyl groups are deuterated before methyl groups. The second observation is that a given ethyl group is completely deuterated before another one incorporates *any* deuterium. Let's consider the first observation first. The mechanism of deuterium exchange is probably related to the reverse of the last two reactions in Figure 17.11, which shows a schematic mechanism for the hydrogenation of an olefin by D_2. The steps necessary for deuterium substitution into an alkane are shown below, and include the dissociative chemisorption of an R–H bond, the dissociative chemisorption of D_2, and the dissociation of R–D. This can occur many times with the same alkane molecule to effect complete deuteration.

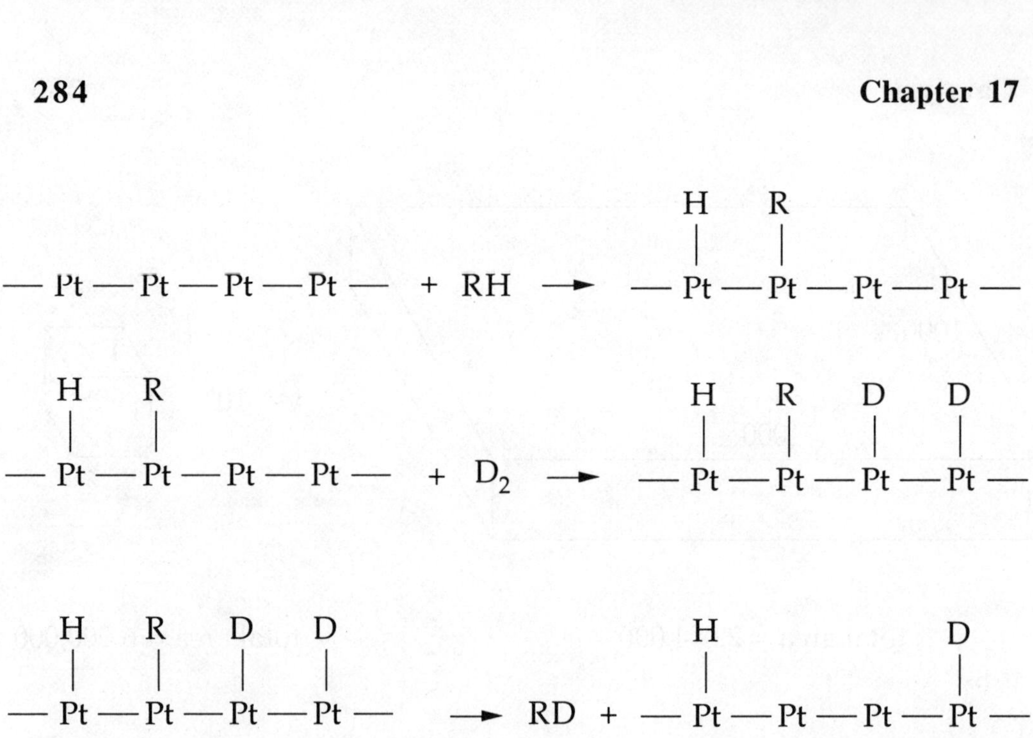

$$\text{— Pt — Pt — Pt — Pt —} \ + \ RH \ \longrightarrow \ \overset{\displaystyle H \quad\quad R}{\underset{\displaystyle \text{— Pt — Pt — Pt — Pt —}}{|\quad\quad\quad|}}$$

$$\overset{\displaystyle H \quad\quad R}{\underset{\displaystyle \text{— Pt — Pt — Pt — Pt —}}{|\quad\quad\quad|}} \ + \ D_2 \ \longrightarrow \ \overset{\displaystyle H \quad\quad R \quad\quad D \quad\quad D}{\underset{\displaystyle \text{— Pt — Pt — Pt — Pt —}}{|\quad\quad|\quad\quad|\quad\quad|}}$$

$$\overset{\displaystyle H \quad\quad R \quad\quad D \quad\quad D}{\underset{\displaystyle \text{— Pt — Pt — Pt — Pt —}}{|\quad\quad|\quad\quad|\quad\quad|}} \ \longrightarrow \ RD \ + \ \overset{\displaystyle H \quad\quad\quad\quad\quad D}{\underset{\displaystyle \text{— Pt — Pt — Pt — Pt —}}{|\quad\quad\quad\quad\quad|}}$$

Since the $-CH_2CH_3$ groups are deuterated before the $-CH_3$ groups, one possibility is that dissociative chemisorption of a C–H bond from a $-CH_2-$ group is faster than dissociative chemisorption of a C–H bond from a $-CH_3$ group. The second observation can be explained by invoking a mechanism for rapid deuterium exchange of the methyl group in the chemisorbed $-CHR(CH_3)$ group ($R = C(CH_3)_2(C_2H_5)$. The scheme below shows such a mechanism. It involves the successive application of the equilibrium shown in Figure 17.11. If this equilibrium is maintained more rapidly than the dissociation of the alkane from the metal surface, the methyl group in question will be completely deuterated before dissociation takes place.

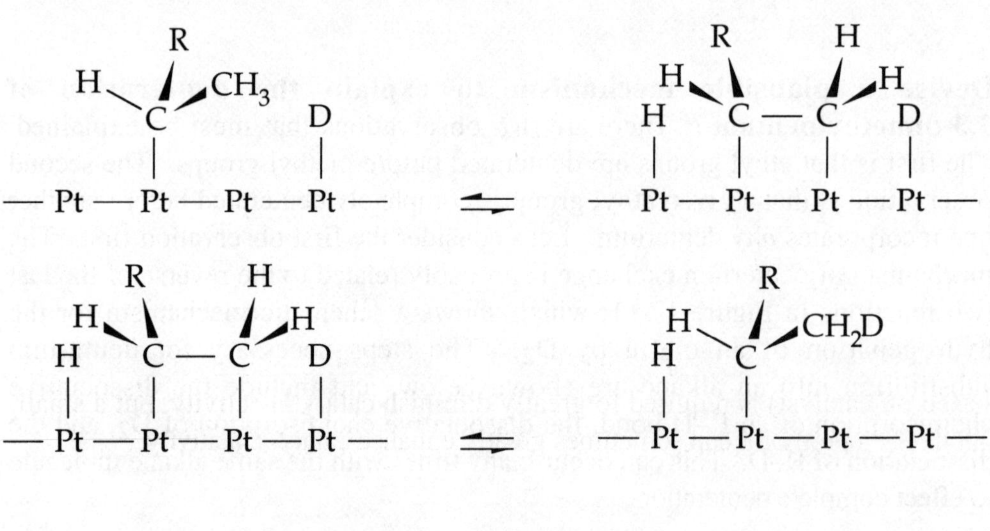

Another possibility, not shown in the scheme above, is that the terminal $-CH_3$ groups undergo more rapid dissociative chemisorption than the sterically more hindered internal $-CH_3$ groups.

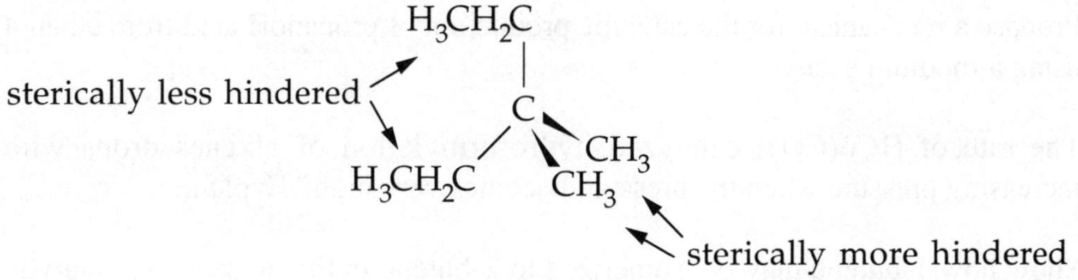

17.16   **Why does CO decrease the effectiveness of Pt in catalyzing the reaction $2\,H^+(aq) + 2\,e^- \rightarrow H_2(g)$?**   The reduction of hydrogen ions to H_2 probably involves the formation of surface hydride species similar to the ones shown in the figure on the opening page of this chapter in *Guide to Solutions*. Dissociation of H_2 by reductive elimination would complete the catalytic cycle. According to Table 17.4, platinum not only has a strong tendency to chemisorb H_2, but it also has a strong tendency to chemisorb CO. If the surface of platinum is covered with CO, the number of catalytic sites available for H^+ reduction will be greatly diminished and the rate of H_2 production will decrease.

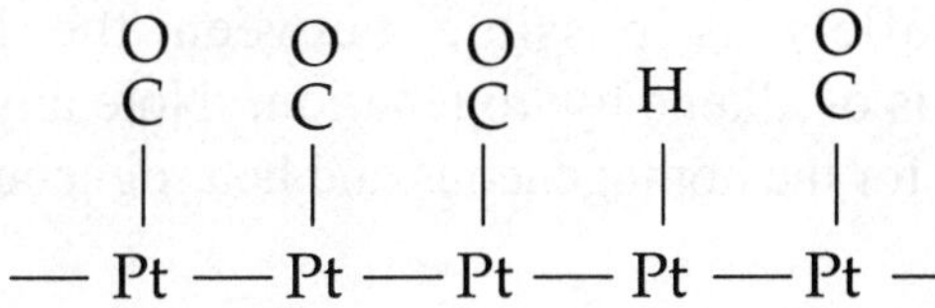

Guide to Solutions Quiz

1   Explain the difference between a catalyst and a catalyst precursor.

2   Explain the observation that addition of a stoichiometric amount of an inhibitor (based on catalyst) is required to greatly diminish catalytic activity, but a small amount of a promotor can sometimes greatly enhance catalytic acitivity.

3   Explain the following premise. If you can isolate a metal complex from a catalytic mixture and structurally characterize it, it is not likely to be a catalytic intermediate.

4   Propose a mechanism for the catalytic production of propanoic acid from ethanol using a rhodium catalyst.

5   The rate of $HCo(CO)_4$-catalyzed hydroformylation of alkenes drops with increasing pressure when the pressure becomes very high. Explain.

6   Show how 1-butene may be isomerized to 2-butene in the presence of catalytic amounts of $HCo(CO)_4$.

7   When the complex $RhH(P(OC_6H_5)_3)_4$ is heated in solution in the presence of D_2, the phenyl group *ortho* hydrogen atoms are exchanged with deuterium to form the product $RhD(P(OC_6H_3D_2)_3)_4$. Write a catalytic mechanism for this reaction.

8   Propose a mechanism for the $Pd(P(OC_6H_5)_3)_4$-catalyzed addition of HCN to alkenes, as shown in the following example (reactions such as this have been used in the production of Nylon):

$$CH_2=CH_2 + HCN \rightarrow CH_3CH_2CN$$

9   Draw as many parallels as possible between the homogeneous and heterogeneous catalysis of alkene hydrogenation. Note any reaction types that are similar chemically for the homogeneous and heterogneous reactions but have different names.

10   By "dealuminating" certain zeolites (Section 10.14), it is possible to make heterogeneous catalysts that approach SiO_2 in composition while still retaining zeolite architecture. What industrial value might such solids have? Explain your answer.

18 Structures and properties of solids

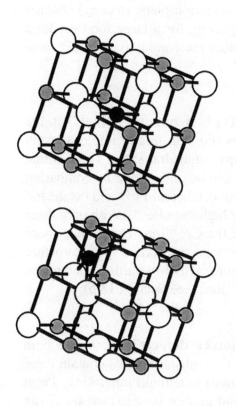

All solids contain defects, which influence properties such as mechanical strength and conductivity. The top figure shows the ideal structure of AgCl with no defects (the Cl^- ions are the larger open spheres). All of the Ag^+ ions, including the highlighted black one, are in octahedral holes of Cl^- ions. In a real sample of AgCl, some of the Ag^+ ions are displaced from octahedral sites to tetrahedral sites, as shown for the black Ag^+ ion in the bottom figure. This is an example of a Frenkel defect, which is a type of intrinsic point defect.

S18.1 Does the measured density of VO indicate the presence of vacancies or interstitials? As in the example, a measured density that is lower than the calculated density is indicative of vacancies (Schottky defects). This is because the volume of the crystal remains essentially constant, but some of the mass is missing. If the measured and calculated densities are nearly the same, the defects are probably interstitial (Frenkel defects). In the case of VO, the measured density is only about 91% of the calculated density. For the 1:1 ratio to be maintained, the vacancies must be on both V and O sites.

S18.2 **Why does increased pressure reduce the conductivity of K⁺ more than that of Na⁺ in β-alumina?** Increasing the pressure on a crystal compresses it, reducing the spacings between ions (for example, subjecting NaCl to a pressure of 24,000 atm reduces the Na–Cl distance from 2.82 Å to 2.75 Å). In a rigid lattice such as β-alumina, larger ions migrate more slowly than smaller ions with the same charge (as discussed in the example). At higher pressures, with smaller conduction plane spacings, *all* ions will migrate more slowly than they do at atmospheric pressure. However, larger ions will be impeded to a greater extent by smaller spacings than will smaller ions. This is because the ratio (radius of migrating ion)/(conduction plane spacing) changes more, per unit change in conduction plane spacing, for a large ion than for a small ion. Therefore, increased pressure reduces the conductivity of K⁺ more than that of Na⁺ because K⁺ is larger than Na⁺.

S18.3 **Rationalize the observation that $FeCr_2O_4$ is a normal spinel?** In the normal AB_2O_4 spinel structure, the A^{2+} ions (Fe^{2+} in this example) occupy tetrahedral sites and the B^{3+} ions (Cr^{3+}) occupy octahedral sites. The fact that $FeCr_2O_4$ exhibits the normal spinel structure can be understood by comparing the ligand field stabilization energy of high-spin octahedral Fe^{2+} and octahedral Cr^{3+}. With six d electrons, LFSE = 0.4 Δ_o for high-spin Fe^{2+}. With only three d electrons, LFSE = 1.2 Δ_o for Cr^{3+}. Since the Cr^{3+} ions experience more stabilization in octahedral sites than do the Fe^{2+} ions, the normal spinel structure is more stable than the inverse spinel structure, which would exchange the positions of the Fe^{2+} ions with half of the Cr^{3+} ions (see Section 18.5).

18.1 **Describe the nature of Frenkel and Schottky defects?** These are both called intrinsic point defects. Intrinsic means that substances that contain these defects may be pure substances; they do not have to contain impurities. Point defects occur at single sites in the lattice and cannot be detected by X-ray crystallography or by electron microscopy, in contrast with extended defects. Frenkel and Schottky defects differ in the way the single sites are affected. A Frenkel defect involves an atom or ion moving from its usual site to an interstitial site (see the figure on the opening page of this chapter). Frenkel defects are difficult to verify experimentally; usually their presence is inferred from ionic conductivity data. A Schottky defect is a vacancy in an otherwise perfect lattice. At low vacancy concentrations, the displaced atoms can take up sites on the crystal surface. Since a substance with Schottky defects contains vacancies, the measured density is lower than the calculated density. The stoichiometry of substances that contain Frenkel and Schottky defects is not changed by the presence of the defects.

18.2 **Pick the compound most likely to have a high concentration of defects?** **(a) NaCl or NiO?** To answer this question, the two things you want to consider are the openness of the structure (open structures provide larger sites that can accommodate interstitial atoms) and the possibility of variable oxidation states. In this case the two compounds both have the rock salt structure (see Table 9.8). Furthermore, inspection of Table 1.5 should lead to the conclusion that ionic radii of Na^+ and Ni^{2+} do not differ a great deal (the table contains the radii for Na^+ and Ca^{2+}, and Ni^{2+} will be somewhat smaller than Ca^{2+} due to imperfect shielding by d electrons). Therefore, the openness of the structure is not an issue. Variable oxidation states are not possible for sodium, but they are for nickel. NiO could contain some Ni^{2+} ion vacancies, with two remaining Ni^{2+} ions oxidized to Ni^{3+} to compensate for each Ni^{2+} vacancy (see Section 18.4). Thus, you should expect NiO to contain more defects than NaCl. The defects will be Ni^{3+} ions occupying Ni^{2+} sites and vacancies on the Ni^{2+} sublattice.

(b) CaF$_2$ or PbF$_2$? In this case too, the two compounds have the same structure (the fluorite structure). However, unlike Na^+ vs. Ni^{2+}, the two cations in question here have very different ionic radii, and Pb^{2+} is much more polarizable. Therefore, you should expect PbF_2 to contain more interstitial defects (Frenkel defects) than CaF_2 (see Section 18.5). The F^- ions in a perfect crystal of PbF_2 would occupy all of the tetrahedral interstitial sites in the close-packed Pb^{2+} ion lattice, with all of the octahedral interstitial sites vacant. Therefore, F^- ions in PbF_2 might be expected to vacate some tetrahedral sites and to occupy some octahedral sites. You might also be alert to the fact that lead can exhibit two stable oxidation states, Pb^{2+} and Pb^{4+}. However, the displacement of F^- from tetrahedral to octahedral sites appears to account for most of the defects in PbF_2, rather than the alternative possibility of Pb^{4+} occupying some Pb^{2+} sites.

(c) Al$_2$O$_3$ or Fe$_2$O$_3$? Both of these compounds have the corundum structure, and both Al^{3+} and Fe^{3+} have approximately the same ionic radius. The important difference in this case is the possibility of variable oxidation states for iron. Fe_2O_3 might contain O^{2-} ion vacancies that are charge compensated by the presence of some Fe^{2+} ions.

18.3   Distinguish intrinsic from extrinsic defects?   Intrinsic defects do not alter the stoichiometry of the compound.   They may involve charge compensating vacancies of cations and anions or the movement of ions from their normal sites to interstitial sites.   Examples are Schottky and Frenkel defects. Extrinsic defects change the elemental composition of the compound in one of two ways.   Sometimes another element is added to the compound (this can be called an impurity if the new element is added accidentally, or a dopant if it is added intentionally).   In other cases, extrinsic defects result from vacancies that are charge compensated by oxidation or reduction of some of the remaining ions. In these cases, no new elements are present, but the original elements are not necessarily present in the ratio of whole numbers (i.e. a nonstoichiometric compound results from this type of extrinsic defect).   Examples are As-doped Si, color centers, and $VO_{0.8}$.

18.4   Where might intercalated Na^+ ions reside in the ReO_3 structure? The unit cell for ReO_3, which is shown in Figure 18.22(a), is reproduced in the figure on the left below (the O^{2-} ions are the open spheres).   As you can see, the structure is very open, with a very large hole in the center of the cell (if the Re–O distance is a, the distance from an O^{2-} ion to the center of the cell is $1.414a$).   A ReO_3 unit cell containing a Na^+ ion (large, heavily shaded sphere) at its center is shown on the right below.

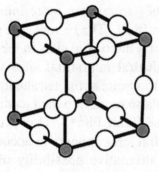

ReO₃

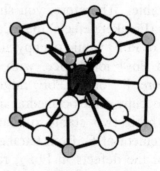

NaₓReO₃

18.5   What is a crystallographic shear plane?   A crystallographic shear plane may be considered a defect or a way of describing a new structure (see the end of Section 18.1).   When crystallographic shear planes are distributed randomly throughout the solid, they are called Wadsley defects.   A continuous range of composition is possible, because a new, discrete phase has not been formed. For example, tungsten oxide can have a composition ranging from WO_3 to $WO_{2.93}$. However, at W/O ratios higher than $1/(2.93)$, the shear planes are distributed in a non-random, periodic manner (i.e. a new stoichiometric phase has been formed, for example $W_{20}O_{58}$ (W/O = $1/(2.90)$))). Thus, crystallographic shear planes are defects when disordered and give a new phase with a new structure when ordered.

18.6   **Interstitial sites in the rock salt structure?**   A figure showing the rock salt structure is shown on the left below. The anions are the larger spheres, most of which are not shaded and three of which are heavily shaded. One cation is black and the others are lightly shaded. Since the cations in the rock salt structure occupy *all* of the octahedral interstices in the close-packed anion lattice, the only available sites for cation migration are tetrahedral interstices. In the figure below, the black cation is shown migrating to a tetrahedral site, through "the bottleneck," a trigonal planar array of anions (the three heavily shaded spheres). Note that there are eight equivalent trigonal planar arrays leading to eight equivalent tetrahedral sites for the migrating black cation.

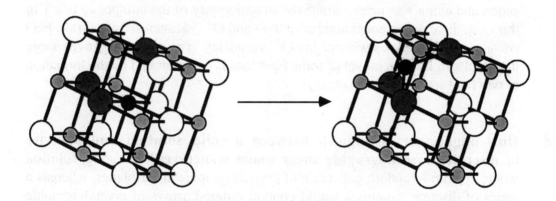

A simpler way to represent this situation is shown below. The two close-packed layers are drawn separately and superimposed. In the separate layers, the position of the black cation is indicated by the letter O (it is in an octahedral hole between the two layers), while the six tetrahedral holes between these two layers are indicated by the letter T. In addition to migrating to any one of the six T positions, the black cation can migrate to a tetrahedral hole above O (between the nonshaded layer of anions and a third layer) or below O (between the shaded layer of anions and a fourth layer).

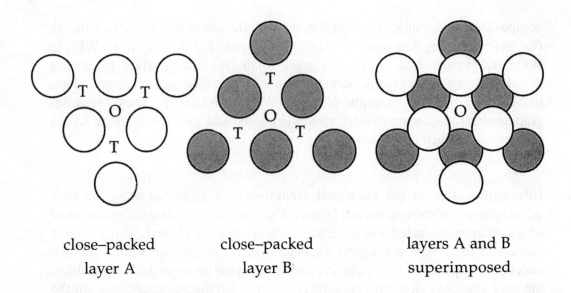

close–packed   close–packed   layers A and B
   layer A   layer B   superimposed

18.7   Contrast the nature of the defects in TiO with those in FeO?
Titanium monoxide contains a very high concentration (~12 M) of Schottky
defects (see Section 18.1). Remember that this type of defect results in both
cation and anion vacancies. Since the stoichiometry of the compound is 1:1 in
this case, there are an equal number of Ti^{2+} and O^{2-} vacancies. In contrast, FeO
contains Fe^{2+} vacancies without any O^{2-} vacancies. The excess negative charge
that results from the removal of some Fe^{2+} ions is compensated by the formation
of two Fe^{3+} ions per Fe^{2+} vacancy.

**18.8   How might you distinguish between a solid solution and a series
of discrete crystallographic shear plane structures?** A solid solution
would contain a random collection of crystallographic shear planes, whereas a
series of discrete structures would contain ordered arrays of crystallographic
shear planes. These two possibilities are represented in the figures below.

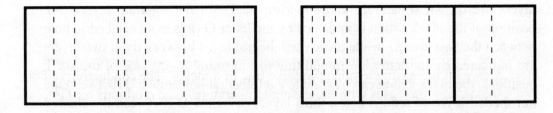

solid solution (a random   discrete phases (a series of
  array of shear planes)   ordered arrays of shear planes)

Due to the lack of long range order, the solid solution would give rise to an electron micrograph showing a random distribution of shear planes. In addition, the solid solution would not give rise to new X-ray diffraction peaks. In contrast, the ordered phases of the solid on the right would be detectable by electron microscopy and by the presence of a series of new peaks in the X-ray diffraction pattern, arising from the evenly spaced shear planes.

18.9   **The ease of oxidation and reduction of TiO, MnO, and NiO?** The standard potentials for oxidation and reduction can be found in Appendix 2. Since $\Delta G° = -nFE°$, the more positive (or less negative) the potential, the more favorable the process. Consider the following oxidations:

$$Ti^{2+}(aq) \rightarrow Ti^{3+}(aq) + e^- \qquad E° = 0.37 \text{ V}$$

$$Mn^{2+}(aq) \rightarrow Mn^{3+}(aq) + e^- \qquad E° = -1.5 \text{ V}$$

$$Ni^{2+}(aq) + 2H_2O(l) \rightarrow NiO_2(s) + 2e^- + 4H^+(aq) \qquad E° = -1.593 \text{ V}$$

According to the criterion mentioned in the exercise, NiO will be the most difficult of these three compounds to oxidize, since Ni^{2+} is the most difficult of the three aqua ions to oxidize. Now consider the following reductions:

$$Ti^{2+}(aq) + 2e^- \rightarrow Ti(s) \qquad E° = -1.63 \text{ V}$$

$$Mn^{2+}(aq) + 2e^- \rightarrow Mn(s) \qquad E° = -1.18 \text{ V}$$

$$Ni^{2+}(aq) + 2e^- \rightarrow Ni(s) \qquad E° = -0.257 \text{ V}$$

Since Ni^{2+} is the easiest of the three aqua ions to reduce, NiO is likely to be the easiest of the three monoxides to reduce. In general, you expect gross similarities between redox chemistry of ions in aqueous solution and of ions in metal oxides.

18.10   **The electrical conductivity of TiO and NiO?** TiO is a metallic conductor while NiO is a semiconductor (see Section 18.4). The conductivity of TiO decreases with increasing temperature, like that of a metal. The conductivity of NiO, in contrast, increases with increasing temperature. The d orbitals on Ti^{2+}, which is early in period 4, are larger than those on Ni^{2+}, which is in the middle of period 4 (see figures below). The d orbitals on the Ti^{2+} ions in TiO are large enough to overlap and form bands that are partly filled, a characteristic of metallic conductors. This cannot occur in NiO.

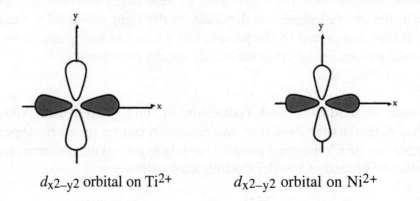

$d_{x^2-y^2}$ orbital on Ti^{2+} $d_{x^2-y^2}$ orbital on Ni^{2+}

18.11 **The Ag^+ ion conductivity of AgI?** The diagram for this experiment is shown on the right. The shaded area between the two silver electrodes is a pellet of AgI that is heated to 165°C. When the two electrodes are connected to a 0.1 V battery, a current passes through what has become a complete circuit.

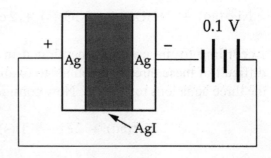

The negative pole of the battery (the left side) is a source of electrons. As current passes, Ag^+ ions in the pellet of AgI migrate to the right through the lattice and combine with electrons to form Ag atoms, which plate out on the electrode (i.e. the Ag^+ ions are reduced). At the electrode on the left side of the apparatus, Ag atoms are oxidized to Ag^+ ions and free electrons. The Ag^+ ions migrate through the lattice and the electrons pass through the external wire to the positive pole of the battery (the right side). The mass of the left electrode decreases while the mass of the right electrode increases.

18.12 **The changes in the silver anode and cathode in a Ag^+ ion conductance experiment?** In the Ag/AgI/Ag cell shown above, the cathode is the silver electrode on the right side (i.e. the electrode at which reduction of Ag^+ ions to Ag atoms occurs). The anode is the left electrode. As current passes, Ag atoms in the left electrode are oxidized to Ag^+ ions, which migrate through the hot AgI to the right electrode, where they are reduced back to Ag atoms and plate out on the electrode. Therefore, over a period of time the left electrode loses mass and the right electrode gains mass. After 10^{-3} mole of electrons have passed through the cell, 10^{-3} mole of Ag atoms will have migrated from the left electrode to the right electrode. Therefore, the left electrode will have lost $(10^{-3} \text{ mol})(108 \text{ g mol}^{-1}) = 0.108$ g Ag and the right electrode will have gained 0.108 g Ag. There will not be any change in the mass of the AgI pellet.

18.13 Contrast the mechanism of interaction between unpaired spins in a ferromagnetic material with that for an antiferromagnet? As discussed in Box 18.1, in a ferromagnetic substance the spins on different metal centers (atoms or ions) are coupled into a *parallel* alignment. In the diagram on the left below, each arrow represents the spin on one metal center. Ordered

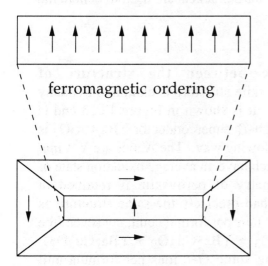

ferromagnetic ordering   antiferromagnetic ordering

One possible domain structure for a ferromagnetic crystal. The arrows represent the net magnetic moment for each domain. The crystal as a whole has a net magnetic moment.

The direction of the net moment =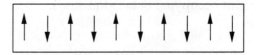

regions of a sample, called domains, are shown in the bottom figure. In an antiferromagnetic substance, the spins on different metal centers are coupled into an *antiparallel* alignment. As the temperature approaches absolute zero, the net magnetic moment of a ferromagnetic domain becomes very large, while that of an antiferromagnetic domain goes to zero.

18.14 Suggest a distribution of cations between octahedral and tetrahedral sites for CoFe$_2$O$_4$? This compound is an example of an AB$_2$O$_4$

oxide. These compounds either exhibit the spinel ($MgAl_2O_4$) or inverse spinel (e.g. Fe_3O_4) structure. When paramagnetic ions are present, both types of structures exhibit antiferromagnetism: the tetrahedral ions have their spins aligned antiparallel to the spins of the octahedral ions. The cations in $CoFe_2O_4$ are Co^{2+} (d^7) and Fe^{3+} (d^5). In the spinel structure, the +2 cations occupy tetrahedral sites and the +3 cations occupy octahedral sites. If $CoFe_2O_4$ exhibited the spinel structure, there would be 7 spins per formula unit (3 spins "up" from one Co^{2+} ion and 10 spins "down" from two Fe^{3+} ions). In the inverse spinel structure, half of the +3 ions occupy tetrahedral sites and the other ions occupy octahedral sites. If $CoFe_2O_4$ exhibited the inverse spinel structure, there would be 3 spins per formula unit (5 spins "up" from one Fe^{3+} ion and 8 spins "down" from one Co^{2+} ion and one Fe^{3+} ion). Since the observed number of spins is 3.4, it appears as though $CoFe_2O_4$ has the inverse spinel structure. Using the notation introduced in Section 18.5, you could write the formula of $CoFe_2O_4$ as $Fe[CoFe]O_4$ (recall that the square brackets are used to denote the ions occupying octahedral sites).

18.15   Describe the difference between the structure of $YBa_2Cu_3O_7$ and perovskite? The perovskite structure is exhibited by many ABO_3 oxides (the prototype is $CaTiO_3$). It is shown in Figure 18.23 and is reproduced below. The structure of the high-T_c superconductor $YBa_2Cu_3O_7$ is related to the perovskite structure in the following way. The A ions are Y^{3+} *and* Ba^{2+} ions. The B ions are copper ions, which have an average oxidation state of +2.33. Thus, the 1:1 cation stoichiometry of perovskite is retained in $YBa_2Cu_3O_7$. However, if $YBa_2Cu_3O_7$ had precisely the same structure as perovskite, there would have to be nine O^{2-} ions per formula unit, not seven (the formula unit would be $(Y_{1/3}Ba_{2/3}CuO_3)_3 = YBa_2Cu_3O_9 \neq YBa_2Cu_3O_7$). Therefore, this superconductor is missing some O^{2-} ions per formula unit relative to perovskite. For this reason, none of the copper ions are six-coordinate. Instead, some are five-coordinate and some are four-coordinate. Furthermore, the Y^{3+} and Ba^{2+} ions do not have the twelve-coordinate dodecahedral coordination of the Ca^{2+} ions in perovskite.

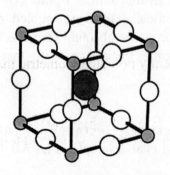

The structure of perovskite, $CaTiO_3$. The open spheres are O^{2-} ions, the small shaded spheres are Ti^{4+} ions, and the large shaded sphere in the center of the cell is the Ca^{2+} ion.

Guide to Solutions Quiz

1 Can an antiferromagnetic substance be distinguished from a paramagnetic substance by measurement of the magnetic susceptibility of both at a given temperature? Explain. Hint: Draw plots of χ vs. T for both types of substances on the graph below.

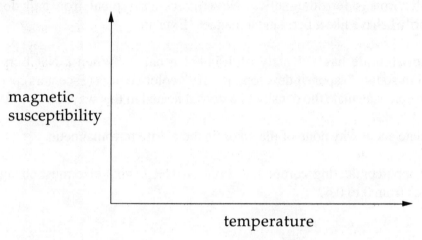

magnetic susceptibility

temperature

2 Explain why Fe_3O_4 has the inverse spinel structure and Mn_3O_4 has the normal spinel structure.

3 What relationship between the radii of the A and B cations favors the inverse spinel structure over the normal spinel structure for AB_2O_4 oxides, all other things being equal?

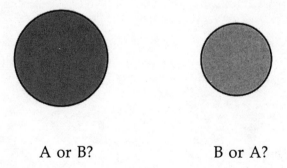

A or B?   B or A?

4 Why are transition metal oxides much more frequently non-stoichiometric than main group metal oxides?

5 When ZnO is heated, it forms the non-stoichiometric compound $Zn_{1.00007}O$. In contrast, when CdO is heated, it forms $Cd_{0.9995}O$. Suggest an explanation.

Hint: ZnO has the zinc-blende structure and CdO has the sodium chloride structure.

6   If lithium is substituted for sodium in a modified silica glass, what should be the effect on the glass transition temperature?

7   Metallic iron is ferromagnetic. Nevertheless, a typical iron nail does not ordinarily behave like a permanent magnet. Explain.

8   Sodium chloride has a density of 1.54424 g cm^{-3}. When a NaCl crystal is heated in sodium vapor, it develops 1×10^{19} color centers (F-centers) per cubic centimeter. Calculate the density of a crystal heated in this way.

9   Speculate about why none of the $4d$ or $5d$ metals are ferromagnetic.

10   In the superconducting compound YBa$_2$Cu$_3$O$_{7-d}$, what else must change as d changes from 0 to 0.4?

19 Bioinorganic chemistry

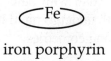

iron porphyrin

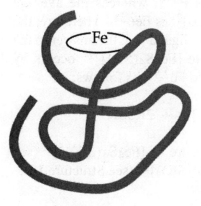

heme protein

Bioinorganic chemistry is concerned with the roles of metal ions in biological systems. Many of the spectroscopic tools that inorganic chemists use to study the kinds of complexes discussed in earlier chapters can be used to study metalloproteins and other metal-containing biomolecules. However, the structure of the active sites in proteins cannot be determined very accurately using X-ray diffraction. Therefore, bioinorganic chemists often study the structures of low molecular weight "model" complexes and make the following inference: if there is a spectral congruence between the model and the protein, it is assumed that there is a structural congruence as well.

S19.1 **Comment on the extent of Co to O_2 electron transfer?** The oxygen species present in KO_2 and BaO_2 are the superoxide ion (O_2^-) and the peroxide ion (O_2^{2-}). The O–O bond distance in $[Co(CN)_5O_2]^{3-}$ is just a little longer than the distance in O_2, so very little electron transfer is evident. The O–O distances in $[Co(bzacen)(py)O_2]$ and $[(NH_3)_5CoO_2Co(NH_3)_5]^{5+}$ are similar to the distance in KO_2, so they are thought to contain an oxygen species similar in electronic structure to the superoxide ion. The O–O distance in $[(NH_3)_5CoO_2Co(NH_3)_5]^{4+}$ is nearly the same as in BaO_2, so the oxygen species is inferred to be similar in electronic structure to peroxide ion. If you calculate the apparent oxidation state of cobalt in each of these four complexes assuming that the above charges on the oxygen species are correct, you will find that it is Co(II) in $[Co(CN)_5O_2]^{3-}$ and Co(III) for the other three.

S19.2 Reasons Nature might select Cu^{2+} for enzymology? You should consider the various differences between Cu^{2+} and, for example, Zn^{2+}. The most obvious difference is that copper is a redox-active metal so, if the catalysis involved electron transfer steps, copper might be selected over zinc. Another difference is that Cu^{2+} binds ligands more strongly than Zn^{2+} (as well as more strongly than most of the other divalent d-block metal ions), so substrate binding to the enzyme active site would be tighter with copper than with zinc. Two advantages that both Cu^{2+} and Zn^{2+} share is that they both undergo rapid ligand exchange reactions and they both have flexible coordination geometries.

S19.3 Difficulties in the preparation of $[Fe_2S_2(SR)_4]^{2-}$? The oxidation state of the two Fe atoms in this dinuclear cluster is Fe^{3+}, whereas the average oxidation state of the four Fe atoms in $[Fe_4S_4(SR)_4]^{2-}$ is $Fe^{2.5+}$. Therefore, the difficulty is that the Fe^{3+} is too oxidizing for a SR^- ligand (see the answer to the example). The conversion of $[Fe_2S_2(SR)_4]^{2-}$ to $[Fe_4S_4(SR)_4]^{2-}$ occurs by oxidation of SR^- to RSSR and reduction of Fe^{3+} to $Fe^{2.5+}$, as follows:

$$2[Fe_2S_2(SR)_4]^{2-} \rightarrow [Fe_4S_4(SR)_4]^{2-} + 2RSSR$$

Notice how each of the six faces of the Fe_4S_4 core of $[Fe_4S_4(SR)_4]^{2-}$ (see Structure 13) is similar to the Fe_2S_2 core of $[Fe_2S_2(SR)_4]^{2-}$ (see Structure **12**).

S19.4 Testing the Marcus theory? Below is an abbreviated table:

entry	$E°$ (V)	distance (Å)	ratio of k_r's
first	0.2	8	
sixth	0.2	10	600
third	0.4	8	
eighth	0.4	16	40,000
fourth	1.1	8	
ninth	1.0	16	22

The two members of each pair of entries have the same net potential. In all three cases, the redox couple with the shorter distance has the faster rate, consistent with the argument that the probability of finding the electron on the other side of a high barrier decreases with increasing width of the barrier. If you compare the first pair of entries with the second pair, you can see that a doubling of the distance produces a much larger decrease in the rate constant than simply changing the distance from 8 Å to 10 Å. However, the third pair of entries shows that the distance change is not the only parameter that is important. A doubling of the distance here produces a relatively small change in k_r, so other factors must come into play.

S19.5 The electron configuration of the resting state of P-450? The Fe(III) oxidation state corresponds to a d^5 configuration. Most five-coordinate Fe(III) porphyrin complexes are high-spin, not low spin. Therefore, each d orbital has one electron. The d-orbital splitting pattern for a C_{4v} complex is shown in Figure 13.14. For the resting state of P-450, it would look like the figure on the right.

$$\uparrow \quad b_1 (x^2-y^2)$$
$$\uparrow \quad a_1 (z^2)$$
$$\uparrow\downarrow \quad b_2 (xy)$$
$$\uparrow\downarrow\;\uparrow\downarrow \quad e (xz, yz)$$
$$C_{4v} \quad Fe^{3+}$$

19.1 Identify where the following elements are concentrated in animals and describe a major function? **(a) Oxygen?** The most important concentration of oxygen in animals is in water, which makes up a significant fraction of the mass of any animal. Oxygen is also an important component of proteins and nucleic acids and of phosphate ion in bones.

(b) Nitrogen? The most important sources of nitrogen in animals is in proteins and nucleic acids. In many cases its Brønsted basicity is its important function, whether it is acting as a catalyst or as a hydrogen bond acceptor to influence the three-dimensional structure of proteins and nucleic acids.

(c) Potassium? Potassium ions are concentrated inside the cells of organisms. An important function of K^+ is to promote the hydrolysis of phosphorylated biomolecules.

(d) Calcium? Calcium is concentrated in bones, shells, and other structural materials in animals. It is also found in a large variety of proteins, where it is normally bound by carboxylate groups of amino acid side-chains. Its role in many of these proteins is to induce conformational changes.

19.2   The ligands that bind Ca^{2+} and Fe^{2+}?   The usual ligands for calcium ions in proteins are carboxylate side-chains of selected amino acids. In contrast, the ligands used to bind divalent iron in hemoglobin are nitrogen heterocycles such as the porphyrin macrocycle and an imidazole side-chain of the amino acid histidine. This difference is sensible, since Ca^{2+} is a hard Lewis acid and Fe^{2+} is a moderate Lewis acid (i.e. softer than Ca^{2+}). The carboxylate oxygen atoms are hard, and the heterocyclic nitrogen atoms are relatively soft, so the typical preference of hard acids for hard bases and soft acids for soft bases is observed.

19.3   Bond length and net spin for O$_2$, O$_2^-$, and O$_2^{2-}$ as ligands?   According to Figure 3.17, O$_2$, O$_2^-$, and O$_2^{2-}$ would have 2, 3, and 4 electrons, respectively, in the $2\pi_g$ set of antibonding orbitals. O$_2$, with the fewest anti-bonding electrons, will have the strongest and shortest bond, while O$_2^{2-}$, with the greatest number of antibonding electrons, will have the weakest and longest bond. With two unpaired electrons, O$_2$ will have a net spin of 1. With one unpaired electron and no unpaired electrons, respectively, O$_2^-$ and O$_2^{2-}$ will have net spins of 1/2 and 0, respectively.

$2\pi_g$ O$_2$ O$_2^-$ O$_2^{2-}$

19.4   Oxygen affinity of hemoglobin and myoglobin?   The net reaction Hb + Hb(O$_2$)$_4$ $\rightleftharpoons$ 2 Hb(O$_2$)$_2$ is the sum of the two equilibria:

$$Hb + 2O_2 \rightleftharpoons Hb(O_2)_2$$

$$Hb(O_2)_4 \rightleftharpoons Hb(O_2)_2 + 2O_2$$

The first equilibrium lies to the left (i.e. hemoglobin that has no O$_2$ bound to it has a relatively low affinity for O$_2$). The second equilibrium also lies to the left (in this case, consider the reverse reaction — hemoglobin that has some O$_2$ bound to it has a high affinity for O$_2$). Therefore, the net reaction lies to the left. The position of the equilibrium Hb(O$_2$)$_4$ + 4 Mb $\rightleftharpoons$ Hb + 4 Mb(O$_2$) depends on the partial pressure of O$_2$. At low partial pressures, where the Hb curve is well below the Mb curve (see Figure 19.8), this reaction lies to the right (myoglobin has a much higher affinity for O_2 than hemoglobin at low partial pressures). At high partial pressures of O_2, the affinity of hemoglobin for O_2 approaches that of myoglobin, and the reaction lies farther to the left.

19.5 A mechanism for CO poisoning? Since O_2 and CO are similar electronically as well as sterically, they can both bind to the same sites in metalloproteins such as hemoglobin. In fact, hemoglobin and myoglobin (and model iron porphyrin complexes) have a higher affinity for CO than for O_2. Therefore, if CO is present, it can bind to, and block, the oxygen transport sites of hemoglobin, thus preventing oxygen from being distributed to various tissues.

19.6 Peptide synthesis at Co(III) centers? The ester carbonyl C atom is the electrophilic center that is subject to nucleophilic attack. Coordination of the carbonyl O atom to Co(III) increases the partial positive charge on this C atom, enhancing its electrophilicity. The nucleophile is the amino N atom of the peptide being added. The leaving group is the ester OCH_3^- group, which is probably protonated before the C–O bond cleavage occurs.

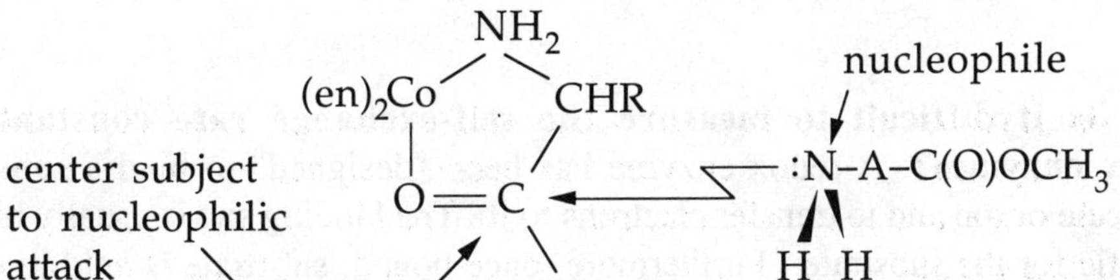

$$
\begin{array}{c}
\text{NH}_2 \\
\text{(en)}_2\text{Co} \diagdown \diagup \text{CHR} \qquad \text{nucleophile} \\
\end{array}
$$

center subject to nucleophilic attack

$$O=C \longleftarrow \ :N\text{--A--C(O)OCH}_3$$

$$OCH_3 \qquad \overset{H\ \ H}{}$$

19.7 High-spin and low-spin Fe(II) porphyrin complexes? The electron configurations for the two spin states in question are $t_{2g}^4 e_g^2$ (high-spin) and t_{2g}^6 (low-spin), as shown below:

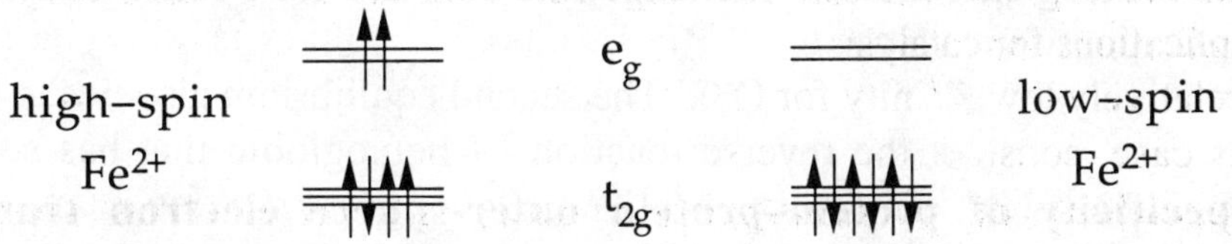

high–spin
Fe^{2+}

low–spin
Fe^{2+}

e_g

t_{2g}

High-spin Fe(II) is the larger of the two because of the two e_g electrons that it possesses. Section 7.4 described the metal–ligand bonding in octahedral complexes. The metal t_{2g} orbitals are either nonbonding or weakly π-bonding MOs, but the metal e_g orbitals are always metal–ligand σ-antibonding MOs (see Figure 7.18). Population of σ-antibonding MOs leads to longer metal–ligand bond distances and an apparently larger metal ion radius. Only weak ligands such as H_2O and ethers induce a small enough Δ_o to yield six-coordinate high-spin Fe(II) porphyrin complexes (one that has been studied by X-ray crystallography is [Fe(porphyrin)(thf)$_2$]). More strongly basic ligands such as pyridine (py) and the imidazole group of histidine, or π-acid ligands such as CO, yield low-spin complexes (a structurally characterized one is [Fe(porphyrin)(py)(CO)]).

19.8   Why are d metals such as Mn, Fe, Co, and Cu used in redox enzymes in preference to Zn, Ga, and Ca? The first group of metals occur naturally in redox enzymes because they each have at least two stable oxidation states. Redox catalysis involves the cyclic oxidation and reduction of the metal ion (for example, from Fe(II) to Fe(III) and back again for the cytochromes). The metals in the second group have only one stable oxidation state (Zn(II), Ga(III), and Ca(II)), so they cannot be oxidized or reduced at physiological potentials (i.e. the potentials of the most oxidizing and most reducing proteins in a cell).

19.9   Why is it difficult to measure the self-exchange rate constants of redox enzymes? A redox enzyme has been "designed" to bind a substrate molecule or ion and to transfer electrons to it. The binding site is usually highly specific for the substrate. Furthermore, once bound, substrate is held in close proximity to the metal center. The self-exchange rate constant is defined as the rate constant for electron transfer between the oxidized form of the enzyme and the reduced form. In the physiological redox reaction, the two forms of the redox enzyme do not transfer electrons with one another. In fact, the self-exchange electron transfer may be very slow, even if the enzyme–substrate electron transfer is fast. Therefore, since the site(s) at which the two forms of the enzyme bind to each other may not be the same as (or even close to) the substrate binding site, the self-exchange rate constant for a redox enzyme has few implications for catalysis.

19.10   The specificity of protein–protein outer-sphere electron transfer? One way that the protein parts of an enzyme contribute to making outer-sphere electron transfer highly specific is by controlling the "docking" of the enzyme with its redox partner. This controls the distance between the metal centers as well as their relative orientations, as illustrated in the figure below. Another way is by holding the ligands for each metal center in a particular geometry, thereby controlling inner-sphere reorganization energies.

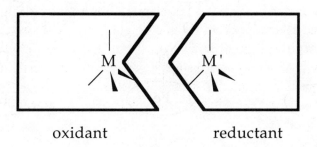

oxidant   reductant

19.11 Identify one significant role in biological processes for the elements Fe, Mn, Mo, Cu, and Zn. See Table 19.2:

    Fe: oxygen transport (hemoglobin, hemerythrin)
             oxygen storage in tissue (myoglobin)
             electron transfer (cytochromes, rubredoxin, ferredoxins)
             oxygenation of hydrocarbons (cytochrome P-450)

    Mn: oxidation of water to O_2 (photosystem II)

    Mo: reduction of N_2 to NH_3, or nitrogen fixation (nitrogenase)

    Cu: oxygen transport (hemocyanin)
             electron transfer (blue copper proteins)

    Zn: conversion of CO_2 to HCO_3^- (carbonic anhydrase)
             hydrolysis of peptide linkages (carboxypeptidase)
             decarboxylation of oxaloacetic acid (oxaloacetate decarboxylase)

19.12 What prevents simple iron porphyrins from functioning as O_2 carriers? Simple iron porphyrins are not protected from aggregation as are the iron porphyrin prosthetic groups in hemoglobin and myoglobin. Aggregation of the two complexes [Fe(porphyrin)O_2] and [Fe(porphyrin)] leads to a μ-peroxo dimeric complex, [[Fe(porphyrin)]$_2O_2$], and ultimately to a μ-oxo dimeric complex, [[Fe(porphyrin)]$_2$O]. Note that the oxidation state of iron is Fe(II) in [Fe(porphyrin)] and Fe(III) in [[Fe(porphyrin)]$_2$O]. Thus, treatment of simple

[Fe(porphyrin)] complexes with O_2 leads to irreversible *oxidation* instead of reversible *oxygenation*.

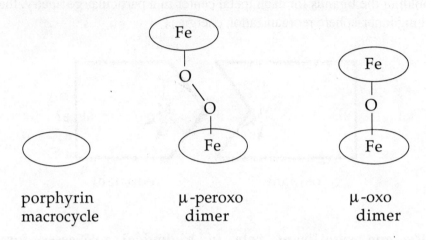

porphyrin   μ-peroxo   μ-oxo
macrocycle   dimer   dimer

19.13   Sketch the steps illustrating a metal complex functioning as each of the following?   (a) A Brønsted acid in an enzyme-catalyzed reaction? A water molecule coordinated to a metal ion is a stronger Brønsted acid than pure water (see Section 5.3). The functioning of carbonic anhydrase is believed to involve a Zn–OH_2 species.

(b) A Lewis acid in an enzyme-catalyzed reaction? The coordination of a basic ligand to an electrophilic metal ion induces a positive charge at the atom adjacent to the point of coordination. For example, coordination of a carbonyl oxygen atom to Zn(II) enhances the partial positive charge at the carbonyl carbon atom. The functioning of carboxypeptidase is believed to involve a Zn–O=C(N–)(R) species.

19.14   What is the average change in oxidation number of the Mn atoms in the S_4 to S_0 step in the PS II cycle? The oxygen atoms in the S_4 (adamantane-like) Mn_4O_6 complex (see Cycle 19.3) and the S_0 (cubane) Mn_4O_4 complex are formally O^{2-} ions. The half-reaction involving oxygen evolution must be $2\,O^{2-} \rightarrow O_2 + 4\,e^-$. Therefore, since the four Mn atoms are reduced by a total of four electrons, each Mn atom experiences a change of –1 in oxidation number.

19.15   Could magnesium be used as the metal ion in a cytochrome? In the case of chlorophyll, the source of the electron is a macrocycle π^* orbital (the electron is promoted to this orbital from the macrocycle HOMO, a π orbital, by absorption of a photon). The Mg atom is involved in the spatial orientation of the chlorophyll molecule relative to other parts of the reaction center, but is not itself involved in electron transfer. In contrast, the source of the electron in a cytochrome is the iron atom itself, as it changes oxidation state from Fe(II) to Fe(III). Magnesium could not be effective in a cytochrome, since Mg(II) cannot be oxidized to a higher oxidation state by any chemical oxidant.

19.16 What potential difference can a 700 nm photon produce? You should use the expressions $\nu = c/\lambda$ and $E = h\nu$. Since you are given $\lambda = 700$ nm $= 700 \times 10^{-9}$ m $= 7.00 \times 10^{-7}$ m, you can use the physical constants $c = 3.00 \times 10^8$ m s^{-1} and $h = 6.63 \times 10^{-34}$ J s to find E:

$$\nu = c/\lambda = (3.00 \times 10^8 \text{ m s}^{-1})/(7.00 \times 10^{-7} \text{ m}) = 4.29 \times 10^{14} \text{ s}^{-1}$$

$$E = h\nu = (6.63 \times 10^{-34} \text{ J s})(4.29 \times 10^{14} \text{ s}^{-1}) = 2.84 \times 10^{-19} \text{ J}$$

since 1 V $= 1.602 \times 10^{19}$ J, the 700 nm photon could potentially produce a potential difference of 1.77 V ($2.84/1.60 = 1.77$). Therefore, the efficiency is (1 V)/(1.77 V) $\times 100\% = 56.5\%$.

19.17 Nature's nitrogen ligands? The most important types of nitrogen ligands that are not directly bonded to the polypeptide chains of proteins are macrocycles such as porphyrins, which bind iron in hemoglobin, myoglobin, and the cytochromes (see Structure **8**), dihydroporphyrins (also called chlorins) such as the chlorophylls, which bind magnesium (see Structure **18**), and corrins such as coenzyme B$_{12}$, which bind cobalt (see Structure **16**). The most important nitrogen ligand that is part of the polypeptide chains of proteins is imidazole, the side group of the amino acid histidine (see Table 19.3, Figures 19.3, 19.4, 19.6, 19.9, and 19.10, Cycle 19.1, and Structures **4** and **5**). Note also that the related benzimidazole ligand is coordinated to the cobalt atom of coenzyme B$_{12}$ (see Structure **16**).

19.18 The emergence of Ca^{2+}, Zn^{2+}, and Cu^{2+} in the control of protein folding relative to Be^{2+}, Al^{3+}, and Cr^{3+}? The control of protein folding by a metal ion involves rapid ligand exchange reactions within the coordination sphere of the metal ion (i.e. water dissociation and reassociation as well as protein side chain dissociation and reassociation; see Section 19.2). Therefore, labile metal ions are required. Inspection of Figure 7.25 shows that Be^{2+}, Al^{3+}, and Cr^{3+} undergo H$_2$O ligand exchange rather slowly, while Ca^{2+}, Zn^{2+}, and

Cu^{2+} undergo rapid H_2O ligand exchange many orders of magnitude more quickly. In addition, a reason for the lack of a role for beryllium in metallobiomolecules is its very low abundance in the universe and the earth's crust; it is not readily available for assimilation by organisms.

19.19   **What is the purpose of the steric hindrance in hemoglobin?** The four polypeptides that surround the four iron(II) porphyrin complexes in a molecule of hemoglobin serve several purposes. From the standpoint of steric hindrance, they are essential for the reversible O_2 binding of hemoglobin. The reaction of O_2 with sterically unencumbered iron(II) porphyrin model complexes is irreversible, and a μ-O^{2-} diporphyrin complex is formed (see Structure **3**). Therefore, steric hindrance prevents the close approach of two iron(II) porphyrin units in hemoglobin, preventing the formation of the inactive μ-O^{2-} complex.

19.20   **Differences between hemoglobin and myoglobin?** There are several important differences. The most obvious is that hemoglobin contains four subunits, each consisting of a polypeptide and an iron(II) porphyrin, while myoglobin consists of a single polypeptide and an iron porphyrin. The four subunits of hemoglobin exhibit cooperative O_2 binding (i.e. the binding of O_2 to the iron porphyrin of one subunit affects the affinity of the other iron porphyrins for O_2), whereas cooperative O_2 binding does not occur for myoglobin. Finally, the binding of O_2 to the subunits of hemoglobin is pH dependent, while the binding of O_2 to myoglobin is not pH dependent.

19.21   **The origin of CO toxicity in mammals?** The binding of π-acid ligands other than O_2 to the iron(II) porphyrins in hemoglobin and myoglobin is discussed in Section 19.3(a). When CO binds to the iron atom, it inhibits the binding of O_2 (CO binds more strongly than O_2 to iron(II) porphyrins). Therefore, CO can prevent the transport of O_2 from the lungs to other tissues in mammals, shutting down the energy-producing respiratory chain and resulting in cell death.

19.22   **Suitable characteristics of Zn^{2+}?** One characteristic of Zn^{2+} that makes it suitable as the catalytic center in many hydrolytic enzymes is its borderline Lewis acid behavior (i.e. it is neither a very hard Lewis acid nor a very soft Lewis acid). For this reason, it can coordinate to a wide variety of biological ligands. Another characteristic is that Zn^{2+} coordinates to ligands more strongly than many other divalent d-block metal ions (see Figure 7.24). A third reason is that

Zn^{2+} undergoes ligand exchange more rapidly than many other divalent d-block metal ions (see Figure 7.25). Lability is a very important parameter in a catalytic cycle (see Figure 17.1).

19.23 Iron–sulfur proteins and redox catalysis? Iron–sulfur proteins are employed in redox catalysis because they are ideally suited for outer-sphere one-electron transfer reactions. The bond distances and angles of iron–sulfur complexes do not change significantly from one redox state to another, and the very small structural reorganization allows for very fast electron-transfer reactions (see Section 14.13).

19.24 The manganese complexes in PSII? According to Cycle 19.3, the manganese ions in PSII are octahedral complexes with a predominance of oxygen-containing ligands such as H$_2$O, OH$^-$, O^{2-}, RCO$_2^-$, and O$_2$.

19.25 Why is manganese better suited for PSII than copper or nickel ? There are two important reasons why manganese is better suited for this task. The first reason is that manganese is harder than either nickel or copper (see Section 5.12), and this property makes it bind more strongly than copper or nickel to the many oxygen-containing ligands in PSII, including H$_2$O, OH$^-$, O^{2-}, RCO$_2^-$, and O$_2$ (see Cycle 19.3). The second reason is that Ni^{2+} is neither readily oxidized nor reduced when complexed by oxygen ligands, making it unsuitable for a role as a redox center in PSII. Although Cu^{2+} is reversibly reduced to Cu$^+$ in many redox proteins, Cu$^+$ is even softer than Cu^{2+}. Therefore, coordination of softer ligands might compete with the coordination of the hard ligand H$_2$O to Cu$^+$.

Guide to Solutions Quiz

1 According to Table 19.1, silicon, aluminum, and titanium are very abundant in the earth's crust but are not very commonly used elements in biomolecules. Suggest a reason for the lack of biological processes involving these three elements. One of the three elements does have a biological function for some organisms. Which element is it, and what is the role?

2 Although synthetic cobalt-containing oxygen carriers have been prepared and studied, Nature does not seem to use cobalt for reversible oxygen-binding

proteins. Propose a possible explanation for this. Hint: consider the implications of Figure 19.7.

3   Alkaline phosphatase is a zinc enzyme that is active in basic solution. It catalyzes reactions such as

$$O_2N \!-\!\!\bigcirc\!\!-\! OPO_3^{2-} \xrightarrow{H_2O} O_2N \!-\!\!\bigcirc\!\!-\! OH \ + \ HPO_4^{2-}$$

Suggest a role for zinc in this type of reaction. Propose, using structures, a catalytic cycle.

4   Do you expect the equilibrium constant for the following reaction to be greater than 1 or less than 1? Explain.

$$2Hb(O_2) \rightleftharpoons Hb + Hb(O_2)_2$$

5   The toxicity of carbon monoxide can be traced to its high affinity for deoxymyoglobin and deoxyhemoglobin. This affinity, however, is much lower than the inherent affinity of CO for model Fe(II) porphyrin complexes. Consider the binding of O_2 and CO to metal ions (see Figure 19.2 and Structure **5**) and suggest how the shape of the heme pocket in oxygen-binding proteins might discriminate against CO in favor of O_2. (This is an important feature of heme proteins: there is a small, naturally occurring concentration of CO in most organisms due to other metabolic processes.)

6   What properties of Fe and Cu make them suitable for redox processes in biological systems? Would Ru and Au be just as suitable?

7   The coordination geometry of Fe^{3+} in nonheme iron proteins is highly variable. Why would you not expect the same to be true for Co^{3+}?

8   Depending on the type of environment of the $Fe_4S_4(SR)_4^{n-}$ cluster in ferredoxin electron-transfer proteins, some proteins shuttle between the $Fe_4S_4(SR)_4^{-}/Fe_4S_4(SR)_4^{2-}$ oxidation levels and some shuttle between the $Fe_4S_4(SR)_4^{2-}/Fe_4S_4(SR)_4^{3-}$ oxidation levels. Determine the average oxidation number for the iron atoms in the three different clusters. When the former group of proteins are denatured (unfolded), only the $Fe_4S_4(SR)_4^{2-}$ and $Fe_4S_4(SR)_4^{3-}$ oxidation levels are accessible. Suggest a reason that involves solvation and hydrogen bonding, and speculate on the two types of protein environments

9 Refer to Figure 19.12. In addition to the heme macrocycle, the iron atoms in many cytochromes b have two imidazole donor ligands (from histidine amino acids) in the axial positions. In constrast, the axial donor ligands in cytochromes c are the functional groups of one histidine amino acid and one methionine amino acid. Explain the difference in redox potentials for these two classes of electron transfer proteins.

10 Consider the complete equation for the fixation of nitrogen by nitrogenase:

$$N_2(aq) + 16\,MgATP^{2-}(aq) + 24\,H_2O + 8\,e^- \rightarrow$$

$$2\,NH_3(aq) + H_2(aq) + 16\,MgADP^-(aq) + 16\,HPO_4^{2-}(aq) + 8\,H_3O^+(aq)$$

Estimate the overall $\Delta G^{\circ\prime}$ (pH 7) for this reduction half-reaction using a Born–Haber-type approach. The free energy of hydration of ammonia is -26.8 kJ mol^{-1}. The free energies of hydration of dinitrogen and dihydrogen are both approximately $+17$ kJ mol^{-1}.